RELIABILITY, YIELD, AND STRESS BURN-IN

A Unified Approach for Microelectronics Systems Manufacturing & Software Development

RELIABILITY, YIELD, AND STRESS BURN-IN

A Unified Approach for Microelectronics Systems Manufacturing & Software Development

by

Way Kuo, Wei-Ting Kary Chien, Taeho Kim

SPRINGER SCIENCE+BUSINESS MEDIA, LLC

Library of Congress Cataloging-in-Publication Data

Kuo, Way, 1951-
Reliability, yield, and stress burn-in : a unified approach for microelectronics systems manufacturing & software development / by Way Kuo, Wei-Ting Kary Chien, Taeho Kim.
p. cm.
Includes bibliographical references and index.
ISBN 978-1-4613-7596-8 ISBN 978-1-4615-5671-8 (eBook)
DOI 10.1007/978-1-4615-5671-8
1. Integrated circuits--Design and construction--Reliability. 2. Microelectronics--Reliability. 3. Computer software--Development--Reliability. 4. Semiconductors--Computer programs--Reliability. I. Chien, Wei-Ting Kary, 1965- II. Kim, Taeho, 1960- III. Title.
TK7874.K867 1998
621.381--dc21 97-39195
CIP

Originally published by Kluwer Academic Publishers in 1998
Softcover reprint of the hardcover 1st edition 1998

Printed on acid-free paper.

To our wives

Chaochou Lee, Ching-Jung Lin, Dongyeon Shin

Contents

List of Tables

List of Figures

PREFACE

The international market is very competitive for high-tech manufacturers today. Achieving competitive quality and reliability for products requires leadership from the top, good management practices, effective and efficient operation and maintenance systems, and use of appropriate up-to-date engineering design tools and methods. Furthermore, manufacturing yield and reliability are interrelated. Manufacturing yield depends on the number of defects found during both the manufacturing process and the warranty period, which in turn determines the reliability.

Since the early 1970's, the production of microelectronics has evolved into one of the world's largest manufacturing industries. As a result, an important agenda is the study of reliability issues in fabricating microelectronic products and consequently the systems that employ these products, particularly, the new generation of microelectronics. Such an agenda should include:

- the economic impact of employing the microelectronics fabricated by industry,
- a study of the relationship between reliability and yield,
- the progression toward miniaturization and higher reliability, and
- the correctness and complexity of new system designs, which include a very significant portion of software.

In the past, most attempts to assure high reliability for microelectronics products included functional testing, life testing or accelerated stress testing of the entire circuit. Integrated circuits (ICs) are extensively used in systems that have high consequence of failures. Critical IC users now are more depending on commercial products when diverse functions of memories, microprocessors, and sensors are being integrated on one chip and new technology in ICs manufacturing is rapidly scaled. Because product testing is getting more expensive, more time consuming, and less capable of effectively identifying the causes for functional failure of microelectronics, the development of new testing strategies and technologies is needed.

This book introduces our research results and development experience from numerous projects related to the reliability of the infant mortality period in modern systems with an emphasis on microelectronics products. This book develops optimal burn-in conditions and analysis procedures for both the component and the system levels for many systems that use evolving technology.

Included are the principles of design for reliability and applications of statistical analysis in those used for stress testing. It intends to provide insight into how to take advantage of quickly-evolving technologies to design a reliable and cost-effective system. The book is also intended for non-microelectronics manufacturing industries that have an interest in designing highly reliable products.

This book will serve students of engineering and statistics, design and system engineers, and reliability and quality managers. It is based on three principles. First, we present the most practical methodologies that can be used to enhance reliability in real manufacturing system design, with illustrations of semiconductor products. Second, we integrate reliability into the manufacturing system analysis. The relationships between reliability and yield, wafer level reliability and stress burn-in, and hardware reliability and software reliability are included in this book. Third, we cover basic reliability, reliability design issues, and applications of statistical methods applied to manufacturing problems, as well as some advanced topics in burn-in analysis. Many illustrations come from our past work experiences with high-tech industries in USA, Taiwan, and Korea and from our research results of projects sponsored by various funding agents.

This book will help designers improve the reliability of their products, help them better specify reliability requirements, and help them design competitive products. Some analysis procedures are presented for the very first time in this book. Using this as a reliability text, readers will gain a better understanding of how reliability theory and optimization are used in design and manufacturing.

Overview of Sections and Chapters

Figure 1 shows the schematic organization of this book, which consists of preface, four sections of twelve chapters, and epilogue.

Chapters 1 and 2 present new techniques needed in manufacturing and reliability improvement in the evolving microelectronics manufacturing arena and an overview of design for reliability. Basic reliability concept and useful distribution functions with illustrations adopted in reliability engineering studies are described in Chapter 3. Because the wearout process under the rapidly advancing manufacturing technology cannot be detected during the normal operation life, the long infant mortality period for sophisticated products deserves special attention. Initially, performance and manufacturing yield are the primary concerns of microelectronics manufacturers. These concerns are compounded by the high initial failure rate of the integrated circuits that are often included in modern systems. Undoubtedly, defects introduced by processing and during manufacturing result in high infant mortality and consequently a concern for low reliability, which should be considered as a yield problem as well.

Some yield models of semiconductor devices which are used to monitor process and to predict the yield of devices before production are presented in Chapter 4. In addition, understanding both failure mechanisms and reliability physics and performing failure analysis are important to a proactive reliability control. Chapter 5 includes environmental stress screening and accelerated life tests. In Chapter 5, stress burn-in and its characteristics are presented. Current

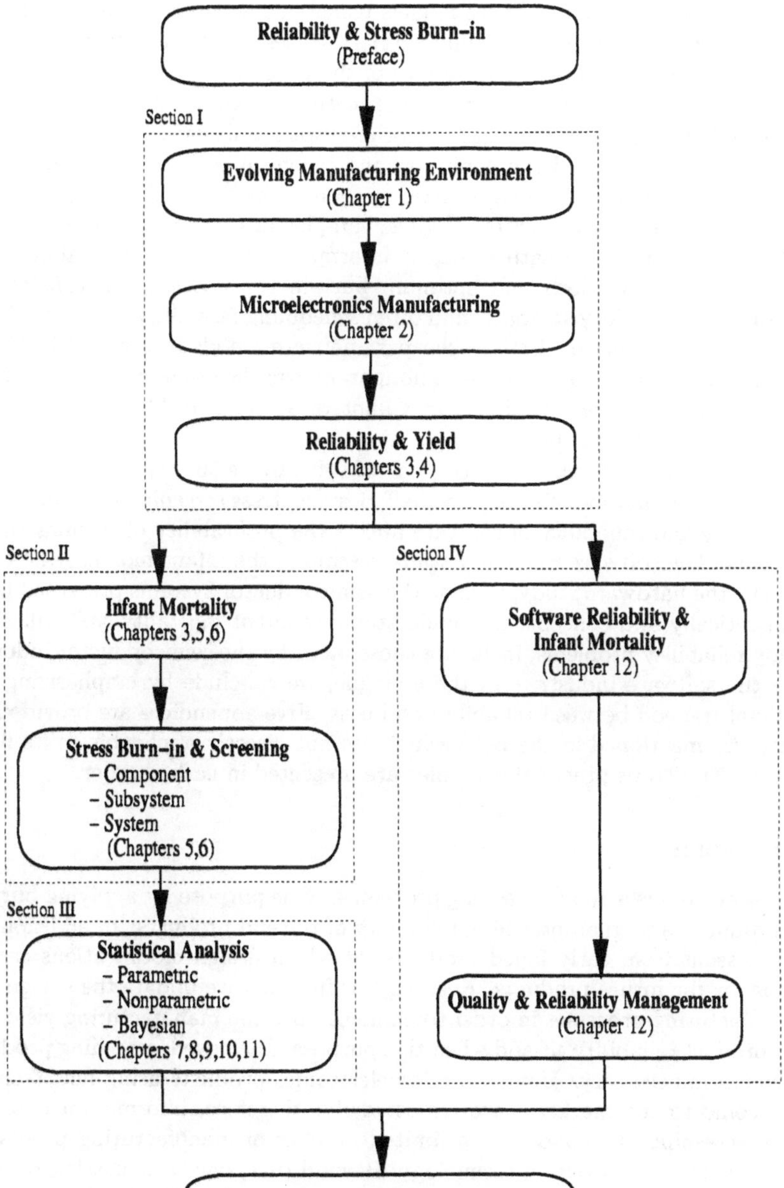

Figure 1: The schematic organization of this book.

reliability prediction models for most semiconductor failure mechanisms are based on empirical data. These models usually suffer over the broad range of stress conditions inherent in highly accelerated tests. Accelerated tests are the best measure of the intrinsic reliability of the product.

Chapter 6 describes the statistical methods used in estimating the functions and their parameters related to the system, subsystem, and component burn-in time. To avoid the confusion and inconvenience of using parametric methods, we introduce nonparametric methods, which provide a model-free estimate of the survival function of given samples in Chapter 7. Stress burn-in needs to be conducted with sufficient information on cost and reliability and with statistical methods and optimum burn-in conditions given constraints. Constraints include cost, space, and often scheduling feasibility, such as a fixed-hour burn-in duration. Besides the parametric methods introduced in Chapter 8, nonparametric approaches, a nonparametric Bayesian approach, and the Dirichlet process are introduced in Chapters 9, 10, and 11, respectively, to determine the optimal burn-in time.

In Chapter 12, the basic concept of software reliability and some useful software reliability models are provided. Besides stress screening tests and burn-in, quality and reliability of software affects the profitability of manufacturing systems. The software reliability study resembles the infant mortality analysis used in the hardware study, such as the semiconductor systems presented here. Theoretically and conceptually, understanding infant mortality well will solve many reliability problems, including those faced by the semiconductor industry and the software industry. In the epilogue, we conclude by emphasizing the optimal tradeoff between reliability and cost. Five appendices are provided for materials mentioned in the book which are not directly applicable to its main theme. Numerous practical examples are presented in each chapter.

Addendum

Stress burn-in is a 100% screening procedure. One purpose for applying burn-in to products is to guarantee high reliability of the end products. In addition, we take lessons from early failed products for which design modifications can be made for the future products. Keeping this in mind, we update the design and manufacturing processes in order to enhance both the manufacturing yield and the product reliability. If and when this purpose is achieved, screening products becomes unnecessary. However, microelectronics products using new technology come to the market place almost daily; therefore, information obtained from screening is valuable for a limited number of manufacturing processing updates using the existing technology. Beyond that, once the existing technology becomes obsolete, the products using new technology need to be evaluated to meet the quality and reliability standards again. Information obtained from burn-in on current products can serve as prior knowledge for burn-in on the design of products due to new technology. Unless we can forecast the exact causes for design and manufacturing flaws of future products, stress burn-in will still serve the screening purpose. In particular, ICs for high consequence of failures, which include applications on communications, medical systems, and others, need to subject themselves to the full screening procedure before they are assembled into a dedicated system.

Perhaps an optimal design methodology in the future will provide us with a fault-free approach, coupled with a flaw-free manufacturing process. At least, we hope that bursting the fault-free design approach converges at a rate no slower than the rate of high reliability required by customers. Until then, burn-in is essential for products that experience high failure rate at the long infant mortality period, such as those using microelectronics or extensive software.

In conclusion, this book presents many new ideas for reliability engineering subjects. Specifically, the book is intended for use by

1. Those who do microelectronics research, design, development, and manufacturing.

2. Those who are interested in stress testing but are not in the semiconductor industry, who will find this book useful and unique for performing reliability analysis.

3. Students and faculty of electrical engineering, industrial engineering, material engineering, and statistics who will find this book to be very useful for courses that address reliability and manufacturing. It elaborates some fundamental theories by providing real examples obtained from the evolving manufacturing industry. Issues addressed are real and not fictitious ones.

4. Those who are interested in real-world reliability problems. Some very interesting research results are available in this publication. We believe this is one of the very few reliability books that apply reliability theory to solve real problems. Challenging problems for future study are also available in this book.

Way Kuo, Texas A&M University, College Station
Wei-Ting Kary Chien, Nan Ya Technology Corp., Taoyuan
Taeho Kim, Korea Telecom, Seoul

ACKNOWLEDGEMENTS

This manuscript grows from our research of the past 12 years. The results reported here come from several research and development projects supported by the National Science Foundation, Army Research Office, National Research Council, Fulbright Foundation, the National Science Council, Taiwan, R.O.C., the IBM Headquarters, Hewlett Packard, and Bell Labs (now Lucent Technologies).

We acknowledge input to this manuscript from William Q. Meeker of Iowa State University; Nozer D. Singpurwalla of the George Washington University; Martin A. Wortman of Texas A&M University; Final Test Engineering Department, the Microcontroller Technologies Group of Motorola, Austin; Yue Kuo of IBM, T.J. Watson Research Center; Henry Hart of Vitro; William A. Golomski of W.A. Golomski Associates; K.C. Kapur of the University of Washington; Dick Moss of Hewlett Packard, Palo Alto; and David Sean O'Kelly, and Kyungmee Oh Kim of Texas A&M University. We acknowledge Mary Ann Dickson for editing this manuscript and Julie Polzer of Texas A&M University for handling correspondence.

Some figures, tables, and statements are obtained or modified from previously published works. We have obtained permissions from *Proceedings of the IEEE* to use Kuo, W. and Kuo, Y., "Facing the headaches of early failures: a state-of-the-art review of burn-in decisions," **71** (11), pp 1257–1266, Nov. 1983, © 1983 IEEE; from *IEEE Transactions on Reliability,* Kuo, W., "Reliability enhancement through optimal burn-in," **R-33** (2), pp 145–156, Jun. 1984, © 1984 IEEE; Chi, D. and Kuo, W., "Burn-in optimization under reliability & capacity restrictions," **38** (2), pp 193–198, Jun. 1989, © 1989 IEEE; Chien, W.T.K. and Kuo, W., "Modeling and maximizing burn-in effectiveness," **44** (1), pp 19–25, Mar. 1995, © 1995 IEEE; from *IEEE Transactions on Semiconductor Manufacturing,* Michalka, T.L., Varshney, R.C., and Meindl, J.D., "A discussion of yield modeling with defect clustering, circuit repair, and circuit redundancy," **3** (3), pp 116–127, Aug. 1990, © 1990 IEEE; Dance, D. and Jarvis, R., "Using yield models to accelerate learning curve progress," **5** (1), pp 41–45, Feb. 1992, © 1992 IEEE; Cunningham, S.P., Spanos, C.J., and and Voros, K., "Semiconductor yield improvement: results, and best practices," **8** (2), pp 103–109, May 1995, © 1995 IEEE; Chien, W.T.K. and Kuo, W., "A nonparametric approach to estimate system burn-in time," **9** (3), pp 461–466, Aug. 1996, © 1996 IEEE; Berglundm C.N., "A unified yield model incorporating both defect and parametric effects," **9** (3), pp 447–454, Aug. 1996, © 1996 IEEE; from *Proceedings of Annual Reliability and Maintainability Symposium,* Hansen, C.K., "Effectiveness of yield-estimation and reliability-prediction based on wafer test-chip measurements", pp 142–148, Philadelohia, Jan. 13–16, 1997, © 1997 IEEE; Kim, T., Kuo, W., and Chien, W.T.K., "A relation model of yield and reliability for the gate oxide failures," Anaheim, CA., Jan. 19–22, 1998, © 1998 IEEE.

The papers, Chien, W.T.K. and Kuo, W., "Optimal burn-in simulation on highly integrated circuit systems," *IIE Transactions*, **24** (5), Nov. 1992, pp 33–43; Chien, W.T.K. and Kuo, W., "Optimization of the burn-in times through redundancy allocation," *Proceedings of the 2nd IERC Conference*, Los Angeles, CA, 1993, pp 579–583; Chien, W.T.K. and Kuo, W., "Determining optimal burn-in using the Dirichlet process," *Proceedings of the 3rd IERC Conference*, Atlanta, GA, 1994, pp 291–295, are reprinted with the permissions of the Institute of Industrial Engineers, 25 Technology Park, Norcross, GA30092, 770-449-0461, Copyright©1992, ©1993, ©1994, respectively.

We have obtained permission to use Chien, W.T.K. and Kuo W., "Extensions of the Kaplan-Meier estimator," *Communications in Statistics-Simulation and Computation*, **24** (4), Marcel Dekker, Inc., N.Y., 1995; from John Wiley and Sons to use Chien, W.T.K. and Kuo, W., "A nonparametric Bayes approach to decide system burn-in time," *Naval Research Logistics*, **44** (7), 1997, pp 655–671.

Copyright 1983 International Business Machines Corporation. Reprinted with permission from IBM Journal of Research and Development, Vol.27, No.6; Stapper, C.H., "Modeling of integrated circuit defects sensitivities," *IBM Journal of Research and Development*, **27**, 1983, pp 549–557.

1. Overview of Design, Manufacture, and Reliability

Competitiveness in international markets is now recognized as an essential element of national economic development. Quality of products, as perceived by the customers, is the most important factor driving this competition. Achieving a competitive level of quality requires leadership from the top, good management practices, effective and efficient operation and maintenance systems, and use of appropriate state-of-the-art engineering design approaches.

The traditional view of reliability holds that reliability problems must be solved in order to eliminate customer complaints. This viewpoint and motivation influence the traditional manufacturing process which includes inspection for detection of defects, statistical process control for process-stability and detection of special causes, and product assurance for failure prevention. From this perspective, it is through assessment and fix that we grow reliability. The process is sequential, and typically takes place downstream. Methodologies developed from this approach have matured in recent years. We have gone from using a few gauges for inspection to an array of sophisticated tools and methods for assessing and tracking reliability.

The special issue of "IEEE Transactions on Reliability" by Kuo and Oh [229] focuses on reliability-related engineering design. The application areas of particular interest are advanced materials-processing and the design and manufacture of electro-mechanical devices. Among the many considerations for designing products that are robust to manufacturing variations and customer use are concurrent engineering, computer-aided simulation, accelerated life testing (ALT), and physical experimentation. Combining these areas with a focus on reliability is a highly complex task which requires the cooperation of engineers, statisticians, and designers working together.

The focus of this chapter is on reliability-related engineering design approaches. This chapter concentrates on important reliability issues from the manufacturing perspective.

1.1. Production and Manufacturing Issues

A modern manufacturing system consists of computers, information, robots, machines, and workers. To raise productivity and simultaneously reduce production cost, we have to apply many new technologies, including computer-aided design and manufacturing (CAD/CAM)* tools, computer-aided process planning (CAPP), group technology (GT), flexible manufacturing system (FMS), Kan-Ban, just-in-time (JIT), human-machine interface analysis, total quality management (TQM), statistical quality control (SQC), experimental design, ALT, and so forth. If we view the whole stream of these new methods, we must assure the reliability of each of the technologies to prevent loss in reliability due to incorporating complex manufacturing process and system configuration. Key concepts are introduced in the following sections.

1.1.1. Concurrent Engineering

Concurrent engineering is the integration of all product life-cycle factors in the design approach. More extensive simulation and physical experimentation are required to reduce the risk of poor quality to levels that will justify the necessary capital investment in tooling and market development. Also, the concurrent engineering approach faces uncertainty early and directly so as to successfully address the needs for rapid prototyping and design qualification. Combining extensive computer aids, experimentation, and concurrent engineering with a focus on quality is a highly complex task; former technical and analytical tools have to be modified or redirected in order to identify and characterize the uncertainties involved in design. As indicated by Kuo [227], increasingly engineers, computer scientists, statisticians, and operations research analysts recognize the need for research in developing and applying approaches for this task. One hopes that the results of such research will add greater power to the now widely accepted CAD systems.

The word concurrent in our context means that the design of the product and its manufacturing process are carried out more or less simultaneously. When companies recognize the potential for an integrated approach, the result may be called the strategic approach to product design (SAPD). The rewards for adapting SAPD can be spectacular because it allows the entire system to be rationalized. Furthermore, SAPD identifies the need for computerized design tools to support this tightly integrated activity and gives it a scientific base for the future. Today SAPD can be accomplished only with teams of highly trained people. To grow in the competitive global economy, a company must prepare in four major areas: marketing, finance, product design, and manufacturing. We will focus on the manufacturing department, where rapid advances in product line flows, fast design updating, and quick changes in production schedules are required. The Toyota production system is a good example of recent advanced production systems [382]. With the objective of achieving efficiencies in a flow line while manufacturing products in small batches, Toyota's system is designed to incorporate the right material in the right place at the right time, continuous learning and improvement, and respect for workers.

By absorbing the philosophies of JIT, Kan-Ban, and low work-in-process inventory [299], we have to bear in mind that the basic steps in manufacturing system design are to:

*Notation and nomenclature used in this book are summarized in Appendix A.

- use either intuitive techniques or computerized methods and select a set of equipment or people that can make the product at the required rate at a reasonable cost,
- determine the production capacity required for the system by taking yield into account,
- tabulate feasible fabrication and assembly techniques for each operation and estimate the cost and time for each,
- select an assembly sequence for use in the design of the assembly system,
- either make preliminary economic analyses or proceed to detailed workstation designs and then perform economic analyses, or
- analyze the product, the necessary fabrication, and the assembly operations.

Design processes are needed to do specific jobs and to be operable after they have been installed. The existence of good process models is a requirement for automating the process. Two types of process models can be implemented: technical models which deal with physical parameters and economic models which deal with time and money. Several design factors must be taken into account. These are capacity planning, resource choice, task assignment, workstation design, floor layout, material-handling equipment choice, part provisioning, and economic analysis. Among them, economic analysis might be the most important from the managerial point of view. Alternative ways of making or assembling parts have to be identified first. Then, we determine the manufacturing or assembly cost for each alternative and derive the investment required to acquire and install each alternative. Finally, we identify the investment that makes the most economic sense in terms of the identified costs and other strategic benefits. Analysts should be familiar with types of manufacturing costs (including fixed costs, variable costs, materials costs, and institutional costs) and must apply various criteria to evaluate alternative investment, such as return on investment, net profit, maximum investment funds available, and reduction in rework and warranty costs.

Typical manufacturing systems and factories strive to achieve a balance among throughput, flexibility, quality, utilization, inventory, and economic efficiency. Concurrent engineering consolidates these factors. The advantages of this process consolidation include reductions of fixture design, fabrication, storage, maintenance costs, refixturing, part movement, the number of part loads and unloads, the number of individual operations required, work-in-process inventory, and manufacturing lead times.

1.1.2. Agile Manufacturing Systems

Because of changeable customer preferences and the competitive global market, the need is urgent for manufacturing systems that can quickly and effectively react to unpredictable environments. Therefore, agile manufacturing systems (AMS) are proposed to achieve flexibility and to resolve uncertainty.

Consumers have learned to take quality for granted, and they expect to buy things at low cost with features that exactly match their needs. To survive, enterprises have to build a reliable and efficient network which links suppliers, producers, and customers. The network provides direct and fast feedback on service diagnostics, sales statistics, consumer reports, government policies, advanced production technologies, and raw material suppliers. Using communication standards, the information is analyzed, filtered, digested, and automatically decoded into computer formats to be used by product designers and marketing personnel for prompt responses on these incoming messages. After being modeled and simulated in a computerized system, products are manufactured on a small scale by a modular and flexible production system with a short development cycle and introduced to the market. All designs and sales records are logged into a distributed data base for easy access among all the joint ventures.

Among those factors, knowledgeable workforce, joint ventures, and all-level consensus are three key factors to achieve agility. The greatest asset in an enterprise is the workforce. When kept informed about the overall goals of the enterprise, the workforce can make suggestions on possible options to achieve the objective using the available technologies and affordable resources. Also, through continuous education, a sense of achievement is a by-product which binds the workforce together with the enterprise. Joint ventures reduce risks and permit the sharing of costs, knowledge, technologies, and resources with other companies. Mutual trust is also important; one must have a trustworthy partner. The workforce's initiative in thinking of appropriate ways to benefit the enterprise is a good way to build all-level consensus.

1.1.3. System Design

System design usually includes performing preliminary system feasibility analysis and detailed work content analysis, where one has to define alternative fabrication system configurations that consider process technology alternatives; to evaluate material-handling alternatives, process control alternatives, process system control alternatives, human resource roles; and to predict system reliability. We also have to evaluate the technical and economic performance of each alternative solution in order to select the strategic criteria to determine the best performing alternative and develop an implementation plan for its installation.

The technical synthesis of system factors to be considered include the modern technology available today, optimal system design strategies, such as the maximum number of stations per worker, quality processes needed for manufacturing, quality and reliable product design, and product life testing. The economic synthesis of system factors considered for each resource include resource price, tool and material-handling price per part, uptime expected, minimum expected task cycle time, tool changing time, operating and maintenance rate, and the ratio of total cost and hardware cost.

A successful shop should respond rapidly to customer demand for a wide variety of parts in small quantities, provide very high-quality interchangeable parts at a low cost, and be able to accommodate rapid increases in part mix and volume on a short-notice surge capacity.

Simulation is an important technique in practical applications because one can use it to evaluate the performance of systems without having to physically build them. The impact of different operational strategies and major external uncontrollable events such as component failures can also be compared without requiring them to occur. In addition, time can be either expanded or compressed to facilitate the phenomena so that observation is possible. Finally, simulation can be applied to manufacturing systems to determine resource utilization and inventory levels and to investigate scheduling strategies and the impact of different batching strategies for batch-process systems. Some of the commonly used simulation languages (or software packages) are ECSL (Extended Control and Simulation Language), GPSS (General Purpose Simulation System), SIMAN, SIMSCRIPT II (an event-oriented simulation package), and SLAM (Simulation Language for Alternative Modeling). Virtual factory is another way to mimic real system operation which can be either at the physical operational level or performance evaluation level.

1.2. Taguchi Method in Quality Engineering

The quality improvement philosophy advanced by W. Edwards Deming has proven critical to Japanese industrial development. Deming points out that approximately 85% of problems can be traced back to the system; thus, two enumerative and analytic studies are emphasized in his method to inspect the production system. Genichi Taguchi shifts Deming's principle back to the design stage. Taguchi method seeks to minimize variability through experimental design and factor classifications combined with social philosophies, such as customers' preferences and their expectations relating to merchandise. Certain losses, however, will necessarily be generated from variability of inspection, production, and manufacturing. The smaller the loss, the more desirable the product. Investments in quality improvement activities appear much more attractive from a long-term point of view. An effective quality control program saves society more than it costs and benefits everybody since a defective product can cause a loss of market share. In a competitive economy where price is determined by the market and profit is determined by the units sold, continuous quality improvement and cost reduction are necessary for staying in business. Customers have ever-rising expectations of products, and all have the notion that quality is never high enough and cost is never low enough.

A continuous quality improvement program includes continual reduction in the variation of product performance characteristics from the target values. A product's quality cannot be improved unless the product's quality characteristics can be identified and measured. However, it is neither economical nor necessary to improve all quality characteristics because not all quality characteristics are equally important. A product's performance characteristics, whose ideal values are called target values (which need not be the midpoint of the tolerance interval), are the primary quality characteristics that determine the product's performance in satisfying the customer's requirements. Continuous scale is better for measuring small changes in quality. If continuous scale is not possible for a parameter, then an ordered categorical scale such as the rating of poor, fair, good and excellent can be used.

1.2.1. Minimizing Variability

The customer's loss due to a product's performance variation is often approximately proportional to the square of the deviation of the performance characteristic from its target value. Because loss may be too large or too small, using squared loss can avoid negative values. The general form of the loss is the cost to the customer of repairing or discarding the product which is expressed as $k\Delta^2$ where k is the constant to be determined, and Δ is the variation. One attribute of the loss function is to help determine an acceptable factory tolerance. Let y be the measurement and refer to Figure 1.1, the Δ used in different cases follows:

- $\Delta = \frac{1}{y}$, when larger is better (curve A)
- $\Delta = y$, when smaller is better (curve B)
- $\Delta = (y - y_s)^2$, when closer to pre-specified value y_s is better (curve C).

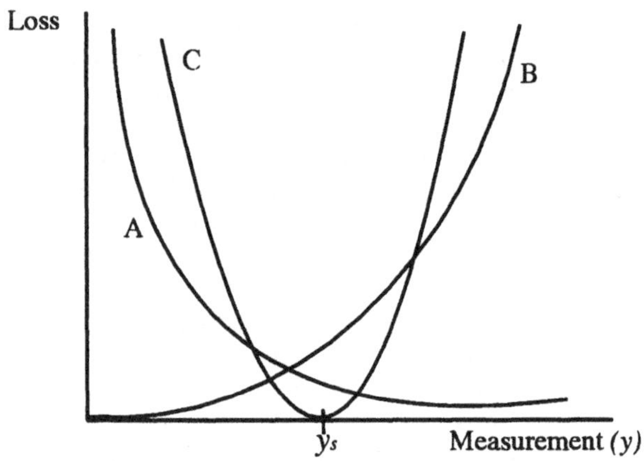

Figure 1.1: Loss function curves.

Similar modification can be made according to the physical interpretation of its model.

The final quality and cost of a manufactured product are determined to a large extent by the design of the product and its manufacturing process. There are three overlapping stages of a product development cycle: product design, process design, and manufacturing. Each stage has many steps, and the output of one step is the input for the next step. Since the complexity of modern products continues to increase, product and process design play a crucial role.

To reduce manufacturing imperfections, we first have to improve product and process design.

Product or process performance variation can be reduced by exploiting the nonlinear effects of the product or process parameters of the performance characteristics. Quality control has to begin with the first step in the product development cycle, and it must continue through all subsequent steps.

1.2.2. On-line and Off-line Quality Control

Quality control methods can be categorized in two ways. Off-line quality control methods are technical aids for quality and cost control in product and process design, while on-line quality control methods, such as sensitivity tests, prototype tests, accelerated life tests and reliability tests, are technical aids for quality and cost control in manufacturing.

1.2.3. Design Sequence

Taguchi introduces three methods to assign nominal values and tolerances to product and process parameters. Namely,

1. the system design is used to produce a basic functional prototype design which accommodates both customers' needs and the manufacturing environment,
2. the parameter design identifies the settings of product or process parameters, and
3. the tolerance design determines tolerances around the nominal settings identified by the parameter design.

At early research and development stages, system design may be used to improve quality with respect to all noise, which is defined as the uncontrollable factors such as the chance of having a surge of electric power to a microelectronic circuit. Several system alternatives may be proposed for testing for robustness against different noises. Once a system is chosen, parameter design can be used to fight the noise effects, to improve the uniformity, and to make the performance less sensitive to causes of variation under pre-determined tolerances. Quality can be further enhanced by tightening tolerances on product or process parameters; tolerance design tries to balance the trade-off between cost and quality. There are debates about the best order for applying parameter and tolerance designs. An engineer may jump to alter the tolerance design when a problem is detected in product development; however, this may inflate cost. That is, though there is a vague sequence of these three design methods, an interactive and closed-form application is advised to make the best use of experience and information, as shown in Figure 1.2.

1.2.4. Experimental Design

Statistically experimental design (SED) is the core of Taguchi method. Taguchi uses SED to identify

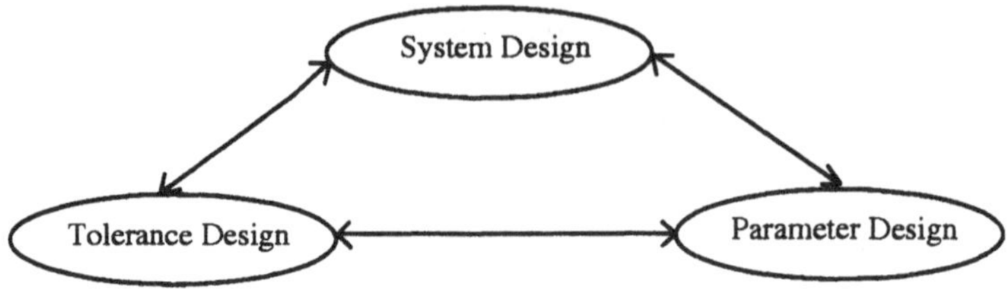

Figure 1.2: A closed-form design sequence.

- the settings of design parameters at which the effects of the sources of noise on the performance characteristic are at minimum,
- the settings of design parameters that reduce cost without hurting quality,
- those design parameters that have a large influence on the mean value of the performance characteristic but have no effect on its variation, and
- those design parameters that have no detectable influence on performance characteristics.

Two types of variables, the design parameters and the sources of noise, affect the performance characteristics of a product. The design parameters, which are also called control parameters, are the product or process parameters whose nominal settings can be chosen by the engineers. The sources of noise come from variables that cause the performance characteristics to deviate from their target values. Not all sources of noise can be included in a parameter design experiment: for example, lack of knowledge and physical constraints. The noise factors or parameters are those sources of noise and their surrogates that can be systematically varied in a parameter design experiment.

Taguchi-type parameter design experiments consist of two parts: a design parameter matrix or control array and a noise factor matrix or noise array. The design parameter matrix specifies the test settings of design parameters. Its columns represent the design parameters, and its rows represent the different combinations of noise levels. Taguchi recommends the use of orthogonal arrays for constructing the design parameter and the noise factor matrices. All common factorial and fractional factorial plans are orthogonal arrays. But not all orthogonal arrays are common fractional factorial plans. Please refer to related literature about experimental design such as Montgomery [288, 289].

Taguchi also proposes the use of signal-to-noise (s/n) ratios as performance statistics to compare different settings of the design parameters. For non-negative continuous performance characteristics with a fixed target, he defines three s/n ratios, depending on the three loss functions in Section 1.2.1:

- $-10\ \log(\frac{1}{n}\sum 1/y_i^2)$ for curve A in Figure 1.1
- $-10\ \log(\frac{1}{n}\sum y_i^2)$ for curve B in Figure 1.1
- $10\ \log(\overline{y}^2/s^2)$ for curve C in Figure 1.1

where n is the sample size, $\overline{y}$ is the sample mean of the n observations $y_1, \ldots, y_n$, and s is the standard deviation of the sample.

The accumulation analysis is applied for the data analysis of ordered categorical measurements. When the performance characteristic is measured on a binary scale, such as good or bad, Taguchi recommends

$$10 \log(\frac{p}{1-p})$$

where p is the proportion of good products. Interested readers are encouraged to refer to specific literature for Taguchi method (see [196, 208, 355, 413]).

1.3. Manufacturing Design and Reliability

A computerized manufacturing procedure is shown in Figure 1.3. Each project should be verified according to the market's needs, attainable manufacturing techniques, and potential profits. These three issues must be evaluated at all times. Once a project is approved after careful examination by the engineering, marketing, and financial departments, computer designs can then be commenced followed by the CAPP, where many techniques are proposed and applied in practice, such as the generative process planning system and the variant approach.†

Prototypes will be manufactured for life-testing. If the quality is acceptable, production will then proceed; otherwise, re-design might be needed. The popularity of a product is justification for possible re-modeling or modification of selling strategies. If it is found that the product is not competitive, development profiles as well as field reports will be put into a database for future reference. After being in the market for some time, a non-profitable product may be terminated and the production equipment may be either sold or used for making other items.

This section emphasizes manufacturing related issues. The life-cycle approach will first be introduced in Section 1.3.1; important issues for CAD/CAM systems will follow in Sections 1.3.2 to 1.3.9, including hardware (mechanical parts and electronic components) failure modes and computer tools, and integrated system and software designs. Software development and reliability modeling will be studied in detail in Chapter 12.

1.3.1. Life-Cycle Approach

Powerful computers have been applied for prediction, production scheduling, manufacturing, and cost estimation. Life-cycle engineering (LCE) is a combined technique using CAD/CAM, and computer-aided support systems (CAS) [56]. LCE can be built by a well-informed system design; an interactive process review by specialists in various disciplines can identify deficiencies that are outside the designer's experience. A detailed description containing cost, maintenance

†The discussion of the CAPP is beyond the scope of this book and will not be included here. Readers can refer to [69] for more information.

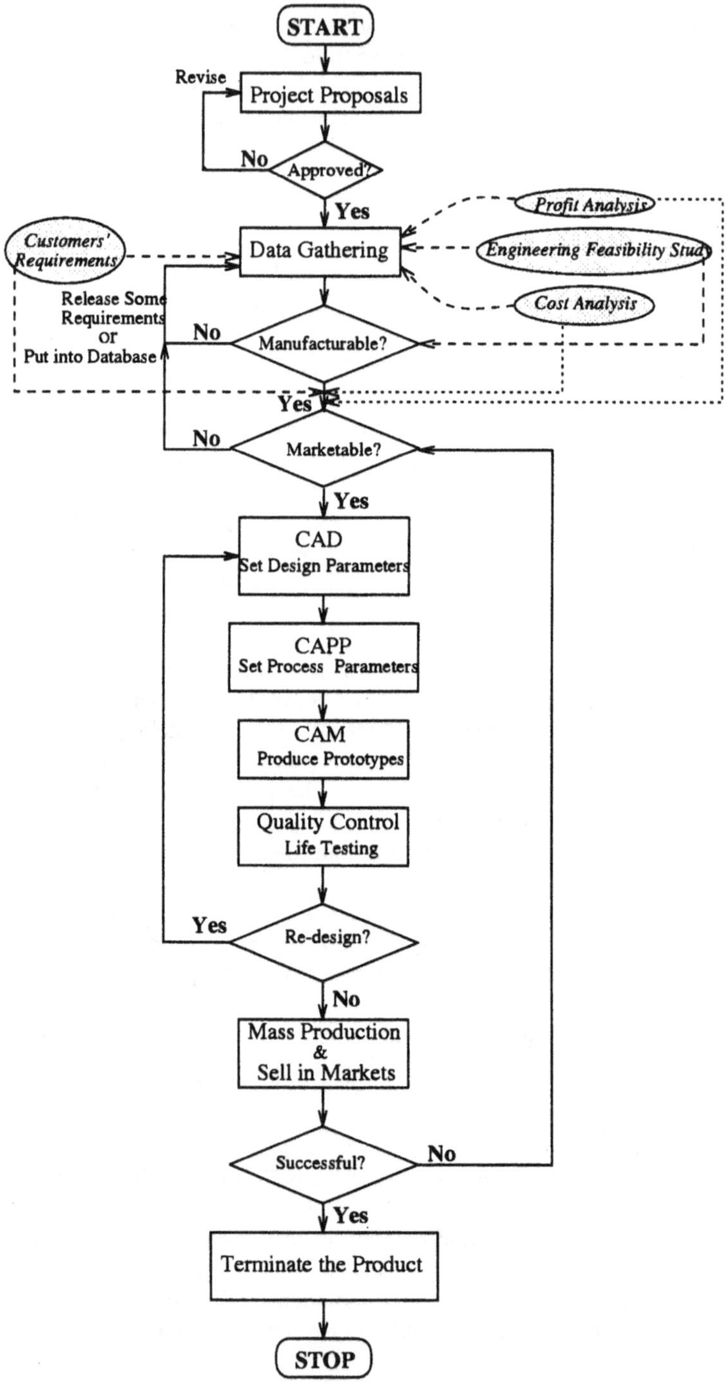

Figure 1.3: A modern manufacturing sequence.

planning, and reliability of a product can provide all necessary information such as reliability prediction; reliability apportionment; mechanical, electric, and electronic failure modes of analysis; sneak circuit analysis; maintainability allocation; accessibility and testability evaluation; test point selections; and test sequencing.

Merchandise can be designed in a more flexible way to fit market demands and to fulfill cost requirements, which are usually the most critical factor. Because no actual production is needed, considerable amounts of resources can be saved. Research and development costs can be decreased to make products more competitive. Burte and Copolla [56] model the life of a product as the function of definable characteristics of the service conditions and environment and measurable characteristics of the state of the hardware, multiplied by a constant, which is related to detailed design.

Similar to the hardware LCE mentioned above, Lin and Kuo [250] categorize the life-cycle cost (LCC) of a piece of software into the cost-estimation model, the resource-allocation model, and the program evaluation model. All three models deal with reliability and describe different aspects of software life-cycle cost.

Carrubba [66] considers integrating life-cycle cost and cost-of-ownership to close the gap between the optimizations of the two models so that a "win-win" situation–which is the saddle point strategy– for both the manufacturers and customers can be found. Kolarik, Davenport, and McCoun [218] introduce a life-cycle reliability model to simulate a system's reliability characteristics in the early design stages. These two studies once again point out the importance of a well-informed design environment. The LCC models can also be found in MIL-HDBK-259 and MIL-HDBK-276.

1.3.2. CAD/CAM Systems

The declining costs of computers and the increasing costs of labor are changing our manufacturing society from human-dependent to machine-dependent systems. Electro-mechanical systems are often developed for automation of internal manufacturing processes in industry. The application of these systems can range from simple assembly operation to large-scale material-handling processes. Machine designs are very much application dependent, and a continuous need exists for new electro-mechanical system designs either for new product development or existing process improvement. System design for an electro-mechanical machine involves mechanism and electrical control. Because of the increasingly significant role of computers, software design is also receiving a new level of attention.

1.3.3. Mechanical Design

A large volume of commercial and public domain software is available to assist mechanical engineers with design and analysis processes.

Solid Modeling

Solid modeling can create a realistic visual display by producing a shaded visible surface image of the solid. Another application involves the motion of the component and its interference with other solids or surfaces. Data generation programs frequently accept the solid model data base as partial input for the analysis required as part of the design process. The most common example is automatic generation of a mesh for a finite-element or finite-difference analysis

The goal behind solid modeling is to integrate all design, analysis, manufacturing, and documentation operations. Almost all information needed for part generation is contained in the solid model. Fabrication and manufacturing errors are significantly reduced if all the design engineers access the same current part description. It thus becomes important to restrict changes and revisions to the parts so that all design and manufacturing groups have the same model.

Finite-element Analysis (FEA)

The practical advantages of FEA in stress analysis and structural dynamics have made it the accepted design tool for the 1990s. It is widely applied in thermal analysis, especially in connection with thermal stress analysis. The greatest advantages of FEA are its ability to handle truly arbitrary geometry and to include nonhomogeneous materials. The designer can treat systems of arbitrary shape that are made up of numerous different material regions. Each material can have constant properties, or the properties can vary with spatial location. The designers, therefore, have a large degree of freedom in prescribing the loading conditions, support conditions, and post-processing items such as stresses and strains. The FEA also offers significant computational and storage efficiencies for elliptic boundary value problems, including electrical fields, ideal fluid flow, magnetic fields, stress analysis, and heat conduction.

Kinematic and Dynamic Analysis

Kinematic analysis is used to calculate the motion of the various bodies in the mechanism, disregarding both their mass properties and any forces in the system. It can solve position, velocity, and acceleration equations without any mass, inertia, or force information. In dynamic analysis, the model may have any number of degrees of freedom, from one up to the machine's total possible degrees of freedom. The mechanism motion may be either planar or spatial. Planar motion is characterized by components that move along the curves that lie in parallel planes and are not necessarily limited to a single plane. Spatial motion takes place on paths that do not lie in parallel planes.

Testing

The purpose of testing is to determine product performance. Computer simulation offers a more economical approach than performing expensive, time-consuming tests. Simulation's capabilities range from structural dynamics, rotating machinery analysis, and fatigue life estimation from strain data, to advanced spectral analysis. Moreover, simulation can create functions (including random numbers) or read them from data acquisition equipment for the Monte Carlo simulation. The results can be expressed in terms of mathematical and statistical diagrams and plots in different forms, such as time histories, histograms, statistics, XY plots, and XYZ plots.

Simulation and Optimization

Simulation involves experimentation with computerized models of a system or process. It allows the designer to consider a range of coefficients, or forcing functions, to test the design before it goes into production; thus, simulation is an important part of the CAD process. The process of mechanical design, or synthesis, may often be viewed as an optimization task. The CAD environment should give the analyst the ability to investigate several "what if" questions during the synthesis process. Such questions can involve many variables which are often subject to various constraints or design rules.

Conceptual Design

Many researchers recognize that intelligent CAD systems for mechanical engineers of the future must include a knowledge-based system that assists rather than impedes the designers. Most of the current CAD tools have emerged from the early drafting systems. The cognitive aspects of mechanical design have rarely been studied. Waldron and Waldron [434] explore the cognitive dimension and the types of information and knowledge that the mechanical designer uses at the conceptual level. This information environment must be included in the CAD system.

1.3.4. Electrical Design

Many CAD packages for electronic circuits have been developed [204, 305]. Some of them are user-oriented analysis programs, such as SCEPTRE and ECAP (Electronic Circuit Analysis Program); the latter was developed by IBM and the Norden Division of United Aircraft. Packages for integrated circuit design, such as SPICE (Simulation Program with Integrated Circuit Emphasis), have been used for component layout and routing for years. However, much effort has been put into computer components [152]. The U.S. Navy Standard Electronic Module (SEM) program (described in MIL-M-28787) provides a range of highly reliable standard modules designed to reduce logistics costs.

1.3.5. Integrated Design

Integrated Design System (IDS)

IDS is a concurrent design system developed in the early 1980s and inspired by a project at Case Western Reserve University named LOGOS for digital systems designers. There are three levels in IDS:

Level 1 – the architecture level specifies the algorithm to be implemented.

Level 2 – **the implementation level** uses the set of HDL (Hardware Description Language) [152] descriptions to form the specification of a register transfer machine.

Level 3 – **the realization level** applies the HDL in level 2 to construct lists of parts and their interconnections.

IDS uses the concept of an object-oriented database (OODB) to allow several designers to work independently on different parts of the design. As long as the interfaces between modules are well defined, the coordination of the component parts will be relatively straightforward. As in an OODB, a module in IDS can call on modules at lower levels to perform functions without regard to how that function is to be implemented. The correctness of a particular module depends only on the interconnections between its components, assuming that each component is itself correct. That is, modular hierarchical systems provide a way to cope with complexity.

Integrated, Intelligent Design Environment

Colton and Dascanio [88] present the Integrated, Intelligent Design Environment (IIDE) system for mechanical design, consisting of (1) a comprehensive aid scheme, (2) a data and information controller, and (3) a number of evaluative applications. The controller is a software application that acts as liaison between the IDS front end and evaluative support applications. The controller is responsible for extracting information from the graphical environment and from the user's input. The information is then conveyed to the appropriate analysis software. Upon completion of the analysis, the controller passes the information to the appropriate recipient, which may be a message for display to the user or an object attribute in the data base. Another function of the controller is to propagate forces and torques through an assembly to determine if the parts and joints are strong enough. A recursive routine moves through an assembly propagating the applied forces and torques through the parts as mechanical analyses are performed. The controller's responsibility involves extracting data from the graphical environment and manipulating it by either passing it to the appropriate modules or storing it in the data base.

Computer-aided Software Engineering (CASE)

CASE is another popular integrated design system. The three fundamental areas in CASE are dictionary support, modeling support, and documentation support. CASE has developed into three main categories: Computer-Aided Structure Analysis/Programming (CASA/CAP), DP-driven CASE (DP CASE), and User Computer-Aided Strategic Planning (CASP)/CASE. In the last few years, CASE has been one of the most significant developments in the software industry [150]. The three categories provide strategic planning at all levels of management. The plans, which are in days rather than months and years, encourage rapid feedback through strategic planning methods.

Although IDS and CASE are powerful, they lack technical analysis capability for electrical circuits or for mechanical dynamics. We suggest applying ideas in CASE for linking different packages and the OODB design in IDS to enhance their capabilities.

1.3.6. Intelligent Manufacturing

System integration has become important in recent years. Because of the high degree of integration, system reliability has been significantly reduced. An intelligent manufacturing system (IMS) is an integration of an information database, a knowledgebase (an expert system) , and a graphical user interface (GUI) which calls for CAD/CAM, CAPP, and computer-aided quality control (CAQC), or other application tools stored in the system. Figure 1.4 illustrates the configuration: the GUI responds to the users' commands or requests, is able to let the system administrators modify the knowledge-based rules, and can also serve as a database management system (DBMS). The knowledgebase segment contains a combination of data structures (facts) and interactive procedures (rules). The facts are declarative knowledge and may contain data like workpiece (machine surface), machining operations, machining tools, tool holding devices, and tolerance information; the rules may be classified into the following groups: tool selection, jig and fixture selection, machining parameter selection, operation selection, and operation sequence. Much information that the knowledgebase needs is stored in the database, and therefore interactions occur between these two segments. It is important to mention, however, that an IMS is not a panacea

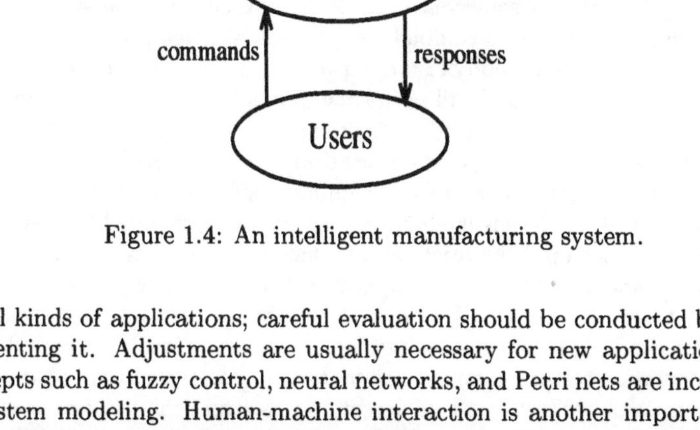

Figure 1.4: An intelligent manufacturing system.

for all kinds of applications; careful evaluation should be conducted before implementing it. Adjustments are usually necessary for new applications. New concepts such as fuzzy control, neural networks, and Petri nets are incorporated in system modeling. Human-machine interaction is another important topic; higher efficiency can be expected from better cognitive systems design.

1.3.7. Component and System Failure Modes

As mentioned by Dasgupta *et al.* [101], both at the early and later stages of the electronic component or system life-cycle, the failure modes are primarily mechanical or electrical rather than statistical phenomenon and thus should be modeled as mechanical failures. Failure modes for mechanical parts and electronic and electrical parts are summarized in Appendix B.

1.3.8. Probabilistic Design

For both mechanical and electrical components, the strength of these components should exceed the stress added to them in order to warrant a safe operation of the systems utilizing them (see Section 3.3 for illustration). A traditional way to design for reliability was to add safety factors to the tolerated stress levels. Rather than being deterministic, however, the strength and stress vary because of the factors that affect strength and stress variations. Kapur and Lamberson [212] consider the probabilistic nature of the design so that the component reliability can be expressed as a function of strength and stress distributions.

1.3.9. Software Design

In today's large systems, software LCC typically exceeds that of hardware. Compiling cost expenditures from all industries, in 1960, about 20% of the system's cost was spent on software. That percentage has risen to 80% in 1985 and 90% in 1996 [385]. For this reason, improving software design has a critical economic impact.

Top-down design defines only the necessary data and control and hides the details of the design at lower levels. It makes a module small enough so that it is within a programmer's intellectual span of control. One major shortcoming of this approach is that a design error will not be discovered until the end of the design stage. A second design method is called structured programming, which uses a single-entry and a single-exit control structure to provide a well-defined, clear, and simple approach to program design. The program structure is often complicated and sometimes has longer execution time since the GO TO statements are eliminated completely. The newly developed object-oriented concept is based on the method of modular design. Many small or modest-sized subprograms are designed and will perform independently for specific functions. These subprograms can be treated as "software-ICs" because they can be used by many programs. The design representation should be used so that a group of techniques are commonly used for designing a software system.

Once a software design is completed, test methods should be specified. Analysts have to set up the statistical inference requirements, software acceptance criteria, and the statistical sampling methods. Bukowski, Johnson, and Goble [54] present 10 problem detection steps: requirements review, design review, source checking tools, module testing, code review/walk-through, library system, integration testing, planned system testing, abusive system testing, and applications/customer testing. This detection procedure can be used as a check list so that problems can be discovered earlier to reduce expensive corrections.

A lot of software testing packages are available on the market. Most of them have functions to determine branch coverage, multiple condition coverage, loop coverage, relational coverage, routine coverage, call coverage, and race coverage. Take Hindsight/C as an example. It is an analysis and graphic toolset that helps unravel large, poorly documented programs, understand logic flow, find invisible segments, plan efficient test suites, trace global and static variable usage, and analyze quality, complexity, and performance times [187]. Hindsight/C uses the structure chart to highlight and isolate subtrees to the lowest level, overlay performance, complexity and test coverage data, edit code, open any diagram for a function, find global and static variable occurrences on the chart, and find global and static variable usage in code [188].

1.4. Reliability Standards

The military standards‡ prepared by the Department of Defense of the United States provide many testing methods for components as well as for systems The military standards for electronic and electrical components, mechanical parts, and for whole systems as well as the ones for software will be briefly introduced in this section.

1.4.1. Hardware

At present, US military standards can be treated as a complete reference set. Items are first classified into–from the simplest to the most complex –part, subassembly, assembly, unit, group, set, subsystem, and system in these standards. The definitions of these item levels, item exchangeability, and models along with the related terms are given in MIL-STD-280A. In addition, MIL-STD-105E (now QSTAG-105) provides a standardized interpretation of quality assurance terms and definitions to be applied throughout the determination of product quality. Some important terms include: acceptability criteria, acceptable quality level (AQL), acceptance, acceptance number, attribute, and average outgoing quality level (AOQL). Standard general requirements for electronic equipment are included in MIL-HDBK-454, which describes the requirements of electronic components such as safety design criteria, inert materials, soldering, bearings, interchangeability, and electrical overload protection.

Component Tests

MIL-STD-202F introduces test methods for electronic and electrical component parts. It establishes uniform methods for testing electronic and electrical component parts, including basic environmental tests to determine resistance to the deleterious effects of natural elements and conditions surrounding military operations and to physical and electrical tests. The settings, criteria, formula,

‡The US military standards may be easily acquired at minimal fee from the Naval Publications and Forms Center, 5801 Tabor Avenue, Philadelphia, PA 19120 (phone: 215-697-2000).

figures, tables and instruments needed are also described. Test methods are categorized into three groups:

1. Environmental tests
   include salt spray (corrosion), temperature cycling, humidity, immersion, barometric pressure, moisture resistance, thermal shock, explosion, sand/dust, flammability, and seal. MIL-STD-810E sets up the guidelines and complete descriptions for the environmental tests.

2. Physical-characteristics tests
   include vibration, shock, random drop, solderability, radiographic inspection, terminal strength, acceleration, and heat.

3. Electrical-characteristics tests
   include dielectric withstanding voltage, insulation and DC resistance, capacitance, contact resistance, current, voltage, and intermediate current switching.

MIL-STD-750D is devoted to the test methods for semiconductor devices. Similar to MIL-STD-202F, there are 12 major test methods: environmental tests, mechanical characteristics tests, circuit performance and thermal resistance measurements, low-frequency tests, high-frequency tests, high-reliability space application tests, electric characteristics for transistors, electric characteristics for FETs, electric characteristics for diodes, electric characteristics for microwave diodes, electric characteristics for thyristors, and electric characteristics for tunnel diodes. MIL-STD-790F is designed particularly for electronic and fiber optic components. More complete and detailed test methods are compiled in MIL-STD-883E.

System Tests

MIL-STD-785B provides general requirements and specific tasks for reliability programs during the development, production, and initial deployment of systems and equipment. It describes which items should be included in a reliability program and requires

- a description of how the reliability program will be conducted to meet the requirements of the statement of work (SOW),
- a detailed description of how each specified reliability accounting and engineering design task(s) will be complied with or performed,
- the procedures (whatever existing procedures are available) to evaluate the status and control each task to identify the organizational unit with the authority and responsibility for executing each task,
- a description of the interrelationships between reliability tasks and activities and a description of how reliability tasks will interface with other system-oriented tasks, (The description will specifically include the procedures to be employed that assure that the applicable reliability data

derived from, and traceable to, the reliability tasks specified are integrated into the logistic support analysis program (LSAP) and reported on appropriate logistic support analysis records (LSAR).)

- a schedule with estimated start and completion points for each reliability program activity or task,
- the identification of known reliability problems to be solved, an assessment of the impact of these problems on meeting specified requirements, and the proposed solutions or the proposed plan to solve these problems,
- the procedures or methods (if procedures do not exist) for recording the status of actions to resolve problems,
- the designation of reliability milestones (includes design and test),
- the method by which the reliability requirements are disseminated to designers and associated personnel,
- identification of key personnel for managing the reliability program,
- description of the management structure, including interrelationship between the line, service, staff, and policy organization,
- statement of what source of reliability design guidelines or reliability design review checklist will be used,
- description of how reliability contributes to the total design and of the level of authorization and constraints on this engineering discipline, and
- identification of inputs that the contractor needs from operation and support experience with a predecessor item or items. Inputs should include measured basic reliability and mission reliability values, measured environmental stresses, typical failure modes, and critical failure modes.

MIL-STD-785B also describes the issues that should be reviewed at specific times to assure that the reliability program is proceeding in accordance with the contractual milestones and that the quantitative reliability requirements for the system, subsystem, equipment, or component will be achieved. The program should be separated into the following reviews: the preliminary design review (PDR), the critical design review (CDR), the reliability program review, the test readiness review, and the production readiness review. Several analyses are included in MIL-STD-785B: failure-mode effect and criticality analysis (FMECA), sneak circuit analysis (SCA), and electronic parts/circuits tolerance analysis.

In defining critical reliability criteria, we consider the following scenarios:

- A failure of the item critically affects system safety, causes the system to become unavailable or unable to achieve mission objectives, or causes extensive or expensive maintenance and repair. For example, LSI and VLSI systems are usually treated as critical parts as defined in MIL-HDBK-454.

- A failure of the item prevents obtaining data to evaluate system safety, availability, mission success, or need for maintenance or repair.
- The item has stringent performance requirement(s) in its intended application relative to state-of-the-art techniques for the item.
- The sole failure of the item causes system failure.
- The item is stressed in excess of specified derating criteria.
- The item has a known operating life, shelf life, or environmental exposure, such as vibrations, thermal susceptibility, or propellant damage; or a limitation which warrants controlled surveillance under specified conditions.
- The item is known to require special handling, transportation, storage, or test precautions.
- The item is difficult to procure or manufacture relative to the state-of-the-art techniques.
- The item has exhibited an unsatisfactory operating history.
- The item does not have sufficient history of its own, or similarity to other items having demonstrated high reliability, to provide confidence in its reliability.
- The item's past history, nature, function, or processing has a deficiency warranting total traceability.
- The item is used in large quantities (typically, at least 10% of the configured items' electronic parts count).

MIL-STD-1629A provides some industry guidelines for performing criticality analysis. The reliability qualification test (RQT) and production reliability acceptance test (PRAT) program are in tasks 303 and 304 of MIL-STD-1629A, respectively. Another standard which provides the system level view point is MIL-HDBK-781, which indicates that a useful test could be the probability ratio sequential test (PRST). As described in the examples, one advantage of applying PRST is the requirement of a smaller sample size; this merit will become significant when the cost of a test is high.

As technology progresses and the manufacturing process becomes more complicated, a successful design is expected to be "robust" so that the system will remain functional if some components are damaged. Since the importance of risk analysis is growing, risk-based design is worthy of consideration. MIL-STD-882C defines the range of failure rates that expresses the range of uncertainty, which has to be considered when designing a system. The system safety program requirements listed in MIL-STD-882C have to be combined with opinions from field reports and engineers as risk analyst relies more on engineering judgment than statistics; previous experience makes precise forecasting possible and, thus, robust products can be designed to suit different kinds of needs.

1.4.2. Software

In the procurement of individual items of software, standards can help ensure that they are fit for the purposes that are intended. Standards can also ensure the successful inter-operation of independent pieces of equipment and, within markets, a free market for third-party suppliers [172].

Requirements of a software quality assurance program can be found in DOD-STD-2168. Software support environment requirements are in MIL-HDBK-782. Some standards produced by the British Standards Institute (BSI) [§] provide similar techniques and procedures for the collection of data for measurement prediction of software reliability as the US military standards (BS 7165 (1991), BS 5750 (Part 13, 1991), and DD 198 (1991)) .

The Institute of Electrical and Electronics Engineers (IEEE) also provides the following standards for software testing and quality assurance [¶]:

610.12-1990	Standards for Software Quality Assurance Plans
828-1983	Standards for Software Configuration Plans
829-1983	Standards for Software Test Documentation
830-1984	Guide for Software Requirements Specifications
982.1-1988	Standard Dictionary of Measures of Reliable Software
982.2-1988	Guide for the use of Standard Dictionary of Measures to Produce Reliable Software
983-1986	Guide for Software Quality Assurance Plans
990-1987	Recommended Practice for Ada as a Program Design Language
1002-1987	Standard Taxonomy for Software Engineering Standards
1008-1987	Standard for Software Unit Testing
1012-1986	Standard for Software Verification and Validation
1016-1987	Recommended Practice for Software Design Descriptions
1028-1988	Standard for Software Reviews and Audits
1042-1987	Guide to Software Configuration Management
1058.1-1987	Standard for Software Project Management Plans
1063-1987	Standard for Software User Documentation
1074-1991	Standard for Developing Software Life-Cycle Processes

1.4.3. Reminders

Asher [11] points out some shortcomings for MIL-HDBK-781. First, the exponential distribution and the Poisson process are not interchangeable. Second, to apply the standard, one should take into account the type of system the standard is applied to a repairable or a nonrepairable system.

---

§The BSI standards can be acquired from the British Standards Institute, Marketing Department, Linford Wood, Milton Keynes MK14 6LE, UK.

¶Contact IEEE Service Center, P. O. Box 1331, Piscataway, NJ 08855-1331 for information.

Similarly, though MIL-STD-756B is an excellent reference for reliability modeling and prediction, one has to carefully investigate the assumptions made for the models and prediction techniques before applying them.

Luthra [259] also criticizes MIL-HDBK-217F, a widely-used handbook. He notes that no prediction model, whether in reliability, finance, system analysis or in any other area, should be applied blindly. There are so many canned programs available on the market that some engineers just plug in numbers and provide the answers without using common-sense checking.

Some areas that need careful consideration in reliability modeling are thermal stress, cooling factors, card configuration, soldering method, and operation frequency. These factors are available in MIL-HDBK-217F. According to Luthra, they should be applied with consideration for an organization's individual experience which differs from company to company. MIL-HDBK-217F does not have a factor for this adjustment.

Luthra also warns users of the standard to be aware of the application of Parts Count. The failure rate predicted by many organizations for various systems is less than half that seen in the field. One reason for the above mismatch is that MIL-HDBK-217F models use a constant failure rate which does not apply to most systems for the initial 2 or 3 years. Likewise, mechanical parts can make up a good portion of a system, and there is no good data-source for model-source. Thus, a good, simple prediction method for mechanical parts is still lacking. A majority of engineers use snapshot methods for mechanical reliability prediction. This is another reason for the wide gap between the predicted and the observed reliability. Other criticism on MIL-HDBR-217F can be seen in O'Connor [303].

Agarwala [3] criticizes MIL-STD-1629A and indicates that the "Item Criticality Numbers" computations fail to include the contributions from the lower severity classifications of a failure mode.

The brief descriptions of some of the more commonly referenced reliability and maintainability standards are listed in Ireson *et al.* [199] and modified below:

Design	MIL-HDBK-251, MIL-HDBK-338, MIL-HDBK-454, MIL-STD-785B, MIL-STD-1670A, ISO9000
Electrostatic Discharge	MIL-HDBK-263, MIL-STD-1686
Environmental Stress Screening	MIL-HDBK-344A, MIL-STD-810E, MIL-STD-1670A, MIL-HDBK-2164
FMECA	MIL-STD-1629A, RADC-TR-83-72
Human Factors	MIL-STD-1472E, MIL-STD-46855
Logistics	MIL-HDBK-502
Maintainability	MIL-STD-470B, MIL-HDBK-472, MIL-STD-471, RADC-TR-83-29
Quality	ISO9000 (ANSI/ASQC Q90 Series)
Reliability Prediction	MIL-HDBK-217F, MIL-STD-756B, LC-78-2, NPRD-91,

	RADC-TR-73-248, RADC-TR-75-22, RADC-TR-83-91
General Reliability	MIL-HDBK-338, MIL-STD-1543B, MIL-HDBK-2155, RADC-TR-83-29
Reliability Growth	MIL-HDBK-189
Safety	MIL-STD-882C
Sampling	QSTAG-105 (MIL-STD-105E), MIL-STD-690C, MIL-PRF-38535D
Software	MIL-STD-498, DOD-STD-2168
Testing	MIL-HDBK-781, MIL-STD-785B, MIL-STD-790F, MIL-STD-810E, MIL-STD-883E, MIL-HDBK-2165

Note that assuming the required reliability should be at no significant cost or time penalties. Although many standards are introduced here, it is a trend that past approaches to qualifying critical ICs by applying the standards are now less appealing. Recently, critical IC users are depending more on commercial products.

Interest in ISO9000 certification increases immensely since 1990s. In addition to address safety-related issues, reliability techniques are applied to commercial products. Reliability techniques are now part of the ISO9000 certification as the total quality initiatives by manufacturing management teams.

1.5. Conclusions

The purpose of this chapter is to introduce the recent information on computer-based and statistical approaches to the engineering design of quality products and processes. Thus far, CAD/CAM approaches have been very effective in the design and manufacture of electro-mechanical devices, and statistical approaches have been effective in process modeling and optimization. Conceptual frameworks and the use of computer-based approaches and experimentation are also considered in this chapter.

Reliability and quality determine the marketability of manufactured products. In this concurrent engineering era, the key to successful design is to characterize and absorb the uncertainties into early design stages so that a sufficiently robust product design can be achieved in a flexible way to meet the design goals and satisfy the customers' requirements.

Quality is a static measure, but reliability is measured as a function of time. When a system becomes complex and has many small components, the system reliability decreases and is less predictable. Recently, many companies have begun to depend increasingly on TQM, which advocates building quality assurance philosophies and tools into all aspects of an organization's operation. The importance of Taguchi method is emphasized because the goal of TQM is to reduce the Taguchi's loss function. Moreover, the close relationship between reliability and Taguchi method is pointed out by Lee [242]: "Reliability is often a major driver of the Taguchi loss function through cost of unreliability and thus should be a main objective of design of experiment."

In summary, reliability is best handled at the design stage and as part of the design process. Software development is also regarded as a manufacturing process. Some rigorous ways for integrating reliability into the design process are suggested in the chapter.

2. INTEGRATING RELIABILITY INTO MICROELECTRONICS MANUFACTURING

The advent of microelectronics has resulted in the fabrication of one billion transistors on one silicon chip. The oxide thickness has decreased to around 40Å, while the channel length is now below 0.2μm. To attain small-size integrated circuits (ICs), a clean room environment of class 1 is required, which allows no more than 1 particle with 0.1μm diameter per cubic foot. It is now an important agenda to study reliability issues in fabricating microelectronics products and, consequently, the systems that employ high technology such as the new generation of microelectronics. Such an agenda should include:

1. the economic impact of employing the microelectronics fabricated by industry,
2. the progression toward miniaturization and higher reliability,
3. the relationship between reliability and yield, and
4. the correctness and complexity of new system designs, which include a very significant portion of software.

Initially, performance and manufacturing yield are the primary concerns of IC manufacturers. However, these concerns are compounded by the long infant mortality period and the high initial failure rate of ICs, even though their useful lives tend to be very long. Undoubtedly, the defects introduced by processing and manufacturing result in both the infant mortality and the reliability problems, which should be considered as a yield problem as well.

2.1. Microelectronics Manufacturing

2.1.1. Technology Trend

Beginning with the manufacture of discrete transistors in the late 1940s and early 1950s, the microelectronics industry has evolved into the world's largest manufacturing business [411]. Beginning with the historical breakthrough invention of the first IC by Jack Kilby in 1958, the first commercial monolithic IC came on the market in 1961, the metal oxide semiconductor (MOS) IC in 1962, and the complementary MOS (CMOS) IC in 1963. The path of continued advancement of ICs is marked by distinct periods of small-scale integration (SSI), medium-scale integration (MSI), large-scale integration (LSI), very large-scale integration (VLSI), and ultra large-scale integration (ULSI) [358]. Table 2.1 [190] shows the overall trend of IC technology and the number of transistors integrated in a chip of dynamic random access memory (DRAM). After year 2000, the IC industry will enter the super large-scale integration (SLSI) era with over 10^9 transistors for 4G (and over) DRAMs.

Table 2.1: The progressive trend of IC technology.

Integration level	Year	Number of transistors	DRAM integration
SSI	1950s	less than 10^2	
MSI	1960s	$10^2 \sim 10^3$	
LSI	1970s	$10^3 \sim 10^5$	4K, 16K, 64K
VLSI	1980s	$10^5 \sim 10^7$	256K, 1M, 4M
ULSI	1990s	$10^7 \sim 10^9$	16M, 64M, 256M
SLSI	2000s	over 10^9	1G, 4G and above

2.1.2. Semiconductor Devices

IC circuits are based on resistors, diodes, capacitors, and transistors. The three major types of materials making up circuits are metals, insulators, and semiconductors, and the basic difference among them is the magnitude of the energy gap [84]. Metallic materials form the conducting element, while the insulator is the dielectric which provides isolation between transistor elements as well as serves as part of transistor structure.

Some principal features of IC devices and technologies are discussed below, and the projected trend of devices is shown in Table 2.2.

Table 2.2: The projected trend of IC devices.

Device Type	1987	1992	1998
CMOS	39%	73%	82%
BiCMOS	0%	2%	6%
NMOS	24%	4%	less than 1%
PMOS	less than 1%	0%	0%
Bipolar (analog)	20%	14%	9%
TTL	12%	4%	1%
ECL	4%	2%	less than 1%
GaAs	less than 1%	less than 1%	1%

Bipolar Transistor

Most analog circuits of the early days were built using bipolar technologies. A bipolar transistor is made by connecting two PN junctions back-to-back. Its structure has a triple layer of P-type and N-type material. Based on the center layer, there exists an NPN and PNP bipolar transistor. For a bipolar transistor, the center layer is referred to as the base, the outer layer of the forward biased junction as the emitter, and the outer layer of the reverse biased junction as the collector. The bipolar transistor usually has some advantages over CMOS [6]: switching speed, current drive, noise performance, analog capability, and I/O speed.

Field Effect Transistor (FET)

FETs are primarily made by MOS technologies and differ from bipolar transistors since only one charge carrier is active in a single device. The FET employing electrons is referred to as N-channel MOS (NMOS), while the FET employing holes is called P-channel MOS (PMOS). The family of FETs include the junction FET (JFET) and the metal oxide semiconductor FET (MOSFET). There are two basic FET operational modes: the depletion mode and the enhancement mode. In the depletion mode, both the JFET and the MOSFET can be operated, whereas in the enhancement mode, only the MOSFET can be used. MOS is sometimes used as an abbreviation for MOSFET.

CMOS

NMOS and PMOS are used together to form CMOS logic circuits. Due to low power, good noise margin, wider temperature, voltage operation range, and high packing density, CMOS has been frequently used in many VLSI applications. The CMOS families are the original CMOS, the high speed CMOS, and the advanced CMOS [53] and the four basic CMOS technologies are the N-well process, the P-well process, the twin-tub process, and silicon on insulator (SOI) technology.

Bipolar Combined with CMOS (BiCMOS)

BiCMOS combines advanced Schottky bipolar technology and advanced CMOS technology in a single IC in order to get the benefits of both bipolar and CMOS transistors. In general, the input structures and internal logic of BiCMOS come from CMOS and the output structures from bipolar. In some cases, however, BiCMOS devices have both bipolar input and output structures. By combining these two technologies, BiCMOS obtains high switching and I/O speed, high driving current, and better analog capability which are key criteria for implementing high performance systems. One drawback of BiCMOS is relatively high cost due to increased complexity. There exist various BiCMOS logic devices: application specific IC (ASIC), random access memory (RAM), programmable logic devices (PLD), and discrete logic devices. The BiCMOS technology is commercialized by TI in the 1980's and it is especially useful for the processes larger than $0.5\mu m$. However, as the CMOS processes are rapidly shrinked to be below $0.2\mu m$, the high-speed merit we gain from the BiCMOS technology has become insignificant. Therefore, under the cost consideration, the BiCMOS processes are seldom used for most ICs manufactured by the design rules below $0.5\mu m$. However, the BiCMOS technology can still be seen from some high-end products, like the microprocessors.

Transistor Transistor Logic (TTL)

TTLs are the most commercial digital bipolar ICs using saturating logic. They are fabricated with multiple emitter transistors and provide the fastest switching speed. There are some TTL logic families [53]: the original TTL, the low-power TTL, the high-speed TTL, the Schottky TTL, the low-power Schottky TTL, the fast TTL, the advanced low-power Schottky TTL, and the advanced Schottky TTL. For large systems operating at high speed, only the fast TTL and the advanced Schottky TTL can be used.

Emitter Coupled Logic (ECL)

ECL is one of the current mode logic (CML) families that uses nonsaturating bipolar transistors, unlike CMOS and TTL. Its basic concept is a differential pair followed by two emitters, one for direct output and the other for inverted output. Since devices fabricated with the old TTL and CMOS logic technologies are too slow, ECL technology was used as an alternative for high performance applications. However, because of the power dissipation and additional terminating resistors required, advanced CMOS and BiCMOS technologies displace ECL. The basic digital IC families include ECL, TTL, and MOS.

GaAs Transistor

High speed and low power are two distinct advantages of GaAs transistors over silicon transistors. Due to these advantages, GaAs FETs are widely used in digital fiber optic communication systems. Three basic GaAs device structures are metal-semiconductor FET (MESFET), heterojunction FET (HFET), and heterojunction bipolar transistor (HBT). Since MESFET and HFET have similar current-voltage characteristics, some design techniques of MESFET can be applied to HFET. Except for the emitter, HBT is similar to the bipolar transistor.

Application Specific Integrated Circuit (ASIC)

ASIC is a unique custom chip which is usually designed by the customer. Thus, ASICs are different from standard ICs, such as those used in memories and microprocessors. The broad categories of ASICs are the fully customed ASIC, the semicustomed ASIC, the field-programmable logic devices (FPLD), and linear array [117]. The semicustomed ASICs are basically grouped into cell-based custom chips and array-based custom chips, and FPLDs are sometimes considered standard devices.

2.1.3. Manufacturing Processes

The manufacturing process for microcircuits is completely integrated, and it starts with the growth of the crystal and proceeds to packaging. Yield and reliability are the driving forces for the success of any manufacturing scheme for a new technology. Yield must be maximized for each processing step while maintaining reliability in excess of 10^7 hrs [84]. The principal IC manufacturing processes are summarized in Figure 2.1. The front-end fabrication processing sequences proceed from the crystal growth up to the final metallization and the back-end fabrication from the wafer probe to the final test. The front-end fabrication processes are the most complex. Typical manufacturing processes start with the conception of new circuits and end with the final electrical tests.

Crystal Growth

Crystal growth processes include crystal pulling , crystal slicing, crystal shaping, and crystal polishing operations. Czochralski growth process is a prevalent process used in growing crystals. The crystal pulling operations pull a single silicon ingot from a crucible. Since the electrical characteristics of silicon are very sensitive to impurities, a crucible is a critical component in the crystal growth system. The two major stages of the crystal shaping operation are grinding and slicing. In the grinding steps, the precise diameter of ingot and primary and secondary flats are formed. These flats are essential in identifying and orienting IC devices. During the slicing steps, a single crystal ingot is sliced into several dimensioned wafers and four important wafer parameters–surface orientation, thickness, taper, and bow–are determined [412]. After contaminated and damaged regions are removed from the wafer, crystal polishing operations, the final operations of crystal growth, start. Smooth and specular wafers are obtained from polishing operations.

Epitaxy

Epitaxy is a process that grows a thin single crystalline film on the surface of a silicon wafer. Thus, this film (2-20μm) has the same crystal orientation as the substrate [190]. Before epitaxy growth the wafer is mechanical or chemical cleaned. The epitaxial growth operations are typically based on the vapor phase epitaxy (VPE) or molecular beam epitaxy (MBE) [84, 412]. The film properties depend on the device requirement [190]. Although the epitaxy was originally developed for the low-resistivity burier layer of the bipolar transistor, it becomes popular in the MOS circuits now.

Oxidation

If a silicon wafer reacts with oxygen or water vapor, a silicon dioxide film is formed on the surface:

$$\begin{aligned} \mathrm{Si} + \mathrm{O_2} &\longrightarrow \mathrm{SiO_2} \\ \mathrm{Si} + \mathrm{2H_2O} &\longrightarrow \mathrm{SiO_2} + \mathrm{2H_2}. \end{aligned}$$

Oxygen source, thermal energy, and silicon substrate are three required elements for oxidation. The silicon dioxide film serves an important role in the manufacturing processes. Its principal functions are surface passivation, diffusion barrier, and surface dielectric [190, 412]. The techniques developed for oxidation of wafers are thermal oxidation, wet anodization, chemical vapor deposition (CVD), and plasma oxidation [412]. Among them, thermal oxidation

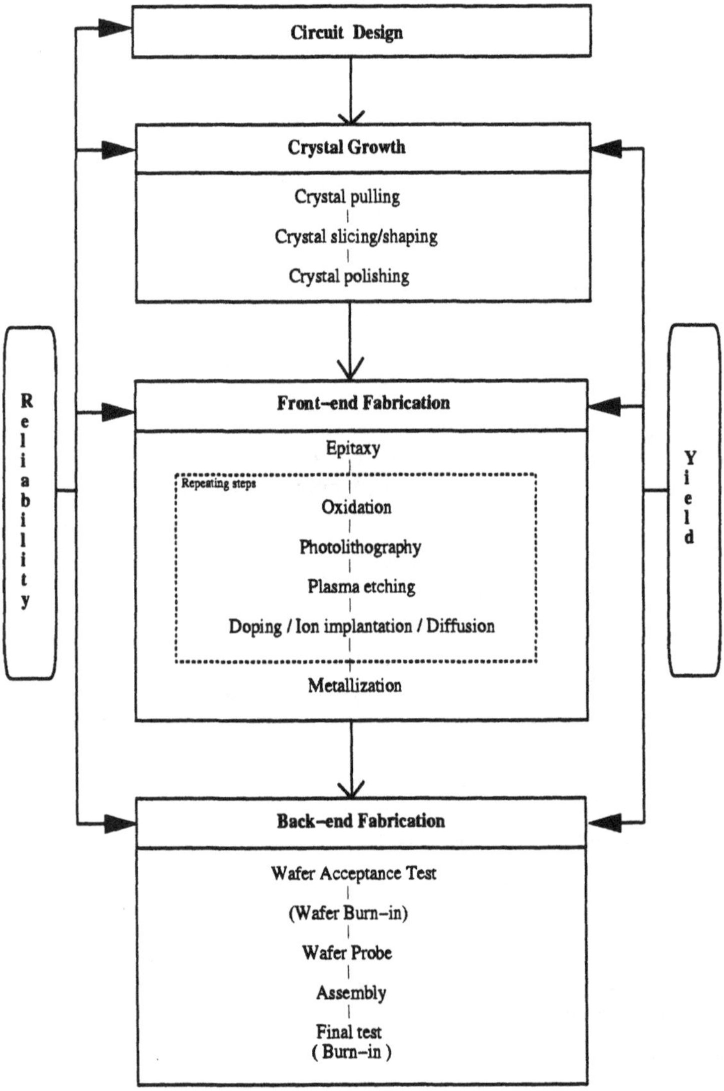

Figure 2.1: Reliability's influence on the IC manufacturing process and yield.

is the most popular technique. Since many wafers are usually oxidized in one batch, cost factors must be considered in the selection of oxidation technique.

Photolithography

Photolithography is a process which transfers the pattern on a photomask (or mask) to the photoresist layer on the wafer. Thus, its goal is to reproduce the circuit patterns accurately. A complete IC circuit is composed of many masking levels. The light source of the photolithography is the ultraviolet (UV) light. The photolithography process consists of following steps: surface preparation, photoresist application, soft bake, mask alignment and exposure, resist development, hard bake, inspection, etch, photoresist removal, and final inspection [84]. These steps are repeated many times during the manufacturing process.

Etching

Etching is a process of removing unmasked portions of materials and transferring the circuit patterns into the top layer of a wafer. Plasma assisted etching (or dry etching) is the most prevalent and suitable technique for VLSI processing [412]. During the plasma assisted etching process, etch gases are excited by the plasma generator and form volatile products with the exposed area on a wafer. The most common plasma assisted etching reactors are plasma etcher, reactive ion etcher (RIE), and down stream etcher.

Doping

Doping is a process of adding dopants (impurities) to the silicon substrate. The electrical property of a semiconductor device is greatly influenced by the doping process. The type and amount of impurities and the depth of doping depend on the requirement of the device. There are two primary doping techniques: diffusion and ion implantation. Diffusion was first developed to attain the designed impurity concentration, whose bulk, grain boundary, and surface diffusion types greatly affect the performance of devices [84]. Most diffusions are composed of two steps: predeposition and drive-in. The predeposition step determines the amount of impurity, and the drive-in step affects the depth profile [190]. Ion implantation was developed to overcome some limitations of the diffusion process. The concentration and profile of an impurity can be accurately controlled by an ion implantation process which projects energized ions to the substrate.

Metallization

Metallization is a process of connecting the fabricated individual component with conductors. The wiring pattern is formed by the deposition and etching of thin metal layer on the wafer [190]. The commonly used metals are aluminum (Al), copper (Cu), and tungsten (W). Sputtering and evaporation are the most common methods for metal deposition. The CVD method is sometimes used to deposit a metal such as tungsten.. In sum, oxidation, epitaxy,

CVD, evaporation, and sputtering are techniques used for thin film deposition on the wafer.

Thin Film Deposition

In addition to metal, dielectrics, such as silicon oxide and silicon nitride, are important films for ICs. They are usually deposited by the CVD method. Plasma enhanced CVD (PECVD) is commonly applied to deposit films at a low temperature such as $< 400^oC$.

Wafer Acceptance Test (WAT)

WAT is an electrical test (E-test) and is done at the wafer level following all of the processes mentioned above and right before the wafer probe (WP). A probe station with probe-pins is connected with a computer system, which is used for operation control and for data analysis. Test engineers design test programs and store them in this computer system. The test samples are the test keys on the wafer. Two kinds of test keys are usually used: the grid line test element group (G/L TEG) and the chip test element group (CHIP TEG). Both TEGs contain test patterns to verify if the important parameters are within the acceptable ranges. The acceptable ranges are sometimes called specifications, which are defined by the R&D and the Technology Development (TD) Department. The CHIP TEG is only made and measured before product release and its area is usually much larger than that of the G/L TEG. Once manufacturing processes become stable and yields are above a pre-specified level, CHIP TEGs are removed to increase the number of chips on a wafer in order to save costs. On the contrary, the G/L TEGs will remain on the wafers throughout a product life cycle and are used as important monitor indices for the wafer quality. The usual test items at WAT are the parameters for transistors, such as T_{ox}, I_{ds}, I_{sub}, V_{th}, G_m, K, L_{eff}, and sheet and contact resistance for certain layers and contacts. Lots with unacceptable WAT data are scrapped to save test and assembly costs and to ensure quality.

The G/L TEG resides on the scribe lines, and they will be destroyed after wafer dicing. From time to time, wafers will be sampled from the production lines and some G/L TEGs will be measured and tested at wafer and package level to monitor the quality of the lots. The lifetimes of the devices, such as NMOS and PMOS transistors and capacitors can be estimated through this kind of test, which is sometimes referred to as the wafer level reliability as shown in Section 2.2.2.

Wafer Probe (WP)

After WAT, wafers are ready for WP, which is also called chip probing (CP) or wafer sorting. Different from WAT, WP tests IC chips instead of the test keys. Similar to WAT, WP tests the ICs by making contacts to the pads. Test programs are also designed and stored in the tester's computer. Basically, two types of test programs are used: direct current (DC) and alternating current (AC) test programs. The WP test time is relatively short [190] (compared with the final tests in the later section); usually less than a minute (depending on the product). DC tests mainly weed out the dice with un-acceptable leakage and driving voltage. AC tests are also called functional tests and contain various test patterns, which are used to verify if the inter-relations of the internal circuits are acceptable. Since the elimination of bad dice at this phase can reduce the assembly, scrap, and material costs, the probe test is very important. Some designs provide a chance to fix the bad dice. This is done by a process called laser repair (L/R), which applies a laser beam to blow fuses on the chip to adjust the internal voltage and current to make the parameters of the dice fall within specifications. The repairable chips will be marked at WP and go to L/R. These repairable chips after L/R will go to the second WP test, which is called the WP2 (or CP2) test, to verify that they are successfully repaired. WP2 can be skipped if the repair yield is above a certain level. For certain ICs, downgraded dice can be sorted with WP tests.

Although full E-tests are not performed at WP, bad dies on the wafer can still be identified and marked. Ink-marking was used in the past and presently most IC makers introduce computer systems to mark the bad dice with different codes (according to the failure modes) to avoid possible reliability deterioration resulting from the marking ink. A wafer map (in paper or in computer files) for each wafer will be passed to the assembly line or to the subcontractors to distinguish good from bad dice.

The WP test equipment is complicated and expensive. Hence, it has become a never-ending endeavor for IC makers to reduce the test time to save costs and to shorten the lead time.

More and more ICs are designed to have wafer burn-in (WBI) testability. WBI is achieved by applying a high bias via a pad before WP to screen the weaker chips in a few seconds. For example, for a 5-V IC, the WBI test conditions may be 8.0V for three seconds. As to the known good die (KGD), manufacturers are required to make full parametric and functional tests with a WP test. The KGD is especially important for the multi-chip module (MCM) makers. Thus, the demand for more comprehensive, effective, efficient, and complete WP tests is increasing as more emphasis is being put on the WP than on the final test [190].

Dicing and Assembly

Dicing is a process that separates a wafer into individual chips. It is consisted of scribing and breaking steps. Before dicing, the wafers are attached on tapes for back-grinding to make wafers thinner. Diamond sawing is the most widely used scribing tool although laser beam scribing is another frequently used [190, 412]. After dicing processes, chip sorting process starts. Bad chips marked in the WP are picked up.

Assembly is a process that puts dice onto lead-frame (L/F) or die pads and seals the ICs as final products. The assembly process includes: mounting, interconnecting (die bond and wire bond), encapsulation (molding), package sealing, molding (with post mold curing), marking, dam-bar cutting, deflash, solder plating, trimming, and forming. Generally, two types of packages are used: hermetic (ceramic or metal) and nonhermetic (plastic or molded). Since plastic packages have many advantages (low cost is the major factor) compared with ceramic packages, they account for more than 90% of the worldwide package volume and will still be popular in the near future. In extreme operation conditions and military, hermetic packages are frequently used. For package mounting, through-hole and surface-mount are widely used. For the interconnecting, tape-automated bonding (TAB), flip chip, and wire bonding are prevalent techniques [190]. The MCM packaging is used to connect different bare dice in a module.

Final Test (FT) and Burn-in

The wafers after WP and assembly will be ready for FT. Usually, the test items in FT are similar but more complicated than the ones in WP. There exist several reasons for this. First, because the samples at FT are the potential final products to sell to customers, more strict test conditions should be applied to ensure that these ICs meet the product specifications. Second, some parameters may shift after the assembly processes and they must be verified by more complex test patterns. Third, the FT may contain more than one stage each at different test temperatures or with different test items. They are called final test 1 (FT1), final test 2 (FT2), final test 3 (FT3), and others. The number of FT stages depends on the product and the production maturity. Even though certain test items are the same in some FT test stages, the test parameters and the passing criteria may be different. Many IC makers arrange burn-in between two FT stages. The FT stage before and after burn-in is sometimes called the pre- and post- burn-in test; these two tests provide important information on the burn-in failure rate. Special test items must be used to detect the burn-in failures. Finally, the first FT stage (i.e., FT1) is usually used to monitor the assembly quality and yield (for example, the open and short failures can be easily detected by the FT1 program).

There are three kinds of E-tests in FT: functional tests, parametric static tests, and parametric dynamic tests. The purpose of functional tests is to verify the designed functions of device under tests (DUTs). Parametric static tests use steady-state voltages and direct currents. Parametric dynamic tests use voltage pulses or alternating current for the verification of more realistic circuit operations. FTs are to identify defects, process changes, and conformity to the .degree of specifications.

Quality Control (QC) Tests

QC tests are usually done on a sampling basis. The sample size depends on the specified AQL level. For example, if the 0.065% AQL is chosen, 200 samples are tested from a production lot. QC tests contain many inspections other than the E-tests. The visual inspection is one important component in the QC tests. The dimension, the marking visibility and durability, and the lead conditions of the ICs are examined either automatically or manually. The E-tests in QC tests are the last gate before products are shipped to customers. The whole lot will be re-tested by QC programs if any failure is found from the samples. Usually, the re-test rate should be controlled below a certain level to reduce test cost.

The procedure to introduce and to quantify a new product is described in the following example.

□ Example 2.1

A new IC is designed by the Research and Development (R&D) group after validating the following issues: customer's and market requirements, design rules, manufacturability, system compatibility, quality, and reliability. Different from quality, reliability is the probability of well-functioning after stress at a certain point of time. A simulation is run to verify the functionality for each part of the IC. The simulation system should be able to link the circuits, the parameter-database of the devices, the fab conditions, and the product specifications so that the reliability can be simulated after releasing this product to field use. The major items to simulate are ESD, latch-up, electromigration (EM), hot carrier injection (HCI), time-dependent dielectric breakdown (TDDB), and the accelerated soft error rate (ASER).

A lot containing 25 wafers is issued for a trial run. Several splits with different recipes may be designed in this pilot lot. The cycle time of the fab production depends heavily on the number of masks used for this product. That is, the lower the number of masks, the shorter the cycle time and the lower the cost. After WAT, WP, and L/R, some wafers are kept for wafer level verifications, such as the ASER test. The TEGs, CHIP TEG and G/L TEG, are measured at this point to see if the TEG data correlate with the WP results. Some TEGs are sent for side-brazing, and they will be used to derive lifetimes for EM, HCI, and TDDB. The WAT and the WP programs are further validated; certain timing parameters may need fine-tuning. The wafers not selected for wafer level measurement will be sent for assembly. The ICs are used for pre-qualification (Pre-Qual), which is mainly for verifying the early failure rate (EFR), the low temperature operating life (LTOL) quality, and the package quality from the environmental test results. Modifications on the assembly processes, technology, and material may be required if failures are found after the environmental tests under certain limits; for example, 96 hrs for HAST, 168 hrs pressure cooker test (PCT), and 300 cycles for temperature cycle (T/C). Detailed descriptions on the environmental tests, such as HAST, PCT, T/C are in Chapter 5.

More than one Pre-Qual is possible if many changes are made. The designs and the production flow are reviewed after each Pre-Qual, and several lots may be issued to find the best production parameters. It is expensive to conduct a qualification, and thus it is better to check every item in detail before each qualification. Usually, for any qualification, Product coordinates among Reliability Engineering (RE), the Integration, R&D, and Quality Assurance (QA) Departments.

Finally, the Process Review Board (PRB) will issue a process-fix notice. The production qualification will not commence unless the process is fixed. It may take as long as 6 months to qualify a product since most reliability tests (e.g., HAST, PCT, THB, T/C, T/S, HTS, HTOL, LTOL) are performed during this period. The samples for every reliability test are verified by E-tests at specified temperatures. For most IC makers, if no failure is found after, say, 500-hr reliability tests, the present product, called the beta version, may be conditionally released to selected customers. If any problem occurs after the conditional release, the products may be called back from the selected customers. Since this may incur high costs, it is a rare case for a matured IC manufacturer. Full release and unlimited production begin if the failure rates are below certain pre-specified levels for the die qualification tests (e.g., HTS, HTOL, and LTOL) and no failure is found at certain test points for the package-related tests (e.g., HAST, PCT, T/C, THB, and T/S). It is customary to guarantee a long-term (10-year field application or 1,000-hr HTOL test) failure rate to be less than 100 FITs (failure in time; please refer to Section 2.2). MIL-STD-883E (military standard), JEDEC (Joint Electron Device Engineering Council Standards) #22, EIAJ (Electronic Industries Association of Japan) ED-4701 and IC-121, IEC (International Electrotechnical Commission) Pub. 68, and JIS (Japan Industrial Commission) C7022 are informative references for IC reliability tests. □

After a qualified new product is released to mass production, continuous and routine process monitoring is still required to ensure that the production is under control. In situ and wafer level measurements and tests can quickly respond to process shift. The level of yield is usually used as a primary index of the chip quality. However, to make precise predictions, efforts must be made to correlate yield and reliability. Example 2.2 describes this concept.

□ **Example 2.2**

Continuous wafer level tests/monitors are able to ensure good correlation between WP and reliability test results. The usual wafer level tests/monitors include EM, TDDB and charge to breakdown (QBD) on capacitors and gate oxides, HCI, metal-line stability, film step coverage, Vt (threshold voltage) stability, and gate oxide integrity (GOI). It is reported that some microelectronic companies have made great advances in reliability prediction based on the WAT results. However, this is difficult when chips are more complicated. That is, the correlation of WP vs. reliability and of WAT vs. reliability for the 64Mb DRAM is much harder to obtain than for the 16Mb DRAM. Similarly, it will not be easy to correlate yield to reliability for BiCMOS and MCM chips.

The yield model should be chosen for a certain fab and process after we believe the process is mature and stable since the model is strongly fab and process dependent. The process must be fixed if the yield model is to be used for reliability prediction. After a model is selected, data has to be gathered to verify the correlation and to estimate the parameters specified in the model. It is a time-consuming process because the chosen model may not work well the first time. Several experiments are needed to see if the parameters shift. Even though a model and the associated parameters are fixed, they still need modification and adjustment if they are to be applied to similar fabs and processes. One method we suggest is to use the nonparametric Bayesian models in Chapter 10.

Today, the life-in-market for a semiconductor product is short (usually no more than 15 months). The model fitting and parameter estimations cannot be commenced unless the process is fixed. Since it takes at least a quarter to fix a process and about 2 quarters or longer to derive a suitable model and parameters, it will take less than 2 quarters for an IC with a good correlation between yield and reliability to be produced if everything is done properly and effectively. That is, after great effort and high cost, the longest duration that a reliable model can benefit the company is 2 quarters. Is this worthwhile? Unless the knowledge gained can be absorbed into the next generation of models, it is doubtful whether one should initiate the development of such a model. One good solution is to apply the Bayesian analysis introduced in Chapters 6 and 10. □

2.2. New Techniques for Reliability Improvement

From the manufacturing standpoint, process technologies for the deep-submicron devices ($< 0.18\mu$m) approach the physical limits. It is difficult to achieve high performance, high packing density, and high reliability [414]. This manufacturing process requires a high initial investment and extremely high operation cost. Thus, cost reduction by developing new techniques or approaches becomes urgent. From a reliability point of view, accelerated life test and end-of-line failure analysis (FA) become less reliable as the chip shrinks and the device structure is very complicated [94]. The simple FA method of sampling the output of a manufacturing line must be shifted to new methods in order to better understand and control the input variables at each point in the manufacturing process [414].

The requirement of new techniques leads to the development of built-in reliability (BIR), wafer level reliability (WLR), and qualified manufacturing line (QML) approaches [84, 194, 380]. In the WLR techniques, both processes and designs affect reliability. Since the traditional reliability approaches may not support enough test time or test parts to resolve failure rates as low as 10 FITs (1 FIT= 1 failure per 10^9 device-hrs), considered the goal for failure rate, the best approach for improving reliability and yield of any manufacturing process must be proactive rather than reactive. Based on the knowledge that anomalous material is generally produced by interaction effects between different process variables, a proactive reliability control obtains the reduction of the process variation and the elimination of some failures that might occur in the future [427]. Table 2.3 [379] shows some important shifts of the reliability paradigm from the traditional to the new techniques.

□ Example 2.3

A company defines the EFR to be the average failure rate within the first year (in FIT) of product release Assume some lots are verified to have quality

Table 2.3: Shift of the reliability paradigm.

Item	Traditional Techniques	New Techniques
Approach	Reactive	Proactive
Test Methodology	Output	Input
Failure Mode	Separated	Integrated
Fab Emphasis	Quality	Reliability
Testing	Test & Predict	Correlation & Input Control

problems after the routine reliability tests (RRT; please see Example 5.6 for details). From failure analyses, fab condition records, and lot history, huge particles are generated inside a thin-film deposition chamber. The number of ICs affected from these lots is about 460,000. One question a Quality Assurance (QA) engineer likes to know is "What is the expected number of returns (rejects) from customers in one year if the estimated EFR is 250 FITs?" Customer returns may greatly damage the company's reputation and, hence, they are handled with the highest priority by IC makers. Provided the information above, we can find the number of failed ICs in a year below:

$$460,000 \times 8760 \times 250\text{FIT} = 1,007.4.$$

This may contribute to 28 parts per million (PPM) of the customer returns if the fab produces 36,000,000 ICs that year. □

2.2.1. BIR Approach

To minimize reliability testing and to achieve target failure rates, reliability structures and high manufacturing yield must be designed into the manufactured products [84]. Hu [194] defines BIR as a methodology or philosophy for manufacturing highly reliable ICs, not by measuring the outputs at the end of production, but by controlling input variables that impact product reliability. The BIR approach thus achieves the reliability goal through the elimination of all defects or weaknesses from the design phase of the product. Although this approach requires high initial cost compared with reliability improvement through enhancing reliability screen tests, reliable products will result in low overall costs. Generally, BIR approaches are effective only beyond the crossover point, since product costs increase due to the large testing costs at that point. The basic ideas of BIR are not new. The systematic use of this approach and the recognition of its benefits have only recently been reported, however. Some useful tools for BIR are statistical process control (SPC), WLR, intelligent burn-in, in-line testing, and circuit reliability simulation [194].

Any process excursions outside the normal variation are regarded as reliability problems; these may be improved through reducing process variation or defects and through reliability screen tests. Elimination of reliability problems from ICs implies that all failure mechanisms are identified and understood and have been eliminated in the subsequent design iterations. The manufacturing trend of the 1990s is toward understanding failure mechanisms of products in order to eliminate them through process improvements. The focus of contemporary manufacturing industry, therefore, is on the failure mechanisms rather than on numbers related to time to failure, whereas the traditional approaches were based on testing and labeling the output product as reliable in terms of their mean time to failure. Traditional approaches were characterized by attention to developing and using methods to measure effects which were then used to predict product reliability. Crook's description of the new approach [94] to reliability centers on the wafer processing aspects of the fabrication of ICs. However, the approach is much more than that. It encompasses all aspects of the fabrication process and applies particularly to any process where it is impractical to directly measure the reliability of the product.

Circuit reliability simulation can be used to simulate circuit reliability at the circuit design stage [194]. One example of reliability simulators is BERT [354] developed at Berkeley. To successfully introduce any reliability simulation tool to production, CAD, CAM, CIM (computer integrated manufacturing), R&D, Product, Integration, QA, and RE engineers have to work together and carefully link the database of design and layout, the simulation tool, the CIM network of the production lines, the reliability test records, and customer claims. Since most reliability simulation tools are very product and process sensitive, a joint task force has to be formed followed by continuous fine-tuning of the simulation system to ensure its profitability.

2.2.2. WLR Approach

Until a few years ago, end-of-line tests were the prevalent methods for most semiconductor industries. Since the reliability of semiconductors was considered as an afterthought at that time, the traditional end-of-line tests for defective devices were sufficient. In recent years, however, semiconductor customers are requiring much higher reliability, so traditional reliability methods, such as operating life tests, are no longer effective. Other reasons for their inapplicability includes cost, shortened lifetime, and increasing customization. The traditional methods, though they are essential during product development, aim only at ensuring specific reliability goals and are not suitable for continuous monitoring [358, 380]. These methods work only if we want to monitor the failure rate every month, and hundreds of thousands of devices would need to be sampled each year to resolve 10 FITs. Another limitation to the end-of-line tests is their inability to assess real time FA [94]. Therefore, a trend definitely driving semiconductor industries is to apply WLR tests to the screening and reliability analysis.

Customer demand for lower failure rates and the limitations of end-of-line tests forced the development of a concept called the WLR approach, which is a transition from the end-of-line concept toward the concept of BIR mentioned in Section 2.2.1. Dellin [103] presents three defining factors of WLR: the wafer level, the test structure, and the highly accelerated stress test. Since the testing is performed at the wafer level to reduce the time and expense of packaging, WLR is significantly different from traditional approaches. According to Turner [427], the purpose of WLR is not to predict a lifetime, but to detect the variation sources that might affect reliability.

To achieve the objectives of the WLR approach, WLR needs fast and highly accelerated wafer level tests (called WLR fast or stressed tests) which are designed to address each specific reliability failure mechanism [94]. They act as an extension of the typical measurements used for semiconductor process control. Generally, WLR fast tests are accelerated tests with durations ranging from milliseconds to one minute (some tests may be as long as two hours to avoid introducing unexpected failure mechanisms) which monitor the degradation rate of performance parameters when the material is subjected to a carefully controlled stress [380]. Very short test times make WLR the only practical technique for monitoring essentially every wafer produced by the fab and providing rapid feedback on every lot produced [380, 427]. The standard wafer level EM acceleration test (SWEAT) [353] is an example of WLR fast tests. The SWEAT lasts for less than a minute and accurately measures the relative rate of electromigration, but not the lifetime of the line. One important issue for SWEAT is to avoid self-heating, which may give analysts false results. In addition to the short test time, the reductions in process variation required for the proactive solution are also obtained because WLR fast tests are very adept at quantifying variations across the wafer or among wafers in a lot. Most common WLR test methods include the gate oxide and capacitor TDDB, QBD, HCI, and the V_t test.

Because current reliability prediction models for most semiconductor failure mechanisms are based on empirical data, they usually describe only the relationship between time and stress conditions. They are not useful for process control. Thus, accelerated tests are best suited as a measure of the degradation rate of the basic materials used in the semiconductor device, providing a relative measure of the intrinsic reliability of the product [380]. Once a process has matured and is well characterized, it is possible to a limited extent to link WLR fast tests to reliability lifetime predictions. This link is only meaningful, however, if the test is not accelerated to the point where the failure mechanism is no longer the one causing parts to fail in the field. In other words, WLR fast tests must be used judiciously to evaluate reliability.

WLR fast tests also allow expedient comparisons of the degree of control between competing process techniques and provide feedback for rapid optimization of the reliability produced by alternative process steps [380]. Therefore, the WLR fast test feedback may be used as an effective process control monitor which quite often will show a statistically significant process drift for reasons not related to reliability concerns. Nevertheless, all process drifts must be analyzed and the cause of the drift determined. WLR fast tests are useful for the identification of statistical process variations, though not for quantification of the effect that these variations have on reliability. Dellin [103] suggests potential application areas of WLR such as process control, qualification of new processes and technology, benchmarking, model development, reliability monitoring, and prediction.

However, Crook [94] and Turner [427] point out disadvantages or limitations of the WLR fast test. Since the WLR fast test is performed at the end of the manufacturing line and is not sensitive enough to detect process drifts, the

WLR fast test is not always an effective process control monitor for detecting variable drifts out of specification and for providing quick feedback [94] and can only be applied with a full understanding of the limitations of the stresses [427]. Another disadvantage is that at higher stress levels, the failure mode may be physically different from what would occur under normal use conditions [103]. In other words, the legitimacy of the most commonly used extrapolation technique in WLR is questionable, which can be seen from Example 2.4.

□ **Example 2.4**

In order to conduct the WLR tests, the test structures, the stressed model, and the pass/fail criteria are required. For example, in the EM test, the shift of resistance, ΔR, is used as an indicator of deterioration. Many IC makers define failure as occurring when $|\Delta R|$ exceeds 10%. Further consider an HCI test. The device is defined to be failed if $\Delta Gm \geq 10\%$ (Gm is called the transconductance). It may take more than 100,000 seconds for a device to reach the 10% ΔGm. Usually, the measurement unit (like HP4155 and HP4156A/B) can be stopped when the test time is longer than a certain value. Then, extrapolation is used to estimate the time for 10% ΔGm. It is important to check the slope of the Gm degradation line on the log to log plot. For example, from experience, the reasonable Gm degradation slope for a particular design is around 0.4. That is, attention must be paid if the slope of the fitted line is significantly different from 0.4.

One obvious drawback of the wafer level measurement is the use of extrapolation. There is no solid proof that the failure mechanisms of normal operating and of the stressed condition are the same. One other shortcoming resides in the applicability of the mathematical models, which are rather process dependent. One last remark is: the DC bias is usually used at the HCI test. Many researchers report that HCI tests with DC-bias give the lifetime at the worst case. From experience, HCI tests with AC-bias give 100 $\sim$ 200 times longer lifetimes than the DC-bias HCI tests. Moreover, the frequency of the AC-bias, the rising time and the falling time of the AC signal all affect the lifetimes estimated from the AC-bias HCI tests. In a word, if the lifetime derived from the DC-bias HCI test is longer than 10 years, then the HCI lifetime is qualified.
□

2.2.3. QML Approach

Recently, under pressure to qualify small quantities of highly reliable circuits, the U.S. Department of Defense (DOD) changed its approach to IC reliability from the qualified product concept to QML [194]. QML is another evolutionary step, with the purpose of developing new technologies for earlier marketing, improving circuit reliability and quality, and doing so at reduced costs. The approach has evolved over the last five years through close working relationships between consumer (government) and supplier (industry) [84].

In QML, the manufacturing line is characterized by running test circuits and standard circuit types [391]. Similar to BIR, understanding failure mechanisms and performing failure analysis are critical elements to implement the QML concept. Therefore, the QML approach places a heavy emphasis on documentation. It also depends on monitoring and controlling the processing steps and on tests to monitor critical failure mechanisms [84]. Soden [391] presents steps of performing FA for ICs:

Step 1 Verify if there is a failure.

Step 2 Characterize the symptoms of the failure.

Step 3 Verify the symptoms causing the observed failure.

Step 4 Determine the root cause(s) of the failure.

Step 5 Suggest corrective action.

Step 6 Document the results of the FA.

QML is another resolution for the recognition of the impracticality of qualifying individual products and the belief that reliability can be built into all products by a qualified manufacturing line [194].

2.3. Manufacturing Yield and Reliability

In the past, most attempts to assure high IC reliability used product testing, life testing or accelerated stress tests of the entire circuit. Because the approach of product testing is getting more expensive, more time consuming, and less capable of effectively identifying the causes of parametric and functional failures of ICs, the development of new technologies is needed. These new technologies make it possible to remove wearout failures due to operational life.

It is reported that there is a strong and measurable relationship between the number of failures in the field as well as in life tests, the yield due to the adoption of the WLR, and the use of reliability related design rules [431]. Kuper *et al.* [232] and Vander Pol *et al.* [431] present models for the yield-reliability relation and the experimental data to show the correlation. Thus, the root causes of reliability failures are the same as those of yield failures, and the manufacturing yield depends on the number of defects found during the manufacturing process, which in turn determines reliability.

The degree of manufacturing success is measured by yield, which is defined as the average ratio of devices on a wafer that pass the tests. Since the yield is a statistical parameter and implies a probability function, yield functions are multiplied in order to attain the total yield. The total wafer yield is a measure of good chips per wafer normalized by the number of chip sites per wafer. Generally, since yields can be measured in various ways, the overall yield is calculated by the product of elements of yield such as the line yield, the WP yield, the assembly yield, the FT yield, and the burn-in yield.

Parameters that affect yield are defects, and the number of defects produced during the manufacturing process can be effectively controlled by introducing test points at crucial times rather than throughout the assembly line [380]. This not only improves the reliability of the outgoing product but also significantly enhances the yield of the manufacturing process, thus increasing the quality of the overall system. Test points are effective only at the critical areas, and their random distribution in the process was observed not to yield the desired results of high quality and minimal defects density. There is another way to control the yield. Since IC device yields are not only a function of chip area but also a function of circuit design and layout, by determining the probabilities of failure and critical areas for different defect types, it is possible to control and manage the yield of ICs [405].

Schroen [369] suggests a new system for studying reliability by utilizing test structures sensitive to specific failure mechanisms. By stressing these structures, more accurate information about the reliability of the circuit could be obtained in a shorter time than by the use of traditional methods. Schroen also regarded it as a means of reducing the dependence on the costly and time consuming burn-in testing.

□ Example 2.5

There are many types of yields: WAT yield, WP yield, assembly yield, FT yield, burn-in yield, QC tests yield, and others. Among them, only the WP yield itself cannot precisely predict the reliability at a certain time for most complicated ICs unless the manufacturing processes are matured. A matured process contains stress screenings in the manufacturing processes so that weaker dice are found and marked before WP. Effective WP programs are important for a matured production line so that defects can be weeded out before assembly. To screen as many defective dice at WP as possible, higher voltages or currents are sometimes required. However, this is not practically true in most WP programs. The main reason for this is the parameters of the circuit (especially timing sensitive parameters) will somewhat shift after the assembly processes; applying high stresses early at WP may over-kill the chips.

From practical experience, we have learned that some potential reliability problems cannot be discovered at WP, such as the reliability failures from the gate oxide thickness, which greatly affects the speed and the HCI lifetime of the devices. The devices with thicker gate oxide may still pass WP. Very possibly, however, they will fail at speed sorting during FT, which is performed usually after the chips are assembled. One promising way to increase the WP effectiveness is to enforce the WAT specifications and screen the chips whose parameters are out of specifications (or control). WAT is performed on a sampling basis on the test keys. However, WP is done basically 100% of the production chips. Thus, defective chips may not be detected at WAT. Moreover, even though a powerful WAT program and specifications are used, certain chips will probably pass WP and fail at FT or at other reliability tests afterwards.

One method to increase yield is to put redundant circuits on the chips so that they may still be usable if the defects are minor and can be repaired by replacing the bad circuits with good redundant ones. The L/R is designed for this purpose. However, L/R may introduce unexpected damage to the chips if certain processes are incorrectly performed. Consider the following case. If the uniformity of an etching process is not controlled correctly, the thickness of pad oxide ($T_{p,ox}$) on each chip varies; one possibility is that the $T_{p,ox}$ of the chips at the center of the wafer is different from that of the chips at the edge of the wafer. Since the L/R energy is the same for one wafer, this may result in insufficient or over-stressed repairs. The yield for insufficient repairs will be low. On the other hand, if the pad oxide is damaged at over-stressed repairs, the yield may not be affected and the chips may pass WP and FT. However, these ICs eventually fail during reliability tests (especially the ones with humidity stress, like HAST and PCT).

Another example is the thickness of the inter-metallic dielectric (IMD) layer. Tiny cracks can be found in the dice from the thick IMD process, and the cracks will become severe after T/C test. Some ICs with tiny cracks may pass WP and FT. However, it is found that the high temperature with bias (HTB) test results in metal lines stringer on these ICs.

In a word, the relationship between yield and reliability is very difficult to define. Even if it is obtained, it is only reliable under the assumptions that the processes are matured and free from particles, contaminations, and mis-operations. □

2.4. Conclusions

Based on projections, the traditional reliability methodologies will no longer be effective to ensure the required failure rate of less than 10 FITs. Some reliability techniques such as, BIR, WLR, or QML, presented in this chapter can resolve this problem. These techniques, however, also have limitations in meeting the challenge of the failure rate goal. To reduce test time and cost, we can implement an equivalent or modified technique, such as one of those introduced in Chapter 10, that is less time consuming and more cost effective.

3. Basic Reliability Concept

Reliability is defined as the probability of a device performing its purpose adequately for the period of time intended under the operating conditions encountered. Therefore, the probability that a system successfully performs as designed is called system reliability. Such probability is also referred to as the probability of being survival. In most cases the probability with which a device will perform its function is not known at first. Also, true reliability is never exactly known, which means the exact numerical value of the probability of adequate performance is not known. But numerical estimates quite close to this value can be obtained by the use of statistical methods.

The elements of reliability definitions, such as "successful operation," "specified period of time," "intended operating conditions," and "survival," are, to a great extent, subjective measures. They are defined, and/or specified, by the user. Mathematically, if $f(t)$ represents the probability density function (pdf) of failure times of a system or component, then the reliability function $R(t)$ is given by

$$R(t) = \Pr\{T > t\} = \int_t^{\infty} f(x)dx \tag{3.1}$$

where T is a continuous random variable with nonnegative values of time to failure or life length.

A device is assumed to be working properly at time $t = 0$. No device can work forever without failure. These statements are expressed as

$$R(0) = 1, \qquad R(\infty) = 0$$

and $R(t)$ is a nonincreasing function between these two extremes. For t less than zero, reliability has no meaning. We define it as unity, however, since the probability of failure for $t < 0$ is zero.

A device becomes unreliable because of failure. We usually recognize a failure as the event, when after its occurrence, the output characteristics of the device shift outside the permissible limits. In a complex system the instances of appearance of failure are usually random events, and so are the failure times. As random events, failures can be either independent or dependent.

Given $R(t)$, from Eq. (3.1), the unreliability function, $F(t)$, is

$$F(t) = 1 - R(t) = \int_0^t f(x)dx. \tag{3.2}$$

Differentiating Eq. (3.2) gives

$$\frac{\mathrm{d}R(t)}{\mathrm{d}t} = -\frac{\mathrm{d}F(t)}{\mathrm{d}t} = -f(t)$$

which says the marginal change in reliability function with respect to time t is the negative of the probability density function of the system failure time.

3.1. Elements of Reliability

3.1.1. Failure Rate

If we express the probability of failure in a given time interval, t_1 to $t_1 + \Delta t$, in terms of the reliability function, the relationship is as follows:

$$\int_{t_1}^{t_1+\Delta t} f(x)dx = \int_{t_1}^{\infty} f(x)dx - \int_{t_1+\Delta t}^{\infty} f(x)dx = R(t_1) - R(t_1 + \Delta t)$$

where Δt is the interval length.

The failure rate can be defined as the rate of the probability that failure occurs in the interval t_1 to $t_1 + \Delta t$, given that it has not occurred prior to t_1. The failure rate can be mathematically stated as

$$\frac{R(t_1) - R(t_1 + \Delta t)}{\Delta t R(t_1)}. \tag{3.3}$$

We now let $t = t_1$ and rewrite Eq. (3.3) as

$$\frac{R(t) - R(t + \Delta t)}{\Delta t R(t)}. \tag{3.4}$$

□ Example 3.1

Consider the burn-in (see Chapter 6 for burn-in) data of ICs given in Table 3.1. A population of 400,000 devices begins to test under some stresses at time zero. During the first hour, 3500 devices fail; between the first and second hours an additional 700 devices fail; and so on until a total of 5980 devices fail after 9-hr burn-in. The question is how failure rates can be calculated. The time interval Δt in which the failures are observed is 1 hr. For the sake of convenience let us assume that the failures have occurred exactly on the hour. That is, the 3500 devices failed after surviving for 1 hr; 700 devices failed after surviving for 2 hrs, and so on. The failure rate is calculated by Eq. (3.4). When the population size is very large and the number of failures is proportionally low, the failure rate can be calculated simply by dividing the number of devices failed in an interval Δt by the total population at time zero. In this case $f(t)$ itself becomes the failure rate. For example, the failure rate at t=5 is

$$\frac{R(5) - R(6)}{1 \times R(5)} = \frac{340}{400000 - 4850} = 0.00086 \text{ hr}^{-1}. \quad \square$$

3.1.2. Hazard Rate

The hazard rate, or instantaneous failure rate, denoted by $h(t)$, is defined as the limit of the failure rate as the interval, Δt, approaches zero. The hazard

Table 3.1: The 9-hr burn-in failure rate calculation.

Time (hrs)	Number of Devices Failed	Number of Cumulative Failures	Failure Rate
1	3,500	3,500	0.00875
2	700	4,200	0.00177
3	350	4,550	0.00088
4	300	4,850	0.00076
5	340	5,190	0.00086
6	240	5,430	0.00061
7	200	5,630	0.00051
8	170	5,800	0.00043
9	180	5,980	0.00046

rate then is

$$h(t) \equiv \lim_{\Delta t \to 0} \frac{R(t) - R(t+\Delta t)}{\Delta t R(t)} = -\frac{1}{R(t)} \frac{\mathrm{d}R(t)}{\mathrm{dt}} = \frac{f(t)}{R(t)} \tag{3.5}$$

which can also be written as

$$h(t) = -\frac{\mathrm{d}\ \ln R(t)}{\mathrm{dt}}. \tag{3.6}$$

The $h(t)dt$ can be explained as the conditional probability that a device will fail in a unit time interval after t, given that it was working at time t. In other words, the instantaneous failure rate at time t equals the failure density function divided by the reliability, with both of the latter evaluated at time t. The hazard rate is also known as force of mortality. For a discrete distribution $\{p_k\}_{k=0}^{\infty}$, the hazard rate is

$$h(k) = p_k / \sum_{j=k}^{\infty} p_j \ \leq\ 1.$$

The hazard rate of a life or failure distribution, $h(t)$, plays much the same role in reliability analysis as the spectral density function plays in the analysis of stationary time series in communication problems. In both cases we are concerned with estimating an unknown function. Note that we are concerned with inferring the function itself and not the parameters of the function. Knowledge of the hazard rate uniquely determines the failure density function, the unreliability function, and the reliability function.

When integrating Eq. (3.6), the reliability function is obtained

$$R(t) = e^{-\int_0^t h(x)dx}. \tag{3.7}$$

Eq. (3.7) is a mathematical description of reliability in the most general form. It is independent of the specific failure distribution involved. Substituting Eq. (3.7) into Eq. (3.5), we have

$$f(t) = h(t)e^{-\int_0^t h(x)dx}.$$

Knowledge of the hazard rate of independent failure distributions also allows us to compute the reliability function of their sum. If X_1 and X_2 are two independent random variables with hazard rates $h_1(t)$ and $h_2(t)$ and unreliability functions $F_1(t)$ and $F_2(t)$, respectively, the sum of their distribution is $X=\min(X_1,X_2)$. The reason for taking the minimum value is that if there is a failure in either of them, the equipment is failed. Thus, the probability of no equipment failure in a time $X > t$ is

$$\Pr\{X > t\} = \Pr\{X_1 > t\}\Pr\{X_2 > t\}.$$

Taking logarithms of both sides yields

$$\ln[1 - F(t)] = \ln[1 - F_1(t)] + \ln[1 - F_2(t)]$$

and since

$$\ln[1 - F(t)] = -\int_0^t h(x)dx$$

then

$$\int_0^t h(x)dx = \int_0^t h_1(x)dx + \int_0^t h_2(x)dx. \tag{3.8}$$

Therefore, we can add the cumulative hazard rates of two independent failure distributions to find the hazard rate of the minimum of these random variables. Also, this operation is independent of the form of the failure distribution; the reliability of the system can be obtained from Eq. (3.7). This additive property of cumulative hazard rates is independent of the functional form of the reliability functions.

Note that hazard rates are equivalent to failure rates only when the interval over which failure rate is computed approaches zero. In spite of this restriction, the term "failure rate" is widely used to describe the measure of reliability which is really hazard rate.

3.1.3. Hazard Rate Family

The hazard rate family includes the following five distinct groups: increasing failure rate (IFR), increasing failure rate average (IFRA), new better than used (NBU), decreasing mean residual life (DMRL), and new better than used in expectation (NBUE). They provide nonparametric tests to determine if a data set falls into a specific category. Buchanan and Singpurwalla [494] provide definitions for most terminologies used in this section.

IFR

Let $R(t) = 1-F(t)$ and $H(t) = -\ln R(t)$. There are three different descriptions for IFR:

1. A life distribution $F(t)$ is IFR if $h(t)$ is monotone nondecreasing in t, or
2. A life distribution $F(t)$ is IFR if and only if $\ln R(t)$ is concave, or

3. A life distribution $F(t)$ is IFR if $R(x \mid t)=R(x+t)/R(t)$ is decreasing in $t, \forall\, x > 0$.

The notion of IFR is equivalent to stating that the residual life of an un-failed item of age t is stochastically decreasing in t. The definition of decreasing failure rate (DFR) can likewise be derived by modifying aforementioned definitions of IFR. IFR distributions have the following useful properties [32]:

- If $F(t)$ is IFR and $F(t) < 1$, then F is absolutely continuous on $(-\infty, t)$ (i.e., there exists a density for F on $(-\infty, t)$). There might exist a jump in F at the right hand end point of its interval of support when this interval is finite.

- If f is a Polya frequency function of order 2 (PF_2, see Karlin [213]) then the corresponding F is IFR.

- Coherent systems of independent IFR components need not have IFR distributions. This is first shown in Esary and Proschan [133].

IFRA

IFRA can be defined in different ways:

1. A life distribution $F(t)$ is IFRA, if $R(t)^{1/t}$ is decreasing in $t > 0$, or
2. A life distribution $F(t)$ is IFRA, if $-\ln R(t)/t = H(t)/t$ is increasing in $t > 0$, or
3. A life distribution $F(t)$ is IFRA, if and only if $R(\alpha t) \geq R^{\alpha}(t)$ for $0 < \alpha < 1$ and $t \geq 0$.

Several properties can be derived:

- IFR forms a subclass of the IFRA distributions.

- Coherent systems with IFRA components have an IFRA distribution [356].

- The IFRA class of distributions is the smallest class containing the exponential distribution, closed under formation of coherent systems and taking limits in distribution.

- The IFRA distributions also arise naturally when one considers cumulative damage shock models. Shocks occur in time according to a device, the damages accumulate until a certain tolerance level is exceeded, causing the device to fail. The time of failure was shown to be a random variable with an IFRA distribution [135], in which the shock that occurs is assumed to be a homogeneous Poisson process. The nonhomogeneous Poisson process model (NHPP) is in [4, 5]; the latter one also considers a pure birth process.

See [22, 30, 47] for more properties.

NBU

A life distribution $F(t)$ is NBU if $R(x)R(t) \geq R(x+t)$, $\forall\, x$ and $t \geq 0$. In other words, the life of a new item is stochastically greater than the residual life of an unfailed item of age t.

DMRL

A life distribution $F(t)$ is DMRL if $\int_0^\infty R(x+t)dx/R(t)$ is decreasing in t. That is, the residual life of an unfailed item of age t has a mean that is decreasing in time t.

NBUE

A life distribution $F(t)$ is NBUE if $\int_0^\infty R(x)dx \geq \int_0^\infty R(t+x)dx/R(t)$, $\forall\, t \geq 0$. That is, the expected life of a new item is greater than the expected residual life of an unfailed item of age t.

Based on the definitions of the aforementioned five groups, the following implications can be derived:

$$\text{IFR} \subset \text{IFRA} \subset \text{NBU} \subset \text{NBUE} \quad \text{and} \quad \text{IFR} \subset \text{DMRL} \subset \text{NBUE}.$$

Refer to Barlow [21] and Barlow and Marshall [29, 30] for the different bounds on reliability and other parameters when one moment or one percentile is known, assuming the underlying life distribution is IFR, IFRA, DFR, or DFRA.

If F_1 (x_1) and $F_2(x_2)$ are both IFR, then their convolution is IFR. However, convolution does not preserve DFR or DFRA [33, pp 100–101]. Let $F(x)$ be the mixture of distributions, $F_\alpha(x)$ and $G(\alpha)$,

$$F(x) = \int_{-\infty}^{\infty} F_\alpha(x)dG(\alpha).$$

If each F_α is DFR, then F is DFR, and if each F_α is DFRA, then F is DFRA. However, this does not apply to the IFR and the IFRA classes, as outlined in Table 3.2, which is explained in detail in Barlow *et al.* [31], Esary *et al.* [134], and Marshall and Proschan [265] (see Appendix A for the full descriptions of abbreviations used in Table 3.2).

3.1.4. Mean Time to Failure

The mean time to failure (MTTF) is sometimes used as a figure of merit for a system, and is defined as the expected time for a system to reach a failed state, given that all equipment was initially operating. MTTF is ordinarily used in explanation of system availability. The expression for the MTTF can either be developed from conventional statistical theory or from some properties of Markov matrices.

Table 3.2: The preservation of life distributions for reliability operations.

Life Distribution	Formulation of Coherent Systems	Addition of Life Distributions	Mixture of Distributions
IFR	Not preserved	Preserved	Not preserved
IFRA	Preserved	Preserved	Not preserved
DMRL	Not known	Not known	Not known
NBU	Preserved	Preserved	Not preserved
NBUE	Not preserved	Preserved	Not preserved
DFR	Not preserved	Not preserved	Preserved
DFRA	Not preserved	Not preserved	Preserved
IMRL	Not known	Not known	Not known
NWU	Not preserved	Not preserved	Not preserved
NWUE	Not preserved	Not preserved	Not preserved

Applying the conventional statistical definition for the MTTF, we define

$$\text{MTTF} \equiv \text{E}[T] = \int_0^\infty t f(t)dt = \int_0^\infty R(t)dt \tag{3.9}$$

since for a physically reliable system it must fail after a finite amount of operating time, and so we want to have $tR(t) \to 0$ as $t \to 0$.

It is important to note that the MTTF of a system denotes the expected or average time to failure and is neither the failure time which could be expected 50% of the time, nor is it the most probable time of system failure. The former interpretation applies only if the failure time density function is symmetric about the MTTF.

□ Example 3.2

Consider two densities:

$$\begin{aligned} f_1(t) &= 0.001e^{-0.001t} && 0 \le t \\ f_2(t) &= 1/2000 && 0 \le t \le 2000. \end{aligned}$$

By using Eq. (3.1), the reliability functions are given by

$$\begin{aligned} R_1(t) &= e^{-0.001t} && 0 \le t \\ R_2(t) &= (2000 - t)/2000 && 0 \le t \le 2000. \end{aligned}$$

Through Eq. (3.9), both of two distributions have the same MTTF=1000 hrs. However, the reliabilities at t=500 are $R_1(500) = e^{-0.5} = 0.607$ and $R_2(500) = (2000 - 500)/2000 = 0.75$. Thus, we can see that the MTTF is not sufficient to determine the characteristics of the failure distribution. □

3.1.5. Mean Residual Life

The mean residual life (MRL) is the expected remaining life (or the additional expected time to failure) given survival to time t. Define $r(t)$ as the MRL at time t, which is given by

$$r(t) = E[T - t | T \geq t] = \frac{1}{R(t)} \int_t^{\infty} R(u) du. \tag{3.10}$$

□ **Example 3.3**

When the hazard rate function is given by

$$h(t) = \frac{3}{t+a} \quad t > 0,$$

the reliability function is calculated as

$$R(t) = \left(\frac{a}{t+a}\right)^3.$$

By using Eq. (3.10) and change of variables, $y = a/(u + a)$, the MRL at time t is

$$r(t) = \left(\frac{t+a}{a}\right)^3 \int_t^{\infty} \left(\frac{a}{u+a}\right)^3 du = \frac{t+a}{2}.$$

In this example, we can see that the relationship between $h(t)$ and $r(t)$,

$$h(t) = \frac{r'(t) + 1}{r(t)},$$

where $r'(t) = dr(t)/dt$, is satisfied. □

3.1.6. Behavior of Failures

Systems and materials start to wear out when they are used, and they can fail due to various failure mechanisms. Often failures need to be defined with a specific bounds given as tolerant limits. Early failures come from poor design, improper manufacture, and inadequate use. It is also known that failures due to the aging process; material fatigue, excessive wearout, environmental corrosion, and undesirable environment can contribute to this process.

A study of many systems during their normal life expectancy has led to the conclusion that failure rates follow a certain basic pattern. It has been found that systems exhibit a high failure rate during their initial period of operation, called the early life, or the infant mortality period or the burn-in period or the debugging stage. The operating period that follows the infant mortality period has a smaller failure rate and is called the useful life period, which tends to remain constant until the beginning of the next phase, called the aging period. Failures during the last period are typically due to aging or cumulative damage. The typical hazard rate behavior is known as the bathtub curve, shown as the solid line in Figure 3.1.

In Figure 3.1, stage A is called the infant mortality (usually less than 1 year for ICs) period, stage B is the useful life period (usually 40 years for ICs), while stage C indicates that the device is almost wearing out and is called the aging period. In Figure 3.1, the bathtub curve is the addition of three curves: the extrinsic failure curve depicted in a, the intrinsic failure curve depicted in b, and the wearout failure curve depicted in c. For semiconductor devices, refer to Section 5.3.1 for discussions on the extrinsic and intrinsic failures.

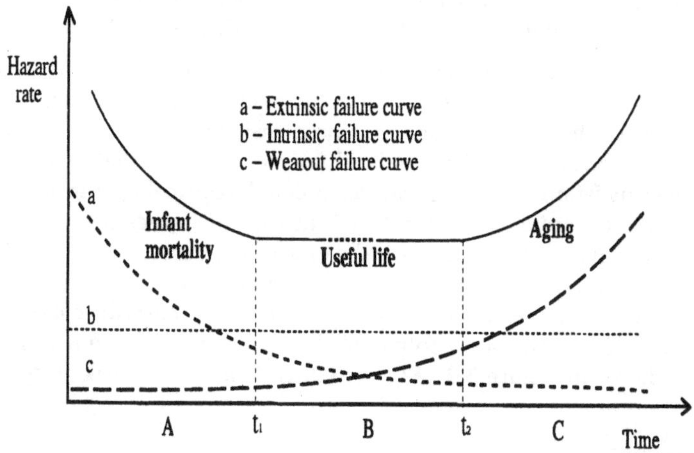

Figure 3.1: A bathtub hazard rate.

Most electronic devices exhibit a DFR in their early life (Bezat and Montagu [44] report that some solid-state electronics also have DFR); this results from the weak individuals that have shorter lives than the normal (stronger) ones. The weak devices (or latent defects [26]) may come from improper operations by workers, a contaminated environment, a power surge of the machines, defective raw materials, ineffective incoming inspection, and inadequate shipping and handling. If the weak devices are released to customers or are used to assemble modules or systems, many of these defects will cause failures in their early lives; from our experience, quite a few failures can be observed in the first year for "immature" products. This early-stage high hazard rate is called infant mortality because the product is in its early life and is not mature enough to be released. Note that infant mortality is viewed for the whole lot instead of for a single device. A device will either fail or pass a test, whereas the failure rate of a lot may follow a decreasing pattern.

In stage B, devices move into the steady-state hazard rate period with an almost constant failure rate. This corresponds to the normal operation of an electronic device and extends well beyond the useful life of most devices. The hazard rate curve, which is related to random events or stresses that cause even good devices to fail, is depicted as constant in time since failures are unrelated to usage. In the context of the failure of devices, if a failure is considered random, it means that every device has the same chance of failure.

Aging occurs when the hazard rate increases and the remaining devices fail. It is reported by Bailey [16] that, due to rapidly evolving manufacturing technology, the wearout process may not be detected during product operational life. However, it is worthwhile to point out that the parts for many nuclear plants, for example, which were built more than 30 years ago, have reached their wearout stages if no proper maintenance plan has been adopted. Extreme precautions have to be taken for this kind of highly hazardous facility to prevent any possible catastrophe. Obviously, it is desirable to have such an item operated upon during stage B, which has a lower hazard rate.

Notice that most, if not all, electronic and non-electronic devices start to wear out once they are put into service. Given a population that these devices are drawn from, the ones that bear with extrinsic failures due to manufacturing defects will wear out and be realized far more quickly that those with the normal wearout failures. In other words, under the normal operating (wearing) conditions, a lot more devices tend to fail at the early stage of their life cycle. Hence the hazard rate decreases until the extrinsic failures are no more a major contributing factor to the wearing. At the aging period, the weraout failure dominates the failure realization. In addition to the bathtub curve given in Figure 3.1, there are two ways to model the hazard rate as a function of time which are shown in Figure 3.2. Figure 3.2(a) depicts the quick trigger of the

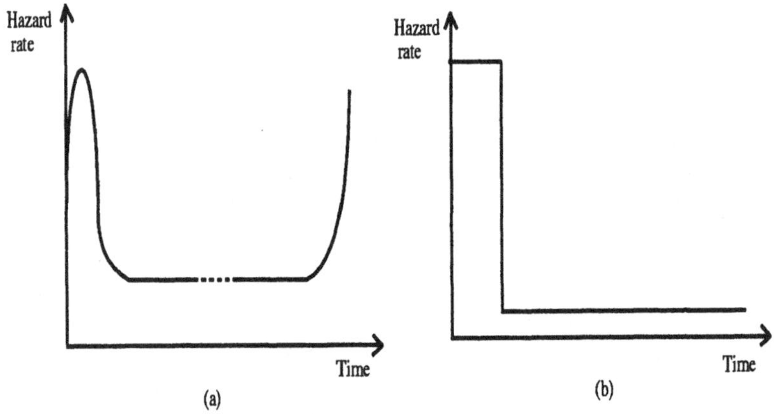

Figure 3.2: Two models for the hazard rate as a function of time.

hazard rate increase once devices are put in use. This may be due to the fact that the incompatible devices are assembled into a system (see Section 6.1.5 for incompatibility), or the exposure of manufactured devices to a new environment. Figure 3.2(b) is a simplified version to show that the hazard rate is significantly higher at the infant mortality period than others. It is rarely that Figure 3.2(b) is used. For the rest of the book, we follow Figure 3.1 unless otherwise state.

Among many statistical distributions, the beta distribution, $\text{BETA}(p,q)$, can be used to model the bathtub hazard rate when both parameters p and q

are less than 1 as shown in Figure 3.3, and

$$f(x;p,q) = \frac{1}{\mathcal{B}(p,q)} x^{p-1}(1-x)^{q-1},\ 0 \le x \le 1, \tag{3.11}$$

where $\mathcal{B}(p,q) = \Gamma(p)\Gamma(q)/\Gamma(p+q)$. The mixture of more than one distribution

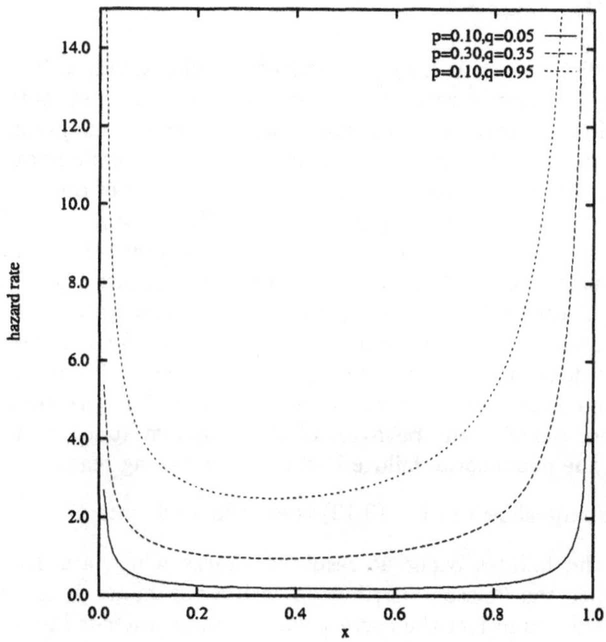

Figure 3.3: Hazard rate functions of beta distribution with various parameters.

may also be used for modeling the bathtub hazard rate, though it may be much more complicated and harder to analyze (see [78, 104, 222]). The determinations for t_1 and t_2 in Figure 3.1 are generally difficult. This is the change point problem. The studies of Balmer [19], Zacks and Brazily [457], and Zacks [458, 459] serve as good references for dealing with this problem. An easier way of finding t_1 and t_2 is via testing the exponential property of the useful life period; this is basically a goodness-of-fit test and will be introduced in Section 7.4.

3.2. Some Useful Life Distributions

Several useful distribution functions are briefly described below. In addition, others used in the book are summarized in Appendix C.

3.2.1. Exponential Distribution

The exponential distribution is one of the most popular distributions used in reliability study. Its pdf is defined by

$$f(t) = \lambda e^{-\lambda t}, \quad t \geq 0, \quad \lambda > 0 \tag{3.12}$$

and the cumulative distribution function(CDF) by

$$F(t) = \int_0^t f(x)dx = 1 - e^{-\lambda t}, \quad t \geq 0, \quad \lambda > 0.$$

Exponential distribution is a simple distribution characterized by a single parameter λ and can also be regarded as a special case of the distribution function from the Weibull distribution with shape parameter=1, scale parameter=λ, or from the gamma distribution with shape parameter=1, scale parameter=$1/\lambda$. It is the right distribution to model the useful life period of the bathtub curve.

From Eqs. (3.1), (3.5), and (3.9), $R(t) = e^{-\lambda t}$, $h(t) = \lambda$, and MTTF=$\frac{1}{\lambda}$. A constant hazard rate means that, in a population of exponential life distribution, the probability of failure of any device in a given time interval is independent of the aging history prior to the beginning of that time interval (memoryless property). Assuming memoryless property to hold true for ICs is equivalent to ignoring time dependent failure rates due to infant mortality and wearout.

Although many distributions modify the pdf of failure times of a system and hence describe the behavior of the reliability function, we are most interested in the exponential failure law for the following reasons:

1. The formula shown in Eq. (3.12) is easy to work with.

2. When the failures occur as result of shocks which are distributed according to the Poisson distribution with a parameter λ, the exponential distribution is in fact the correct one to use for inter-failure times.

3. Some devices really behave in the manner of exponential distribution, such as electronic devices with no wearout properties and no infant mortality.

4. It can sometimes be used to provide bounds on the true reliability. If the hazard rate $h(t)$ is bounded by a and b, $a \leq h(t) \leq b$, then $e^{-at} \geq R(t) \geq e^{-bt}$.

5. Even though $h(t)$ is not bounded above and/or below, we can sometimes set limits on $R(t)$ if the distribution is IFR or DFR.

6. Exponential behavior is observed in complex systems where a large number of devices are repaired instantaneously.

3.2.2. Normal Distribution

The normal distribution is used to model the wearout period and its pdf is

$$f(t) = \frac{1}{\sigma\sqrt{2\pi}} \exp\left[\frac{-(t-\mu)^2}{2\sigma^2}\right], \quad -\infty < t < \infty, \quad -\infty < \mu < \infty, \; \sigma > 0,$$

where μ is the mean failure time and σ is the standard deviation of normal distribution. The reliability and hazard rate expressions for the normal distribution are given by

$$R(t) = \frac{1}{\sigma\sqrt{2\pi}} \int_t^\infty \exp\left[\frac{-(x-\mu)^2}{2\sigma^2}\right] dx,$$

$$h(t) = \exp\left[\frac{-(t-\mu)^2}{2\sigma^2}\right] \left\{\int_t^\infty \exp\left[\frac{-(x-\mu)^2}{2\sigma^2}\right] dx\right\}^{-1},$$

respectively. As the shape of the normal pdf indicates, a normal failure law implies that most of the items fail around the mean failure time. Therefore, in order to achieve a high reliability, the operating time must be considerably less than μ. Because the range of the normal distribution is from $-\infty$ to ∞, this distribution should be used extremely carefully for analyzing life problems.

3.2.3. Lognormal Distribution

The logarithmic normal distribution implies that the logarithms of the lifetimes are normally distributed. The pdf of the lognormal distribution is obtained by replacing times to fail by their logarithms as

$$f(t) = \frac{1}{\beta t\sqrt{2\pi}} \exp\left[\frac{-(\ln t-\alpha)^2}{2\beta^2}\right], \; t > 0, \quad -\infty < \alpha < \infty, \; \beta > 0.$$

The mean and variance are

$$E[T] = \exp[\alpha + \frac{\beta^2}{2}], \quad Var(T) = \exp[2\alpha + \beta^2](\exp[\beta^2] - 1),$$

respectively. The reliability and hazard rate expressions for the above distribution function are given by

$$R(t) = \frac{1}{\beta\sqrt{2\pi}} \int_t^\infty \frac{1}{x} \exp\left[\frac{-(\ln x-\alpha)^2}{2\beta^2}\right] dx,$$

$$h(t) = \frac{1}{t} \exp\left[\frac{-(\ln t-\alpha)^2}{2\beta^2}\right] \left\{\int_t^\infty \frac{1}{x} \exp\left[\frac{-(\ln x-\alpha)^2}{2\beta^2}\right] dx\right\}^{-1},$$

respectively. Ordinarily, the hazard rate of this distribution is an IFR of time followed by a DFR. The hazard rate approaches zero for initial and infinite times [105].

The distribution can be derived fundamentally by considering a physical process wherein failure is due to fatigue cracks. Many wearout failure mechanisms, such as TDDB and electromigration, cause failures that are usually modeled by the lognormal distribution [358]. The lifetimes of semiconductor devices after burn-in are also well described by the lognormal distribution [168].

3.2.4. Gamma Distribution

The gamma distribution describes the infant mortality period well. Its pdf is given as

$$f(t) = \frac{\beta^\alpha}{\Gamma(\alpha)} t^{\alpha-1} e^{-\beta t}, \quad t \geq 0, \ \alpha > 0, \ \beta > 0, \tag{3.13}$$

where

$$\Gamma(\alpha) = \int_0^\infty x^{\alpha-1} e^{-x} dx = (\alpha - 1)\Gamma(\alpha - 1) = (\alpha - 1)!.$$

The mean and variance of the gamma distribution are given by

$$E[T] = \frac{\alpha}{\beta}, \quad Var(T) = \frac{\alpha}{\beta^2},$$

respectively. The reliability and hazard rate expressions are

$$R(t) = \frac{\beta^\alpha}{\Gamma(\alpha)} \int_t^\infty x^{\alpha-1} e^{-\beta x} dx,$$

$$h(t) = t^{\alpha-1} \exp(-\beta t) \left[\int_t^\infty x^{\alpha-1} \exp(-\beta x) dx \right]^{-1},$$

respectively.

The gamma distribution is related to several other distributions. For the special case of β=1/2, we have the chi-square distribution. From the gamma density function, we can obtain the Erlang distribution of:

$$f(t) = \frac{\beta^{\alpha+1}}{\Gamma(\alpha+1)} t^\alpha e^{-\beta t} = \frac{\beta(\beta t)^\alpha e^{-\beta t}}{\alpha!}, \quad \alpha > 0, \ \beta > 0$$

when α is an integer.

3.2.5. Weibull Distribution

The pdf for the Weibull distribution is expressed as

$$f(t) = \lambda\beta(\lambda t)^{\beta-1} e^{-(\lambda t)^\beta}, \quad t \geq 0, \ \lambda > 0, \ \beta > 0 \tag{3.14}$$

and the CDF as

$$F(t) = 1 - e^{-(\lambda t)^\beta} \tag{3.15}$$

where β is the shape parameter and λ is the scale parameter. From Eqs. (3.5) and (3.9),

$$h(t) = \lambda\beta(\lambda t)^{\beta-1}, \quad \text{MTTF} = \frac{1}{\lambda}\Gamma(1 + \frac{1}{\beta}). \tag{3.16}$$

The Weibull distribution models both IFR and DFR by varying the value of β above and below 1, respectively. If $\beta = 1$, then $h(t)$ is equivalent to that of exponential distribution with $h(t) = \lambda$. The Weibull distribution can be used to model all three regions of the bathtub curve.

By setting $\eta=1/\lambda$,

$$f(t) = \frac{\beta}{\eta}(\frac{t}{\eta})^{\beta-1}e^{-(\frac{t}{\eta})^{\beta}}, \quad F(t) = 1 - e^{-(\frac{t}{\eta})^{\beta}}, \quad h(t) = \frac{\beta}{\eta}\left(\frac{t}{\eta}\right)^{\beta-1}. \tag{3.17}$$

The Weibull distribution represents an appropriate model for a failure law whenever the system is composed of a number of independent components and failure is essentially due to the most severe flaw among a large number of flaws in the system.

□ **Example 3.4**

For a given hazard rate function $h(t)$:

$$h(t) = \frac{\lambda p(\lambda t)^{p-1}}{1 + (\lambda t)^p},$$

the reliability function $R(t)$ is calculated, using Eq. (3.7), as

$$R(t) = \frac{1}{1 + (\lambda t)^p}$$

and the pdf is

$$f(t) = h(t)R(t) = \frac{\lambda p(\lambda t)^{p-1}}{(1 + (\lambda t)^p)^2}.$$

This is a log-logistic density of failure time that is obtained by the transformation of logistic density. When $p = 1$ the hazard rate function is monotonically decreasing. However, if $p > 1$, the hazard rate has a peak point at $t = (p-1)^{1/p}/\lambda$. □

3.3. Strength and Stress Analysis

For a component whose strength level X and stress level Y are random variables following some distribution functions, the reliability of the component is the probability that the strength exceeds the stress:

$$R = \Pr\{X > Y\} = \int_0^{\infty} g(y)\left[\int_y^{\infty} f(x)dx\right]dy = \int_0^{\infty} f(x)\left[\int_0^{x} g(y)dy\right]dx$$

where $f(x)$ and $g(y)$ are the pdf's of X and Y, with means μ_x and μ_y, and variance σ_x^2 and σ_y^2, respectively. If we define $Z = X - Y$, then

$$R = \Pr\{Z > 0\} = \int_0^{\infty}\int_0^{\infty} f(z+y)g(y)dydz.$$

Such a relationship is depicted in Figure 3.4.

The traditional safety margin(SM) can be described as

$$SM = \frac{\mu_x - \mu_y}{(\sigma_x^2 + \sigma_y^2)^{1/2}},$$

which separates the means of strength from the stress in a normalized way.

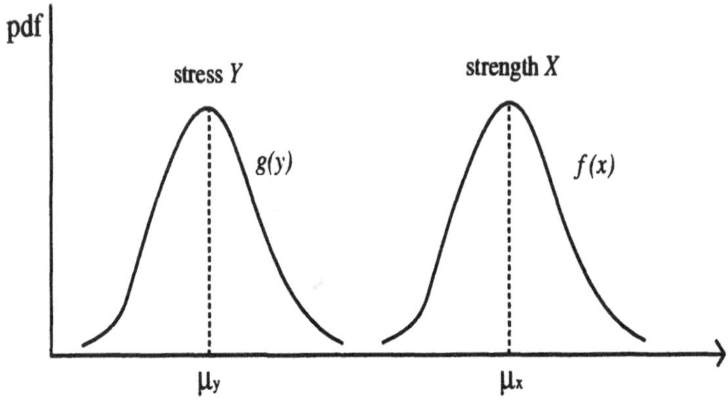

Figure 3.4: The relationship between strength X and stress Y.

□ **Example 3.5**

A new Pentium II MMX PC operates at least 99.9% of the time without a failure caused by being overheated. This computer can tolerate the heat generated inside the PC case due to different power levels. The power tolerance level follows the normal distribution with the mean 38 watts and the standard deviation 2 watts. When an additional memory chip is added to the computer, it moves the heat stress up to a new level. The new power level also follows a normal distribution with the mean 28 watts and the standard deviation 2.5 watts. When the new memory chip is added, do we need to use a more powerful PC fan to warrant at least the same reliable operation of the PC ?

Notice that the CDF's of the strength and the stress as

$$F_X(x) = \Phi\left(\frac{X-38}{2}\right) \text{ and } F_Y(y) = \Phi\left(\frac{Y-28}{2.5}\right),$$

respectively, where $\Phi(w)$ is the CDF of N(0,1).

The reliability can be written as

$$R = \Pr\{X > Y\} = \Pr\{X - Y > 0\}.$$

Since $X \sim N(38, 2^2)$ and $Y \sim N(28, 2.5^2)$,

$$\Pr\{X - Y > 0\} = \Phi\left(\frac{\mu_x - \mu_y}{\sqrt{\sigma_x^2 + \sigma_x^2}}\right) = \Phi\left(\frac{38-28}{\sqrt{4+6.25}}\right) = 0.99911 > 0.999.$$

Therefore, a new powerful PC fan is not needed to cool the PC temperature down in order to achieve the required reliable operation. It is usually not necessary to increase cooling capacity when adding RAM to a PC. Modern SIMM type memory uses only about 5 watts of power per 16MB. When upgrading CPUs, however, there can be considerable heat increase. Replacing a 25MHz 486 using 8MB RAM with a 266MHz Pentium II with 64MB of RAM would make a CPU cooling fan necessary and the power supply fan should be checked for good air flow. All cooling air vents should also be cleaned. □

3.4. Multicomponents Systems

3.4.1. Series System

The simplest and most often used configuration is the series system, where all components must function correctly for the system to function correctly. Let C_i indicate the ith component of system. The reliability for series system after operating for a given time t becomes [168]

$$\begin{aligned} R_{sys}(t) &= \Pr\{\text{system works}\} \\ &= \Pr\{\text{all components work}\} \\ &= \Pr\{C_1 \text{ works}\}\Pr\{C_2 \text{ works}|C_1 \text{ works}\}\ldots \\ &\qquad \ldots\Pr\{C_n \text{ works}|C_1,\ldots,C_{n-1} \text{ work}\}. \end{aligned} \tag{3.18}$$

Assuming independence between components, Eq. (3.18) becomes

$$R_{sys}(t) = \Pr\{C_1 \text{ works}\}\Pr\{C_2 \text{ works}\}\ldots\Pr\{C_n \text{ works}\} = \prod_{i=1}^{n} R_i(t) \tag{3.19}$$

where $R_i(t)$ is the reliability of the ith component at time t. Eq. (3.19) implies that a series system is much less reliable than any of its individual components.

□ **Example 3.6**

A series system has five identical components. How good must each component be for the system to meet the reliability objective of 0.95 ?
Given $R_{sys}(t) = R_i(t)^5 \geq 0.95$, each component must be at least at the reliability level of 0.9898. □

3.4.2. Parallel System

Another simple configuration of components is the parallel system, which performs satisfactorily if any one or more of the components performs satisfactorily. The independence between components is again assumed. Such a system is overdesigned with built-in redundancy, but the added cost is presumably a worthwhile trade-off for increased reliability [168]. The system reliability for parallel system is

$$\begin{aligned} R_{sys}(t) &= \Pr\{\text{system works}\} \\ &= 1 - \Pr\{\text{all components fail}\} \\ &= 1 - \Pr\{C_1 \text{ fails}\}\Pr\{C_2 \text{ fails}\}\ldots\Pr\{C_n \text{ fails}\} \\ &= 1 - \textstyle\prod_{i=1}^{n}[1 - R_i(t)]. \end{aligned} \tag{3.20}$$

From Eq. (3.20), we can find that the parallel system is more reliable than any of its individual components.

□ **Example 3.7**

A parallel system has five identical components. How low may the reliability of a component be for the system to meet the reliability objective of 0.95 ? Given $1-[1-R_i(t)]^5 \geq 0.95$, each component reliability can be as low as 0.4507 for the system to have a reliability of 0.95. □

3.4.3. *k*-out-of-*n* Systems

Sometimes, the pure parallel system can be replaced by the k-out-of-n system, where the system functions satisfactorily if any k of the n components are functioning correctly. The series system is equivalent to the n-out-of-n system and the parallel system to the 1-out-of-n system. The system reliability of the general k-out-of-n system is [168]

$$R_{sys}(t) = \sum_{r=k}^{n} \sum_{j=1}^{nCr} \prod_{i=1}^{n} R_i(t)^{V_{i,j}} [1 - R_i(t)]^{\bar{V}_{i,j}} \tag{3.21}$$

where $\sum_{i=1}^{n} V_{i,j} = r$ and $V_{i,j}$ and $\bar{V}_{i,j}$ are complementary indicator functions for which $V_{i,j}$=1 if the ith component functions and $V_{i,j}$=0 if the ith component fails, respectively. For identical or equivalent components where $R_i(t) = R(t)$, Eq. (3.21) has the cumulative binomial form:

$$R_{sys}(t) = \sum_{r=k}^{n} nCrR(t)^r[1 - R(t)]^{n-r}, \tag{3.22}$$

where nCr=$n!/r!(n-r)!$.

□ **Example 3.8**

Assuming that a 2-out-of-5 system has identical components and their failure rate is constant, then the reliability of each component is $R(t) = e^{-\lambda t}$ and the system reliability and MTTF can be calculated using Eqs. (3.9) and (3.22), respectively:

$$R_{sys}(t) = 10e^{-2\lambda t} - 20e^{-3\lambda t} + 15e^{-4\lambda t} - 4e^{-5\lambda t},$$

$$\text{MTTF} = \int_0^\infty (10e^{-2\lambda t} - 20e^{-3\lambda t} + 15e^{-4\lambda t} - 4e^{-5\lambda t})dt = \frac{77}{60\lambda}.$$

In the case of λ=0.001/hr,

$$R_{sys}(200) = 0.9954, \quad \text{MTTF} = 1283.3 \text{ hrs}.$$

Table 3.3 shows the system reliability at t=200 hrs and MTTF for each k. □

Table 3.3: The reliability and MTTF of k-out-of-5 system for t=200.

System	$R_{sys}(200)$	MTTF (hrs)
1-out-of-5	0.9963	1483.3
2-out-of-5	0.9954	1283.3
3-out-of-5	0.9554	783.3
4-out-of-5	0.7751	450.0
5-out-of-5	0.3679	200.0

3.4.4. Non-series-parallel Systems

There are other configurations that do not fall into the previous types. Two kinds of system structure are considered in this section for illustration: a bridge system in Figure 3.5 and a complex system in Figure 3.6. Grosh [168] explains three standard methods for arriving at the reliability of the non-series-parallel systems. They are enumeration, path tracing, and conditioning on a key element.

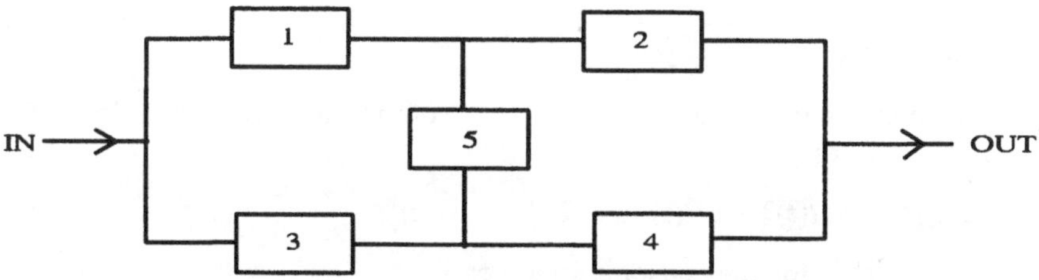

Figure 3.5: Configuration of a bridge system.

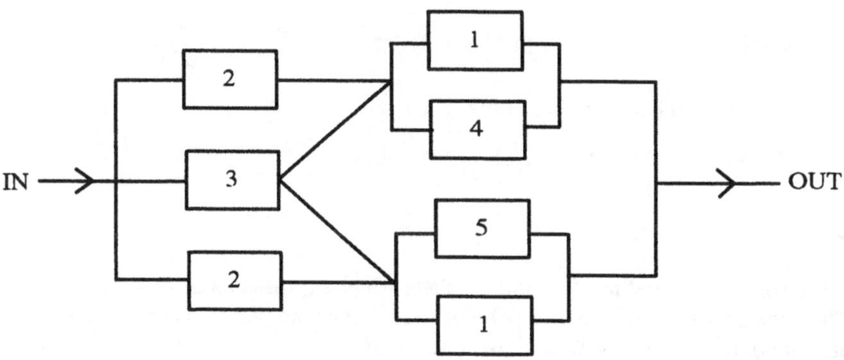

Figure 3.6: Configuration of a complex system.

Among these techniques, we use the technique of conditioning on a key element to calculate the reliability of the systems in Figure 3.5 and Figure 3.6. The component 5 of Figure 3.5 and component 3 of Figure 3.6 are chosen as the key elements since it is probably most fruitful to choose a complicating element as a key element. Using the decomposition rules, the system reliability for Figure 3.5 is

$$\begin{aligned} R_{sys}(t) = & \Pr\{\text{System is good} \mid C_5 \text{ is good}\}\Pr\{C_5 \text{ is good}\} \\ & +\Pr\{\text{System is good} \mid C_5 \text{ is not good}\}\Pr\{C_5 \text{ is not good}\} \\ = & R_1(t)R_2(t) + R_3(t)R_4(t) + R_1(t)R_4(t)R_5(t) \\ & +R_2(t)R_3(t)R_5(t) - R_1(t)R_2(t)R_3(t)R_4(t) \\ & -R_1(t)R_2(t)R_3(t)R_5(t) - R_1(t)R_2(t)R_4(t)R_5(t) \\ & -R_1(t)R_3(t)R_4(t)R_5(t) - R_2(t)R_3(t)R_4(t)R_5(t) \\ & +2\,R_1(t)R_2(t)R_3(t)R_4(t)R_5(t). \end{aligned} \tag{3.23}$$

Reliability in Figure 3.6 can be calculated by

$$\begin{aligned} R_{sys}(t) = & \Pr\{\text{System is good} \mid C_3 \text{ is good}\}\Pr\{C_3 \text{ is good}\} \\ & +\Pr\{\text{System is good} \mid C_3 \text{ is not good}\}\Pr\{C_3 \text{ is not good}\} \\ = & 1 - [1 - R_1(t)]^2[1 - R_4(t)][1 - R_5(t)]R_3(t) \\ & -\{1 - R_2(t)[1 - (1 - R_1(t))(1 - R_4(t))]\} \\ & \times\{1 - R_2(t)[1 - (1 - R_1(t))(1 - R_5(t))]\}(1 - R_3(t)). \end{aligned} \tag{3.24}$$

□ **Example 3.9**

In the case of Figure 3.5, assuming identical components with reliability $R(t)$, Eq. (3.23) reduces to

$$R_{sys}(t) = 2R(t)^5 - 5R(t)^4 + 2R(t)^3 + 2R(t)^2.$$

Since $R(t) = e^{-\lambda t}$ for the constant failure rate,

$$R_{sys}(t) = 2e^{-5\lambda t} - 5e^{-4\lambda t} + 2e^{-3\lambda t} + 2e^{-2\lambda t}.$$

MTTF is obtained by Eq. (3.9),

$$\text{MTTF} = \int_0^\infty (2e^{-5\lambda t} - 5e^{-4\lambda t} + 2e^{-3\lambda t} + 2e^{-2\lambda t})dt = \frac{49}{60\lambda}.$$

Suppose λ=0.001/hr and t=200 hrs, then

$$R_{sys}(200) = 0.9274, \quad \text{MTTF} = 816.7 \text{ hrs.} \qquad \square$$

3.5. Conclusions

We present some definitions and mathematical expressions of basic reliability. The characteristics of the bathtub curve and their applications to the life cycle of semiconductor products are also described.

Another important reliability factor is lifetime distribution. It is known that both the Weibull distribution and the extreme value distribution represent appropriate models for a failure law for multicomponent systems. The Weibull distribution is commonly used to model the infant mortality stage of semiconductor products. Many aging failure mechanisms of semiconductor devices are well described by the lognormal distribution.

4. Yield and Modeling Yield

Among the performance indices for successful IC manufacturing, the yield of manufacturing is regarded as the most important one. Yield is usually defined as the ratio of the number of usable items after the completion of production processes to the number of potentially usable items at the beginning of production [148]. In order to reduce the cycle time and cost, rapid identification of yield losses and early elimination of them are critical. El-Kareh *et al.* [122] emphasize that the process of reducing the chip size should be accompanied by the improvement of yield in order to obtain productivity. They define the chip cost as

$$\text{Chip cost} = \frac{\text{cost per wafer}}{\text{yield} \times \text{number of dice per wafer}}.$$

4.1. Definitions and Concept

4.1.1. Overall Yield

IC device yields depend on many factors such as chip area, circuit design, circuit layout, and others. It is desirable to mathematically explain the overall yield and to effectively control and manage yields by determining the failure probabilities and critical areas for each defect types [405].

The overall yield can be broken down into several components depending on the process grouping or the purpose of application. Here are three key yield components that are commonly used in semiconductor manufacturing:

1. Process yield (Y_{pro}) is the ratio between the number of wafers completing a parametric test to the number of wafers started,

   $$Y_{pro} = \frac{W_o}{W_i},$$

   where W_i and W_o are the numbers of wafers started and completed, respectively.

2. Die yield (Y_{die}) is the ratio between the number of chips that pass specifications to the number of chips that reach the electrical test,

   $$Y_{die} = \frac{CH_o}{CH_i},$$

where CH_o is the number of chips per wafer passing wafer test and CH_i is the number of potentially good chips per wafer. Therefore, the number of chips to be tested is $CH_i \times W_i \times Y_{pro}$.

3. Assembly yield (Y_{asm}) is the ratio between the number of packaged modules that meet the specifications to the number of modules that reach packaging,

$$Y_{asm} = \frac{M_o}{M_i},$$

where M_i and M_o are the numbers of modules to test and to pass the test, respectively. Let CH_m be the number of chips per module. Then,

$$M_i = \frac{Y_{die} \times CH_i \times W_i \times Y_{pro}}{CH_m}.$$

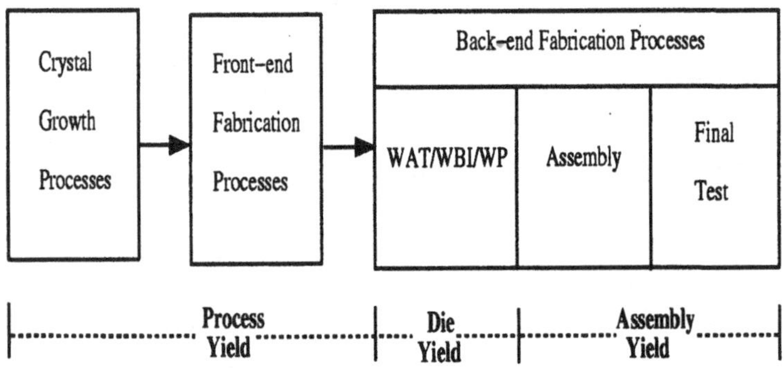

Figure 4.1: IC manufacturing processes and related yield components.

Figure 4.1 shows the relationship between IC manufacturing processes and yield components. Sometimes, the line yield is interchangeably used with the process yield. Define $Y_{line,n}$ as the line yield of the nth process step. Then,

$$Y_{line,n} = \frac{L_{o,n}}{L_{i,n}},$$

where $L_{i,n}$ and $L_{o,n}$ are the numbers of wafers started and completed n^{th} process, respectively. Cunningham [96] describes the line yield over a given production period as

$$Y_{line} = \left[\frac{\text{total wafer moves in the line}}{\text{moves} + \text{ scrap wafers} - \text{ rework wafers}}\right]^N$$

where N is the number of process steps to process a wafer. Usually, the yield models are related to the die yield.

The overall yield can be presented as [122, 148]

$$Y = Y_{pro} \times Y_{die} \times Y_{asm}.$$

Cunningham *et al.* [97] subdivide the yield of a semiconductor into line yield, die yield, and final test yield. This yield categorization is very similar to Ferris-Frabhu's [148]. Generally, wafer fabrication processes directly affect the line yield, and the die yield and packaging process influence the final test yield.

The process yield and die yield are two important factors in managing the productivity of semiconductor manufacturing. Most semiconductor industries focus on the die yield loss especially because the profit obtained from a small investment is relatively large. To remain competitive in the industry, one must attain a high level of die yield.

4.1.2. Defects

For yield prediction, it is useful to categorize defects as nonrandom defects and random defects [122, 148, 404]. Defects which cause circuit failures are called faults or fatal defects [148, 280, 400]. The distinction between defects and faults has an important role in calculating yield based on the defect density and chip area. Another parameter that affects yield is defect clustering. Stapper [400, 401, 403] explains the effects of clustering on yield. For the same average defect density, clustering usually gives higher chip yield [165].

Nonrandom Defects

Nonrandom defects include gross defects and parametric defects. Gross defects are relatively large compared with the size of a circuit or a chip and almost always result in yield loss. Parametric defects affect the device electrical parameters such as gain, breakdown, transconductance, voltage, and resistance, and in most cases lead to reliability failures rather than yield loss.

Random Defects

Random defects are defects that occur by chance. Particles that cause shorts and opens or local crystal defects are random defects. From the point of defect size, random defects are categorized as point defects and spatial defects. Except for lithography defects, the major defects of semiconductor are the point defects. For example, dust particles, defects in the wafers, surface inhomogeneities, pinholes in insulating layers, or crystallographic imperfections are point defects.

4.1.3. Critical Area and Defect Density

The critical area is an area where the center of a defect must fall to create a fault [398, 399]. That is, if a defect occurs in the critical area, then it causes a fault. It is specific to a failure mechanism, a defect size, and layout topology.

Let A_c and $A_c(x)$ be an average critical area and a critical area for defect size x, respectively, and $s(x)$ the pdf of defect size. For the given $s(x)$, A_c is written as

$$A_c = \int_0^\infty A_c(x)s(x)dx. \tag{4.1}$$

The average defect density of all sizes and average defect density of size x are defined as D_0 and $D(x)$, respectively. From the definition, the average defect density is given by

$$D_0 = \int_0^\infty D(x)dx$$

and the relationship between D_0 and $s(x)$ is shown as

$$D(x) = D_0 s(x).$$

For the point defects, if N_D is the total number of random defects in the A_t, the total area on which a defect can fall, then the average defect density is

$$D_0 = \frac{N_D}{A_t}.$$

Next, define μ as the average number of faults caused by defects. It is obtained by

$$\mu = \int_0^\infty A_c(x)D(x)dx = D_0 \int_0^\infty A_c(x)s(x)dx. \tag{4.2}$$

Hence, from Eq. (4.1) and (4.2), we have a direct relationship between D_0 and μ:

$$\mu = A_c D_0. \tag{4.3}$$

The probability of a defect becoming fatal is the fault probability which is denoted as Φ. If a point defect occurs in the critical area, then it will be fatal. However, the probability of a spatial defect being fatal depends on its location and size. Thus, the ratio of the critical area to the total area on which a defect can fall simply gives its fault probability:

$$\Phi = \frac{A_c}{A_t}.$$

4.1.4. Defect Size Distribution

The defect size distribution varies depending on process lines, process time, learning experience gained, and others. Stapper [398] and Ferris-Prabhu [146] assume that there is a certain critical size at which the density function peaks, and then decreases on either side of the peak. Though there exist some distribution functions that have a behavior like this, it is not easy to handle them analytically. Therefore, it is assumed that the defect size pdf is defined by a power law for defects smaller than the critical size and by an inverse power law for defects larger than the critical size [148]. Let x_0 be the critical size of the defect with the highest probability of occurrence. Generally, the defect size pdf is defined as

$$s(x) = \begin{cases} c x_0^{-q-1} x^q, & 0 \le x \le x_0 \\ c x_0^{p-1} x^{-p}, & x_0 \le x \le \infty \end{cases} \tag{4.4}$$

where $p \neq 1$, $q > 0$, and $c = (q+1)(p-1)/(q+p)$. It is experimentally shown that x_0 must be smaller than the minimum width or spacing of the defect monitor, which is denoted as s_0 [398]. Figure 4.2 shows the defect size pdf. Defects smaller than x_0 can not be resolved well by the optics used in

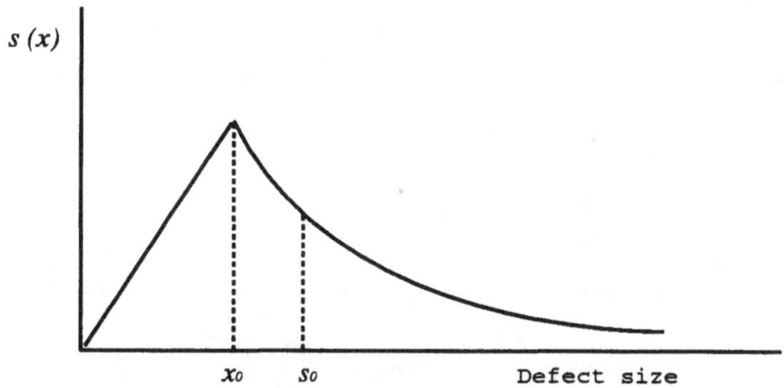

Figure 4.2: The defect size pdf.

the photolithographic process [399]. Since very small defects are assumed to increase linearly with defect size to a point x_0, Stapper [398, 399] indicates that using values of $q = 1$ and $p = 3$ for the spatial distribution agrees reasonably well with experimental data. In his results, p=2.85 for open circuits and p=3.10 for short circuits are obtained. When $q = 1$ and $p = 3$, the defect size distribution of Eq. (4.4) becomes

$$s(x) = \begin{cases} x_0^{-2} x, & 0 \le x \le x_0 \\ x_0^2 x^{-3}, & x_0 \le x \le \infty. \end{cases} \tag{4.5}$$

Assuming that x_0 is the minimum defect size,

$$s(x) = \frac{2x_0^2}{x^3}, \quad x_0 \le x \le \infty. \tag{4.6}$$

It is necessary to keep in mind that although the defect density is a widely used term, it is practically not a measured quantity, but an inferred quantity [147]. The randomness of the defect distribution implies that manufacturing processes do not introduce new defects that may change the distribution function.

□ **Example 4.1**

(1) A Single Conductive Line

To start with a simple case, assume that we have a very long, straight conductor that has a width w and a length L. Open circuits are usually the most likely sources of failure in the conductor and are caused by holes in the conductive material. These holes result in missing photolithographic patterns. If these defects are relatively smaller than the width of a conductor, then it is possible to conduct currents without failure. But, if the width of conductive material left is less than d_m, then the conductor is considered affected. We assume the defect is circular and the defect size is the diameter of that circle, which is denoted as x in Figure 4.3.

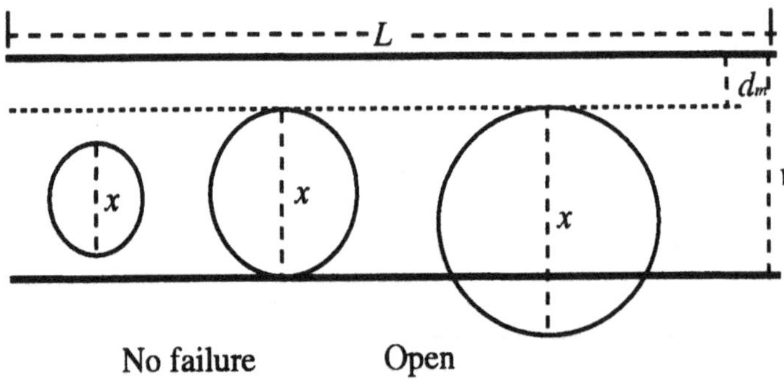

Figure 4.3: Defects in a single conductive line.

Thus, the critical area for a single conductor is mathematically given by

$$A_c(x) = \begin{cases} 0, & 0 \leq x \leq w - d_m, \\ L(x + 2d_m - w), & w - d_m \leq x \leq \infty. \end{cases} \tag{4.7}$$

The average critical area and and the average number of faults are

$$A_c = \int_{w-d_m}^{\infty} L(x + 2d_m - w)\frac{2x_0^2}{x^3}dx = \frac{Lx_0^2}{w}\left(\frac{w}{w - d_m}\right)^2$$

and

$$\mu = D_0\frac{Lx_0^2}{w}\left(\frac{w}{w - d_m}\right)^2,$$

respectively. In the FET memory chip yields, it is convenient to set d_m equal to zero [398]. When d_m=0, Eq. (4.7) becomes

$$A_c(x) = \begin{cases} 0, & 0 \leq x \leq w, \\ L(x - w), & w \leq x \leq \infty. \end{cases} \tag{4.8}$$

In the same way, the average critical area and the average number of faults are obtained as

$$A_c = \int_{w}^{\infty} L(x - w)\frac{2x_0^2}{x^3}dx = \frac{Lx_0^2}{w} \tag{4.9}$$

and

$$\mu = \frac{D_0 L x_0^2}{w},$$

respectively.

Next, consider a chip with length L and width W having a single conductor. The critical area for the conductor on a chip is given by [399]

$$A_c(x) = \begin{cases} 0, & 0 \leq x \leq w, \\ L(x-w), & w \leq x \leq w+W, \\ LW, & w+W \leq x \leq \infty. \end{cases}$$

The average critical area for this $A_c(x)$ is

$$A_c = \int_w^{w+W} L(x-w)\frac{2x_0^2}{x^3}dx + \int_{w+W}^{\infty} LW\frac{2x_0^2}{x^3}dx = \frac{Lx_0^2}{w}\frac{w^2+W^2}{(w+W)^2}.$$

If the chip width W is much longer than the conductor width w, it approaches the average critical area of Eq. (4.9) for a single conductive line.

(2) Multiple Conductive Lines

Figure 4.4 shows open and short failures in two conductive lines. For two very

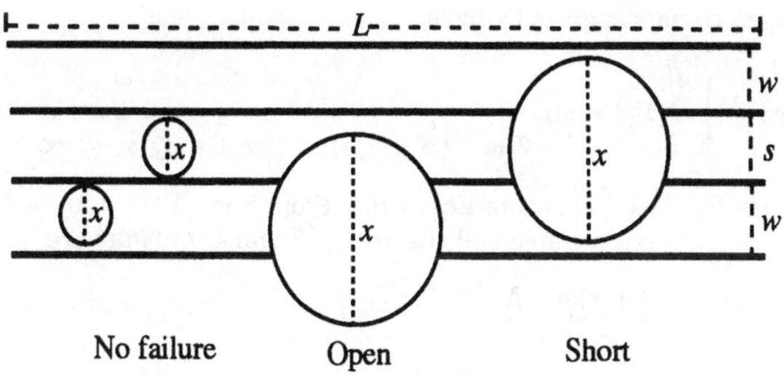

Figure 4.4: Defects in two conductive lines.

large conductors, separated by a narrow slit of width s and length L, the critical area for open circuits is given by [399]

$$A_c(x) = \begin{cases} 0, & 0 \leq x \leq w \\ 2L(x-w), & w \leq x \leq 2w+s \\ L(x+s), & 2w+s \leq x \leq \infty. \end{cases} \tag{4.10}$$

From Eq. (4.1), (4.2), (4.6), and (4.10), the average critical area and the average number of faults can be calculated by

$$A_c = \int_w^{2w+s} 2L(x-w)\frac{2x_0^2}{x^3}dx + \int_{2w+s}^{\infty} L(x+s)\frac{2x_0^2}{x^3}dx = \frac{L(3w+2s)x_0^2}{w(2w+s)}$$

and

$$\mu = \frac{D_0 L(3w+2s)x_0^2}{w(2w+s)},$$

respectively.

We can consider open circuits as well as short circuits in two conductive lines, but the critical area for the short circuits is calculated by using s instead of w in Eq. (4.8):

$$A_c(x) = \begin{cases} 0, & 0 \leq x \leq s, \\ L(x-s), & s \leq x \leq \infty. \end{cases}$$

For a chip of length L and width W with two conductive lines, the critical area is calculated by the same concept:

$$A_c(x) = \begin{cases} 0, & 0 \leq x \leq w \\ 2L(x-w), & w \leq x \leq 2w+s \\ L(x+s), & 2w+s \leq x \leq W-s \\ LW, & W-s \leq x \leq \infty. \end{cases}$$

Since most IC chips have many interconnecting wires, it is useful to find the critical area of N parallel conductors. The space between two conductors is s, and N conductors have the same line length L and width w. The critical area for open circuits is given by [399]

$$A_c(x) = \begin{cases} 0, & 0 \leq x \leq w \\ NL(x-w), & w \leq x \leq 2w+s \\ L(x+(N-2)w+(N-1)s), & 2w+s \leq x \leq \infty. \end{cases} \tag{4.11}$$

If N=2, then Eq. (4.11) equals to Eq. (4.10). From Eqs. (4.1), (4.2), (4.6), and (4.11), the average critical area and the average number of faults are

$$A_c = \frac{Lx_0^2}{w}\left(\frac{(N+1)w+Ns}{2w+s}\right)$$

and

$$\mu = \frac{D_0 L x_0^2}{w}\left(\frac{(N+1)w+Ns}{2w+s}\right),$$

respectively. When the spacing between the lines is very large, we get

$$\lim_{s\to\infty} A_c = N\frac{Lx_0^2}{w},$$

which is equal to N times the critical area of a single conductive line in Eq. (4.9). When, however, the space between the conductive lines is equal to the line width, we obtain

$$A_c = \left(\frac{2N+1}{3}\right)\frac{Lx_0^2}{w}.$$

This suggests that the total critical area is less than the sum of the critical areas of N conductors due to the proximity effect [399].

To calculate the critical area of the short circuits in N conductive lines, we consider that there exist $(N-1)$ spaces. By swapping w and s of Eq. (4.11), we get

$$A_c(x) = \begin{cases} 0, & 0 \leq x \leq s \\ (N-1)L(x-s), & s \leq x \leq 2s+w \\ L(x+(N-3)s+(N-2)w), & 2s+w \leq x \leq \infty. \end{cases}$$

For a chip of length L and width W with N conductive lines, the critical area for the open circuit is given by [399]

$$A_c(x) = \begin{cases} 0, & 0 \leq x \leq w \\ NL(x-w), & w \leq x \leq 2w+s \\ LW, & 2w+s \leq x \leq \infty. \end{cases} \tag{4.12}$$

The average critical area for Eq. (4.12) is calculated as

$$A_c = \frac{Lx_0^2}{w}\left(\frac{N(w+s)^2+wW}{(2w+s)^2}\right).$$

Consequently, the average critical area is determined by the known design parameters. □

4.2. Yield Models

A yield model is used to bridge from monitor to product, to bridge from product to product, or to predict yield before committing to a product [165]. That is, it is used to estimate the future yield of a current or new product and yield loss from each of the process steps. The Wallmark's model [435] is known as one of the earliest yield models. He presented the probability of transistor failure as a ratio of the number of failing transistors in a batch of 100 and used a binomial distribution to estimate the yield, Y, of IC with redundant transistors:

$$Y = \frac{n!}{m!(n-m)!}(x/100)^{n-m}(1-x/100)^m$$

where n is the total number of transistors, m the number of good transistors, and x is the number of failing transistors occurring in a batch of 100. Among the models developed after this, the Poisson yield model and negative binomial yield model are frequently used.

4.2.1. Poisson Yield Model

The Poisson model assumes that the distribution of faults is random and the occurrence of a fault at any location is independent of the occurrence of any other fault. For a given μ, the probability that a chip contains k defects is

$$P_k = \frac{e^{-\mu}\mu^k}{k!}, \quad k = 0, 1, \ldots.$$

Since the yield is equivalent to the probability that the chip contains no defects,

$$Y = P_0 = e^{-\mu} = e^{-A_c D_0}. \tag{4.13}$$

If the defect density is not a constant, instead has its own probability density function, $f(D)$, then the yield model of Eq. (4.13) leads to another expression [294]:

$$Y = \int_0^\infty e^{-A_c D} f(D) dD. \tag{4.14}$$

Using the Dirac function, $\delta(D - D_0)$, which vanishes everywhere except at $D = D_0$, the constant defect density is presented as:

$$f(D) = \delta(D - D_0).$$

From Eq. (4.14),

$$Y = \int_0^\infty e^{-A_c D} \delta(D - D_0) dD = e^{-A_c D_0}.$$

The Poisson yield model is widely used, but it sometimes gives a lower predicted yield than what is observed [148].

□ **Example 4.2**

A design house is looking for a foundry fab to manufacture its product. To have a common comparison criterion, the Poisson yield model of Eq. (4.13) is used. After reviewing the historical records of many candidates, two foundry makers are considered:
(1) Fab A cites an initial defect density, D_0, for process capability between $0.4\mu m$ and $0.6\mu m$ is about $0.6/cm^2$. After reviewing the circuit design and evaluating its own process technology, the process manager of fab A proposes that the defect density can be improved to $0.2/cm^2$ after two quarters using a $0.45\mu m$ production line. Fab A estimates the critical area, A_c, will be around 100 mm^2 and there will be 240 dice in an 8-inch wafer. In other words, by the Poisson yield model, the yield of this $0.45\mu m$ production line is expected to be enhanced from 55% (132 dice/wafer) to 81.9% (196 dice/wafer).
(2) Fab B is able to manufacture the product using $0.35\mu m$ technology in its 6-inch production line. It's estimated that the average defect density in 6 months is $0.4/cm^2$. The critical area is $64mm^2$ and there will be 185 dice in a 6-inch wafer. That is, the yield will be 77.4%, which is equivalent to 143 dice in each 6-inch wafer. Assuming linear increase in the yield for fab A, the average number of dice in an 8-inch wafer is 164, which is still larger than the average good dice per wafer for fab B (143 dice/wafer). However, fab B may win the bid because

- the unit cost per die for fab A may be higher than that for fab B since fab A uses the 8-inch production line,
- generally, the yield of a 6-inch production line is more stable than that of an 8-inch line, and

- the die size of fab B is only 65% of that of fab A.

If the HCI and TDDB lifetimes of both designs are similar, the smaller dice from fab B will usually have better overall reliability because the probability for die crack and reliability problems from assembly for smaller dice is usually lower than that of bigger dice. In a word, yield, cost, and reliability issues all have to be taken into account when choosing a subcontractor as shown above. □

4.2.2. Murphy's Yield Model

If we assume that defect density follows a normal distribution which is approximated by a triangular distribution (i.e. Simpson distribution), then Murphy's yield model is obtained. Define the pdf of the defect density as

$$f(D) = \begin{cases} D/D_0^2, & 0 \leq D \leq D_0 \\ (2D_0 - D)/D_0^2, & D_0 \leq D \leq 2D_0. \end{cases} \tag{4.15}$$

From Eqs. (4.14) and (4.15), the yield model is

$$Y = \left(\frac{1 - e^{-A_c D_0}}{A_c D_0} \right)^2 . \tag{4.16}$$

The predicted yields of this model agree well with actual yields within a tolerance [294].

4.2.3. Seed's Yield Model

The assumption that defect density is exponentially distributed gives another yield model. Define

$$f(D) = \frac{1}{D_0} e^{-D/D_0}.$$

From Eq. (4.14), the yield model is

$$Y = \frac{1}{1 + A_c D_0}. \tag{4.17}$$

Seed's yield model generally gives us higher yields than the actual observations [148]. Price [334] derived the same result by considering the total number of ways indistinguishable defects can be distributed among chips.

4.2.4. Negative Binomial Yield Model

If the defect distribution is random but clustered, the negative binomial yield model is used. The negative binomial model assumes that the likelihood of an event occurring at a given location increases linearly with the number of events that have already occurred at that location [147].

Assuming that the defect density follows a gamma distribution

$$f(D) = \frac{1}{\Gamma(\alpha)\beta^\alpha} D^{\alpha-1} e^{-D/\beta},$$

from Eq. (4.14), the probability that one chip contains k defects follows negative binomial distribution

$$P_k = \int_0^\infty \frac{e^{-A_cD}(A_cD)^k}{k!}\frac{D^{\alpha-1}e^{-D/\beta}}{\beta^\alpha\Gamma(\alpha)}dD = \frac{\Gamma(\alpha+k)(A_c\beta)^k}{k!\Gamma(\alpha)(1+A_c\beta)^{\alpha+k}} \qquad (4.18)$$

where

- k the exact number of faults in given area
- μ the expected number of faults in given area
- α the clustering factor, $(\mu/\sigma)^2$
- μ the average defect density$(\alpha\beta)$
- σ^2 the variance of defect density $(\alpha\beta^2)$.

Since $D_0 = \alpha\beta$, Eq. (4.18) can be rewritten as

$$P_k = \frac{\Gamma(\alpha+k)\left(\frac{A_cD_0}{\alpha}\right)^k}{k!\Gamma(\alpha)\left(1+\frac{A_cD_0}{\alpha}\right)^{\alpha+k}}.$$

Therefore, the yield model is given by

$$Y = \left(1+\frac{A_cD_0}{\alpha}\right)^{-\alpha}. \qquad (4.19)$$

The clustering factor α determines the degree of clustering of the model. If α is equal to 1, then Eq. (4.19) is equivalent to the Seed's yield model of Eq. (4.17). If α goes to ∞, then Eq. (4.19) gives the same result as the Poisson model of Eq. (4.13), implying no clustering. The practical range of α is 0.3 to 5.0. Stapper [717, 402] reports that this model fits well with actual yield data.

4.2.5. Other Yield Models

If the defect density is uniformly distributed over the interval $[0, 2D_0]$, then the yield is given by

$$Y = \int_0^{2D_0} \frac{1}{2D_0}e^{-A_cD}dD = \frac{1-e^{-2A_cD_0}}{2A_cD_0}.$$

This model predicts yield higher than the observed yield [294].

Okabe [307] presents another yield model. When each mask level has the same defect density D_0 and exponential density distribution, the overall density distribution follows an Erlang distribution:

$$f(D) = \frac{x/D_0}{(x-1)!}e^{-xD/D_0}D^{x-1}$$

where x is the number of mask levels. Therefore, the yield is given by

$$Y = \frac{1}{(1+A_cD_0/x)^x}$$

and it is structually similar to the negative binomial yield model of Eq. (4.19), but the derivation is different. Stapper [402] notes that this yield model does not agree well with data.

□ **Example 4.3**

A single circuit has an average of 0.001 faults per circuit. If we use the Poisson yield model of Eq. (4.13), then the yield of a chip with 500 of these circuits is equal to

$$e^{-500\times 0.001} = 60.7\%.$$

If the defects are clustered with α=0.5, then the yield is calculated by the Eq. (4.19) and equals to

$$\left(\frac{1}{1+0.5/0.5}\right)^{0.5} = 70.7\%.$$

Table 4.1 shows estimated yield for α=0.5, 1.0, 2.0, ∞ with different yield models. □

Table 4.1: The estimated yield for different redundancy.

Model	α	D_0=0.001 A_c=1 μ=0.001	D_0=0.001 A_c=500 μ=0.5	D_0=0.001 A_c=10000 μ=10
Negative Binomial	0.5	99.9%	70.7%	21.8%
	1.0	99.9%	66.7%	9.1%
	2.0	99.9%	64.0%	2.8%
Poisson	∞	99.9%	60.7%	0.0%
Murphy	-	99.9%	61.9%	1.0%
Seed	-	99.9%	66.7%	9.1%

4.2.6. Different Approaches to Yield Modeling

A Variable Defect Size Approach

Berglund [43] represents a variable defect size (VDS) yield model including both the conventional small defects and much larger parametric or area defects. To do this, Eq. (4.13) can be modified as:

$$Y = \exp\left[-\int A_c(x)D(x)dx\right] \tag{4.20}$$

where $D(x)$ is the mean density of defects of size x and $A_c(x)$ is the critical area to a defect of size x. Assume that the larger defects are circular in shape with diameter x for a die of length L and width W, the total critical area sensitive to such larger defects are [43]

$$A_c(x) = LW + (L+W)x + \pi x^2/4.$$

Let Y_d be the yield loss due to the defects of small size and Y_x the yield loss due to the defects which are comparable or larger than the die size. Berglund [43] shows that Eq. (4.20) is written in the product form of two exponential terms:

$Y = Y_d \times Y_x$. Therefore, the die-area-independent yield loss term, Y_d, can be viewed as the gross yield and the additional die-area-dependent term, Y_x, accounts for the added yield loss around the edges of the larger parametric defects. Berglund [43] concludes that by selecting appropriate values for some parameters, the VDS model will satisfactorily match most experimental data of yield versus die area that can also be matched by defect clustering models.

A Circuit Count Approach

It is generally believed that yield is a function of chip area and that larger chips give lower yields. However, there are some cases in which the yields scatter in the wide range for chips with the same areas, because of the variation in circuit density. Stapper [405] presents a circuit count approach to yield modeling which includes the number of circuits, n_j, and the average number of random faults, μ_j, per circuit type j. The binomial yield model of this approach is given by

$$Y = Y_0 \left\{ 1 + \sum_j \frac{n_j \mu_j}{\alpha} \right\}^{-\alpha}$$

where Y_0 is the gross yield and α is a cluster factor.

Absolute Yield Model

To analyze and compare the yield of products from different semiconductor manufacturing facilities, Cunningham *et al.* [97] present a common yield model. The first step needed is to select the technological and organizational factors influencing the yields of different manufacturing processes. Table 4.2 shows factors considered by them [97]. After applying a linear regression model to

Table 4.2: Factors for the absolute yield model.

Process/Product	Facility	Practice
die area	cleanroom class	automated SPC
feature size	cleanroom size	commitment to SPC
mask layers	facility age	automated CAM
technology type	facility region	paperless CAM
wafer size	photo make/model	organizational
process age	photo link	yield modeling

a sample of yield data from 72 die types in separate processes, they conclude that die size, process age, and photo link are significant variables. The resulting absolute yield model with R^2 (coefficient of determination)=0.6 is given by [97]

$$\begin{aligned} Y &= e^Z(1+e^Z)^{-1} \\ Z &= 0.33 - 0.80X_1 + 0.34X_2 + 0.39X_3, \end{aligned}$$

where

X_1 the die size variable refers to the area in cm^2

X_2 the process age variable refers to the time span in months between the first and last yield data supplied

X_3 the photo link variable, which is 1 if the photolithography system is linked and -1 otherwise.

Yield Model with Repair Capabilities

Michalka *et al.* [280] suggest a yield model to illustrate the effect of non-fatal defects and repair capabilities on yield calculations. Assume a die having both core and support areas in which defects randomly occur. The support area yield is the probability that there are no fatal defects in the support area:

$$Y_s = \int_0^\infty e^{-DA_s} f(D)dD \tag{4.21}$$

where A_s is the support critical area. The core area yield includes the chance of defects being repaired. To do this, we need one more assumption: that fatal defects can be independently repaired with probability P_{rep}; however, no repair is possible in the support area. Let $Y_c(i)$ be the core yield given that there are i defects in the die. Then, the core area yield is [280]

$$Y_c = \sum_{i=0}^{\infty} \left\{ \int_0^\infty \frac{(DA_{t,c})^i e^{-DA_{t,c}}}{i!} f(D)dD \right\} Y_c(i), \tag{4.22}$$

where $A_{t,c}$ is the core area. From Eqs. (4.21) and (4.22), the die yield is the product form of the support area yield and core area yield:

$$Y = Y_s Y_c.$$

Productivity of Wafer

The productivity of wafer is defined as the number of circuits available per wafer after fabrication [148]. All parameters except defect density are invariant after the design is fixed. The defect density is not a design parameter but results from the processes of fabrication. Based on an existing reference product, Ferris-Prabhu [148] suggests a method to predict the productivity of a new product, q quarters after start of normal production, which is given by

$$P_q(s) = n_s N(s) Y_q(s).$$

where n_s is the number of circuits per square chip of edge s, $N(s)$ is the number of square chips per wafer, and $Y_q(s)$ is the predicted yield for new product after q quarters.

Yield Models to Accelerate Learning

Dance and Jarvis [100] present a performance-price improvement strategy using yield models and an application of yield models to accelerate learning. Using yield models to accelerate the progress of a learning curve reduces learning cycle time to deliver required manufacturing technology within the time frame set by the competition [100]. They present four major improvement techniques to accelerate learning [100]: fine-grain yield models, short-loop defect monitors, equipment particulate characteristic, and yield confidence intervals.

4.3. Yield Prediction

By scaling yield data from an existing product, yield for a new product can be predicted. Since the defect density, $f(D)$, or the average number of faults per chip, μ, is not known in advance, design-related and yield-related information and sources of yield losses are required to predict yield [148]. Ferris-Prabhu [148] mentions three scale factors (area, sensitivity, and complexity) and two characteristic yield prediction models (composite yield model and layered yield model).

4.3.1. Scale Factors

Under the assumption that the average number of faults per chip is not constant across a wafer, the area scale factor represents the chance of finding a good chip as decreasing with the radial distance of the chip from the center of the wafer. Let γ_a be the area scale factor. Then,

$$\gamma_a = \left(\frac{A_n}{A_e}\right)^{1-b}, \quad 0 \le b \le 1$$

where A_n and A_e are the areas of the new and existing products, respectively, and b is the measure of the deviation of the defect density. Stapper [402] finds that b=0.68 is valid for many products.

The second factor that affects the defect density is the sensitivity scale factor, which represents the sensitivity of a product to defects such as the minimum dimension of the pattern. That is, a product with smaller design dimensions is more susceptible to smaller defects, though the size distribution and defect density are the same. Let γ_s be the sensitivity scale factor. Then,

$$\gamma_s = \left(\frac{u_e}{u_n}\right)^{q-1}$$

where u_n and u_e are the minimum design dimensions on new and existing products. Usually, the value of q is taken to be 3.

The complexity scale factor accounts for the difference in the number of mask layers. If the number of mask levels increases, then a chance exists that the number of faults will increase. Let γ_c be the complexity scale factor. The range of γ_c is usually between 0.9 and 1.1 [148].

4.3.2. Prediction Models

Composite yield models predict the yield based on the composite chip and the average number of faults of all types. In the layered yield models, the yield of each layer is predicted, and the total yield is calculated by taking the product of each predicted yield. Since the Poisson yield model satisfies the property that the average of the sum of all faults is equal to the sum over all layers of the average number of faults per layer [148]:

$$\mu = \sum_{i=1}^{N} \mu_i,$$

it is frequently used to predict yield for a new product.

If the yield is predicted using a composite yield model, then the predicted yield for a new product is given by [148]

$$Y = e^{-\mu} = e^{-\gamma \mu_e} = Y_e^{\gamma}$$

where

Y_e the yield for existing product

μ_e the average number of faults of all types for exsiting product:

$$\mu_e = -\ln(Y_e)$$

γ the overall scale factor:

$$\gamma = \frac{M_n}{M_e} \gamma_a \gamma_s \gamma_c$$

M_n the number of mask levels for new product

M_e the number of mask levels for existing product.

If we apply the Poisson yield model to the layered yield model, then the yield of the ith layer is

$$Y_i = e^{-\mu_i} = \prod_{j=1} Y_{ij} = \prod_{j=1} Y_{e,ij}^{\gamma_{ij}}$$

where Y_{ij} and $Y_{e,ij}$ are the predicted and existing yield for j^{th} fault type of the ith layer, respectively. The predicted yield is given by

$$Y = \prod_{i=1}^{M_n} Y_i.$$

It is known that the layered yield model is better than the composite yield model, since the layered model uses more information separately [148].

4.4. Yield Estimation

A typical VLSI manufacturing process generally produces 100 - 200 chips in a wafer and 30 - 40 wafers in a batch, and the cost effectiveness of this process is measured by estimating the yield of each wafer through measurements of test chips on the wafer [174] . Hansen and Thyregod [175] suggest techniques for estimating defect density such as: visual inspection, electrical measurements, outlier inspection, and the normal score method. They conclude that the outlier inspection technique is more effective than others in estimating yields and defect densities.

4.4.1. Test Chips and Structures

The purpose of using test chips is to measure easily the quality of a process because it would be difficult or impossible to measure the fully functional chips. A scaling factor, γ, exists between a fully functional chip and a test chip structure. Hanssen [174] assumes that the scaling factor γ is either known or estimated through a separate experiment. The scaling factor is simply given by [174]

$$\gamma = \frac{A_F}{A_T}$$

where A_F and A_T are the critical area of a fully functional chip and test chip, respectively. If the test chips yield is very close to 1, then the reduction of the width of the metal serpentine of test structure is recommended. In this case, we obtain a scaling factor close to 1 or even less than 1, due to the increased critical defect density and a chance to detect latent defects.

4.4.2. Yield Estimation from Test Chips Measurements

Hansen [174] considers the chips on the wafer as the coordinate system (i,j) with origin in the center of the wafer and the chips as organized in a matrix. Define N_{ij} as the total number of yield defects of a chip at location (i,j), $D(x)$ the defect density of size x, and $P(x)$ the probability of causing a fault. It is convenient to assume that N_{ij} follows a Poisson distribution with mean defect μ_{ij}, which is given by

$$\mu_{ij} = D_{ij} A_T$$

where D_{ij} is defect density of a chip at location (i,j)

$$D_{ij} = \int_0^\infty D(x)P(x)dx.$$

Since the critical area is constant for a given test structure, μ_{ij} is considered as the parameter varying between test chips. However, the μ_{ij} may not be the same for two different wafers and batches. For the given n_T test chips per a wafer, the sample wafer yield of test chips Y_T is [174]

$$Y_T = \frac{\text{Number of good test chips}}{\text{Number of test chips}} = \frac{W_T}{n_T} = \frac{1}{n_T} \sum_{ij(TT)} I(N_{ij} = 0)$$

where $\sum_{ij(TT)}$ is the sum of all test chip (i,j) locations and $I(N_{ij} = 0)$ is an indicator function. Therefore, the expected test chip yield is given by [174]

$$E[Y_T] = \frac{1}{n_T} \sum_{ij(TT)} Pr(N_{ij} = 0) = \frac{1}{n_T} \sum_{ij(TT)} e^{-\mu_{ij}}.$$

If the defect density is constant across the wafer (μ_{ij}=μ, $\forall$ i,j), the number of good chips W_T follows a BIN(n_T,$e^{-\mu}$) and

$$E[Y_T] = e^{-\mu} = e^{-DA_T}.$$

The expected wafer yield of fully functional chips is obtained as [174]

$$E[Y_F] = e^{-DA_F} = e^{-D\gamma A_T} = E[Y_T]^\gamma. \tag{4.23}$$

Therefore, by Eq. (4.23), the estimator for wafer yield of fully functional chips is

$$E[\hat{Y}_F] = Y_T^\gamma. \tag{4.24}$$

However, there is no guarantee that the estimator defined by Eq. (4.24) is unbiased:

$$E[Y_T^\gamma] \neq E[Y_T]^\gamma.$$

We can see that only when γ=1 the unbiased estimate is obtained.

4.4.3. Monte-Carlo Simulation

For demonstration purpose, it is convenient to use Monte-Carlo simulation to study the effectiveness of the yield estimation, due to the complexity in estimating the wafer yield based on scaled test structures [174]. Hansen [174] uses a wafer with 121 chip locations and considers three layouts of test chip locations (n_T=5, n_T=9, n_T=13). He evaluates the effectiveness of an estimator by comparing the mean square error and the standard error. The t-test can be applied in testing the null hypothesis (the estimator is unbiased).

Basic steps of the Hansen's simulation process follow:

Step 1 Create the chip failure probability for each chip location.

Step 2 Simulate each chip as failing or not failing.

Step 3 Calculate the test chip yield (Y_T).

Step 4 Estimate the expected wafer yield ($\hat{y_F}$) using Eq. (4.24).

Step 5 Calculate the observed yield of fully functional chips and the expected value based on the model used.

Three different models are considered for the variation of the defect density across the wafer:

1. Constant defect density.

$$\begin{aligned} \mu_{ij} &= \mu, \quad \forall i,j \\ \Pr\{\text{test chip failure}(i,j)\} &= 1 - e^{-\mu} \\ \Pr\{\text{fully functional chip failure}(i,j)\} &= 1 - e^{-\gamma\mu}. \end{aligned}$$

2. Radial variation of the defect density .
The defect density near the edge of the wafer is higher than the defect density near the center.

$$\begin{aligned} \mu_{ij} &= \mu_0 + (\mu_1 - \mu_0)\frac{(i^2+j^2)}{R^2} \\ \Pr\{\text{test chip failure}(i,j)\} &= 1 - e^{-\mu_{ij}} \\ \Pr\{\text{fully functional chip failure}(i,j)\} &= 1 - e^{-\gamma\mu_{ij}} \end{aligned}$$

where $R^2 = max_{ij}\{i^2 + j^2\}$ and μ_0 and μ_1 are the defect densities near the center and near the edge of the wafer, respectively.

3. Random defect density with autoregressive spatial variation.
Instead of the defect density varying deterministically within the wafer, the defect density is a random variable with autoregressive spatial variation.

$$\Lambda_{ij} = \mu_0 + \psi_{ij}$$

where μ_0 is a constant and ψ_{ij} a random variable. An autoregressive model for ψ is

$$\psi_{ij} = \frac{a}{4}(\psi_{i-1j} + \psi_{i+1j} + \psi_{ij-1} + \psi_{ij+1}) + \epsilon_{ij}, \quad \epsilon_{ij} \in N(0, \sigma^2)$$

where $-1 < a < 1$.

The combined model for the three defect densities is

$$\Lambda_{ij} = \mu_0 + (\mu_1 - \mu_0)\frac{(i^2 + j^2)}{R^2} + \psi_{ij}.$$

4.5. Fault Coverage and Occurrence

The defect level is defined as the percentage of circuits passing all phases of a manufacturing test that are actually defective [91]. Thus, the defect level represents the proportion of product which may fail because of extrinsic failure (or infant mortality) [197]. Let D_L be the defect level of IC. Then it is given by [445]

$$D_L = 1 - Y^{1-T} \tag{4.25}$$

where Y is the yield and T the fault coverage, which is the ratio of the number of detected faults and the number of faults assumed in the fault list. The reliability, in this context, represents the proportion of chips having no reliability defects which could only fail for random or wearout mechanisms. Sometimes, it is called the reliable fraction of a chip, which is given by

$$R_L = 1 - D_L = Y^{1-T}.$$

The basic assumption of Eq. (4.25) is that all faults have equal probability of occurrence, which implies no clustering. That is, the faults are uniformly distributed. Corsi [91] extends this for non-equiprobable faults, using a generalized weighted fault coverage T:

$$T = \frac{\sum_{j=1}^{m} A_{cj} D_{0j}}{\sum_{i=1}^{n} A_{ci} D_{0i}}$$

where m and n are the number of faults tested and the total number of faults assumed, respectively. This relationship is useful to estimate the defect level (or reliable fraction) after a test or to determine how much testing is necessary to obtain an assigned defect level (or reliable fraction).

Seth and Agrawal [371] define the probability of fault occurrence as the probability with which the fault will occur on a chip and a method of determining it from chip test data. Their attempt is to find a probability of fault occurrence for individual faults instead of a distribution for them and to revise the conventional fault coverage by combining the probability of fault occurrence with the probability of fault detection. They call the product of these two probabilities the absolute failure probability of a chip.

It is assumed that a chip failure on an applied test vector is a random event with a density function $f(x)$, and x is the probability that a fault has occurred and is tested by the vector. Thus, $f(x)dx$ is the fraction of chips that fail on a vector with a probability between x and $x + dx$. Since only a fraction of $1 - Y$ of the total chips can fail, we have

$$f(x) = Y\delta(x) + p(x)$$

where $\delta(x)$ is the delta function and $p(x)$ is a partial density function. Since $1 = \int_0^1 f(x)dx = Y + \int_0^1 p(x)dx$,

$$\int_0^1 p(x)dx = 1 - Y.$$

Let N be the total number of test vectors applied. After application of N test vectors, the true yield is given by [371]

$$Y = 1 - \frac{1}{c}\sum_{i=1}^{N} c_i - \frac{2N+1}{N}\frac{1}{c}\sum_{i=1}^{N} c_i \frac{i(i+1)}{(N+i)(N+i+1)}$$

where c is the total number of chips tested and c_i is the number of chips that fail exactly at test vector number i.

4.6. Yield-reliability Relation Model

Yield and reliability are two important factors affecting the profitability of semiconductor manufacturing, and the correlation between them has not been clearly identified. Some reliability prediction models describe the defect level or the reliable fraction of products as a function of yield. Most models are based on the relationship of the device degradation and the long-term reliability. These models can only be used to estimate the defect level after a final test or to interrelate failures with the ultimate reliability [91, 197, 232, 445]. If one wants to identify the effects of stresses or conditions causing the infant mortality failures, it is necessary to relate the reliability model with defect reliability physics and to describe that as the function of yield.

Defects that result in yield loss are called yield defects. Defects that cause reliability problems under wearing conditions are called reliability defects. Huston [197] presents some causes of reliability defects such as random contamination, wafer handling, excessive processes or material variations. Since reliability defects cause near open-circuit or near short-circuit failures, they are very difficult to detect and usually more likely to occur in the infant mortality region.

Various sizes of defects may happen on a chip. If the size of a defect is small enough, it may have a negligible effect on the chip. Whereas, if the size of a defect is very large, it will always cause yield loss no matter where it is. The defects of intermediate size cause yield loss or reliability failure depending on their locations. In Figure 4.5, for example, oxide defects and their relation to failure problems are shown.

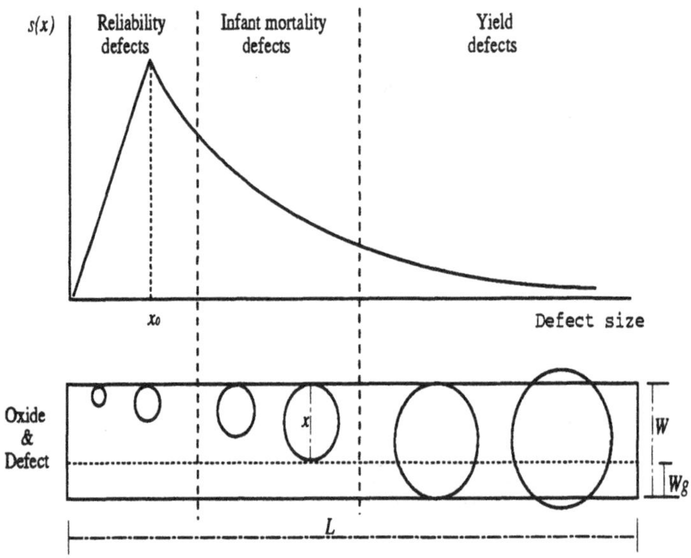

Figure 4.5: The defect size pdf and related oxide problems.

The gate oxide issue is one of the most important problems in manufacturing VLSI CMOS ICs. It is well known that the TDDB of thin oxide structures is one of the major factors limiting the IC reliability [102]. In general, the breakdown is a result of charge trapping in the oxide due to the excessive field and current in the SiO_2. Thus, the lifetime of an oxide is determined by the time required for the charge to reach a critical value [240].

There are two oxide breakdown failures, the intrinsic breakdown failure and the defect-related extrinsic breakdown failure. The intrinsic breakdown is defined as the breakdown that occurs during the wearout period of the oxide lifetime, but the defect-related extrinsic breakdown occurs at the infant mortality region. Defects which shorten the time to breakdown are mostly due to contamination, crystalline defects in the substrate, localized thin regions, or surface roughness at the weak spots. The infant mortality parameters, such as the failure rate and the early failure percentages and the relationship between oxide yield and oxide reliability of VLSI, can be determined only by the defect-related extrinsic breakdown. Since defect-related oxide failures have a DFR in the infant mortality region, voltage accelerated tests or burn-in can effectively screen weak devices having oxide defects.

Assume a thin gate oxide has a length of L and a thickness of W and a defect is circular with diameter x. If the oxide thickness left due to defects or processes is W_g or less, then the oxide is considered to be defective and breakdown occurs when the defect growth exceeds the amount of W_g.

Let A_r and $A_r(x)$ be the average reliability critical area and the reliability critical area for a defect size x in a single gate oxide, respectively. For the threshold thickness W_g, $A_r(x)$ is mathematically given by

$$A_r(x) = \begin{cases} 0, & 0 \le x \le W - W_g \\ L(x + 2W_g - W), & W - W_g \le x \le W \\ L(2W_g), & W \le x \le \infty. \end{cases} \tag{4.26}$$

From Eqs. (4.1), (4.5), and (4.26), the average reliability critical area, A_r, is given by

$$A_r = \frac{Lx_0^2}{2W}\left(\frac{W^2}{(W - W_g)^2} - 1\right).$$

Using Eqs. (4.1), (4.5), and (4.8), the average yield critical area, A_c is given by

$$A_c = \frac{Lx_0^2}{2W}.$$

After comparing A_r and A_c, we can express the oxide reliability as a function of yield Y.

Assuming that the reliability failures occur randomly and follow the Poisson distribution with the average number of reliability failure defined in Eq. (4.3), μ_r , the reliability in the extrinsic failure region, R_1, is given by

$$R_1 = e^{-\mu_r} = e^{-A_r D_0}. \tag{4.27}$$

Using the Poisson yield model of Eq. (4.13), Eq. (4.27) is modified to get the relationship between yield and reliability. The result is

$$R_1 = e^{-A_c D_0 (A_r/A_c)} = Y^{A_r/A_c} = Y^{(1/(1-u)^2 - 1)} \tag{4.28}$$

where $u = W_g/W$ and $0 \le u \le 1$. The relative degree or severity of the defect growth can be expressed by u. At the starting time of operation, W_g is set to 0 and, thus, $u = 0$ and $R = 1$. As time goes on, the value of u will increase with the amount of Eq. (4.30) and then cause the reliability to decrease. Figure 4.6 shows the reliability changes as u increases for some given yields.

The reliability of gate oxide at time t is determined by the defect growth during $[0, t]$. When the original defect size, plus defect growth at time t, exceed a certain threshold in the gate oxide, the oxide breaks down. Let W_g be the expected growth of oxide defect during $[0, t]$. Since defects with various sizes are distributed over the wafer, the time to breakdown is determined by the W_g value of the weakest spot in a chip. Lee *et al.* [240] present a statistical model of silicon dioxide reliability to predict the yield and lifetime of actual

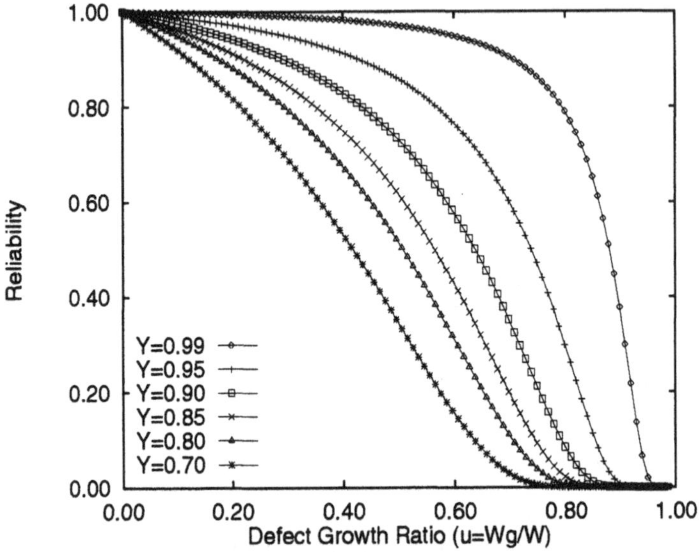

Figure 4.6: The reliability changes for some fixed yields.

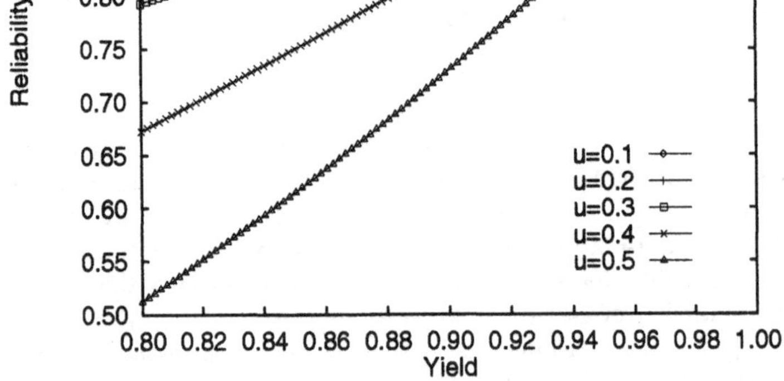

Figure 4.7: The relationship between yield and reliability for some given u values.

capacitors or oxide devices in an IC. Their model to calculate the defect-related time-to-breakdown is given by

$$t_{BD} = \tau \exp(\frac{GW_g}{V}) \tag{4.29}$$

where

t_{BD} breakdown time (sec)

τ a constant($\approx 1 \times 10^{-11}$ sec) determined by the intrinsic breakdown time under an applied voltage of V

V the voltage across the oxide in volt which is generally different from the applied voltage

G a constant ($\approx$ 32,000 MV/m).

From Eq. (4.29), the defect growth for a given breakdown time t_{BD} is calculated as

$$W_g = \frac{V}{G}\ln(\frac{t_{BD}}{\tau}). \tag{4.30}$$

Using Eqs. (4.29) and (4.28), we can predict the oxide reliability in the infant mortality region. The relationship between yield and reliability is presented in Figure 4.7, and it is almost linear in the high yield region ($Y \geq 0.8$). Therefore, the result obtained by Prendergast [333] is well explained. Without any analytical model, he suggests that there exists a strong linear relationship between yield and reliability. When compared with Huston's model [197], Eq. (4.28) gives the reliability at time t based on failure physics of oxide. Huston [197] describes the relationship only through defect sizes and critical areas.

Combining the concept of fault coverage into the relation model and assuming that the average number of faults is proportionally reduced after $100T\%$ fault coverage, the average number of faults can be expressed as

$$\mu_r^{'} = (1 - T)\mu_r.$$

Therefore, the reliability model including the fault coverage is

$$R_2 = e^{-\mu_r^{'}} = e^{-(1-T)\mu_r} = Y^{(1-T)\left(1/(1-u)^2-1\right)}.$$

In a word, after considering all the aforementioned models, reliability can be predicted by the power law of the yield. That is, $R = Y^c$ for a constant c which is time-dependent.

□ Example 4.4

Suppose a product with devices whose W is 120Å($1.2 \times 10^{-6} cm$). The internal bias, V, is 3.3V. From Eq. (4.30), the projected W_g after one year (t=31,536,000 sec) will become 43.93Å. Using Eq. (4.28), the relation model at that time, for a given yield, Y, is shown as

$$R_1 = Y^{1.489}.$$

From previous experience and results of failure analysis, if W_g is set to be 50Å$(0.5 \times 10^{-6} cm)$, the equivalent value of t is 1.139×10^{10} sec by Eq. (4.29), and the relation model for this is

$$R_1 = Y^{1.939}.$$

The reliabilities for Y=0.95 and 0.99 are 0.9053 and 0.9807, respectively. Table 4.3 summarizes four relation models after 3, 6, 12, and 120 months including fault coverage factor. □

Table 4.3: Some relation models for given times.

t (month)	W_g (Å)	Relation model	R_1 (Y=0.95, T=0)	R_2 (Y=0.99, T=0.95)
3	42.49	$Y^{(1-T)\times 1.397}$	0.9309	0.99930
6	43.21	$Y^{(1-T)\times 1.442}$	0.9287	0.99928
12	43.93	$Y^{(1-T)\times 1.489}$	0.9265	0.99925
120	46.30	$Y^{(1-T)\times 1.651}$	0.9188	0.99917

4.7. Cost Model

Christou [84] shows a yield/cost model to quantitatively relate the chip cost to the baseline process variables and to the process yield. In general, the unit total cost, C_{Tot}, is the sum of three components: fabrication cost C_F, assembly cost C_A, and testing cost C_T:

$$C_{Tot} = C_F + C_A + C_T.$$

These three cost components may be expressed in terms of yield, labor, and materials cost:

$$C_F = \frac{C_W}{Y_{pro} C_{MAX}},$$

$$C_A = \frac{C_{AL} + C_{AM}}{Y_{pck}} + (1 - Y_{pck}) C_F,$$

$$C_T = \frac{C_{TL}}{Y_{test}} + (1 - Y_{test})(C_F + C_A).$$

where

C_W	wafer processing cost including the impact of line yield
C_{MAX}	total chips per wafer
C_{AL}	labor cost
C_{AM}	material cost or package cost
C_{TL}	test labor

Y_{pck}	package yield
Y_{test}	final test yield.

In the above yield/cost model, the assembly yield, Y_{asm}, is divided into two yield components, Y_{pck} and Y_{test}, and the die yield, Y_{die}, is included in the process yield, Y_{pro}.

□ **Example 4.5**

Let the wafer processing cost including line yield be \$6000, the C_{MAX} 250, and the process yield 80%. The chip fabrication cost, C_F, is calculated as

$$C_F = \frac{6000}{0.8(250)} = \$30.00.$$

When we assume 80% package yield using \$15 package cost and \$20 labor cost, the assembly cost, C_A, is

$$C_A = \frac{15 + 20}{0.8} + (1 - 0.8)30.0 = \$49.75.$$

If the test yield is 90% and test labor \$20, then the test cost, C_T, becomes

$$C_T = \frac{20}{0.9} + (1 - 0.9)(30.0 + 49.75) = \$30.2.$$

Therefore, the total cost, C_{Tot}, is

$$C_{Tot} = 30.0 + 49.75 + 30.2 = \$109.95.$$

If the package yield increases by 10% and the test yield by 5% without any other cost changes, then the total cost becomes \$93.7. Thus, \$16.25 per unit can be saved through yield improvement. □

4.8. Conclusions

The likelihood of finding a chip with no defect is defined as the yield. If we could assume that defects are equally distributed across the wafer, the Poisson yield model would be a fit. Because the defect density is not always constant over the wafer, however, defect distributions are introduced and integrated into the Poisson yield model. The variety of these yield model depends on the fact that the defects occur unequally across the wafer. In order to build a good yield model, we need information on the defect density distribution $f(D)$ and have to choose a plausible $f(D)$. In addition to this, there is another difficulty; namely, how to build a general yield model. The yield model for a specific manufacturing process or product, no matter how well it agrees with observations, does not guarantee applicability to another process or product. Hence, a flexible model that can be easily modified to a different yield environment is better from a practical standpoint than a specific model. For these reasons the negative binomial yield model gives the best approximation of wafer yield, since it includes parameters which can be adjusted.

5. Reliability Stress Tests

Except in the infant mortality period, semiconductor devices normally have low failure rates and long working lives. Failure rate can be down to 0.005% per 1000 hrs or lower. Because the failure rate can be extremely low under normal working conditions, it is necessary to take actions to force the devices to fail within convenient time scales (generally 0~6 months). Increase of failure rates is achieved by application of controlled stress. Acceleration factors introduced by the stress can be determined, enabling the failure rate of the device to be calculated.

5.1. Accelerated Life Tests

The accelerated life tests (ALTs) that subject devices to higher than usual levels of stress (e.g., voltage, temperature, humidity, corrosion, magnetic field, current, pressure, radiation, vibration, salt, and loading) are techniques used to speed up the deterioration of materials or electrical components so that analysts are able to collect information more promptly. Because of rapidly changing technologies, customer expectations for higher reliability, and more complicated products, ALTs have become very important. Most ALT data analyses include a combination of graphical methods and analytical methods based on maximum likelihood estimation (MLE). Some characteristics of these methods include that [277]:

- the normal distribution is not applicable,
- data are censored,
- users are interested in the lower tail of the life distribution, and
- extrapolation is needed for drawing conclusions from the data, and it should be used with caution since, in some ranges, the errors resulting from assumptions about the modeling may be amplified by improper extrapolation.

To implement an ALT, one has to first design experiments. Not all the statistically optimal experiments are applicable. In some cases, compromise plans are preferable even if they sacrifice some statistical efficiency because they may improve overall properties such as the robustness to misspecification of unknown input. Second, factor level combinations have to be chosen. The number of test units must be decided. Typically, one should allocate more units to lower stress because of limited test time and our primary interest of inferences at lower level of stresses. Temperature is usually the most critical factor for component failures. It is known that about 40% of failures are due to temperature; vibration is the second highest factor, which accounts for 27% of total failures; moisture accounts for 19%; sand and dust, 6%; salt, 4%; altitude, 2%; and shock, 2%. The reliability manual of Bell Labs (now Lucent Technologies) [13] also indicates that since not enough knowledge is available about the effects of electrical stress on infant mortality to model its effect, the discussion will be limited to describing the effect of temperature. Another reason to consider temperature influence is due to an example in avionic equipment; almost 25% of failures can be attributed to temperature, and this percentage might be even higher for ordinary electronic equipment. Therefore, we will focus on the discussion of temperature as a stress factor.

However, one has to bear in mind that there are other ways besides using higher temperature to design an ALT. For example, the machine-level ALT is performed on Xerox copiers [119] by making more copies during tests; if it is estimated that a certain customer will make 1,000 copies per day, then the test operators may make 30,000 copies per day to simulate the operational environment for 30 days. Likewise, if we are testing a printer, and the estimated pages of printing jobs per day are 100, we may print 3,000 pages per testing day, which is equivalent to using the printer in operational profile for 30 days.

Some other practical applications on ALT are thermal verification testing [153], thermal stress screening (TSS) [376], run-in tests [460], and combined environmental testing [108], to name a few. These testing plans, in spite of being given different names, share a common theoretical basis. Epstein ([125] $\sim$ [130]) publishes a series of papers on ALT, which serve as an excellent foundation for advanced studies on ALT. Elsayed [121] classifies ALT models into three types: parametric and nonparametric statistic-based models, physics-statistics-based models, and physics-experimental-based models.

Typical accelerated tests use temperatures between -65^oC and 250^oC, although lower or higher temperatures may be necessary depending on the failure mechanisms. The device could be subjected to either a standard bias or a fully functional operation while being stressed at the elevated temperature. Dynamic life tests range from 168 hrs (1 week) to 2,000 hrs (3 months), depending on the conditions of stress, sample size, and the possible failure mechanisms. Accelerated tests may also use environmental factors, such as humidity, as a stress factor. The highly accelerated stressed test (HAST), which was originally invented by the British Telecommunications (BT) Lab, is conducted at 130^oC and 85%RH (relative humidity) with a bias (usually 10% higher than the normal operating voltage) applied at the pins in an alternating order (i.e., pin 1 at high, pin 2 at ground, pin 3 at high, and so forth). From the relationship among temperature, pressure, and humidity shown in Figure 5.1, the pressure under the HAST test condition is about 2.3 atm. That is, HAST can also be categorized as a pressure-acceleration test and, therefore, the HAST chamber must be able to sustain this high pressure.

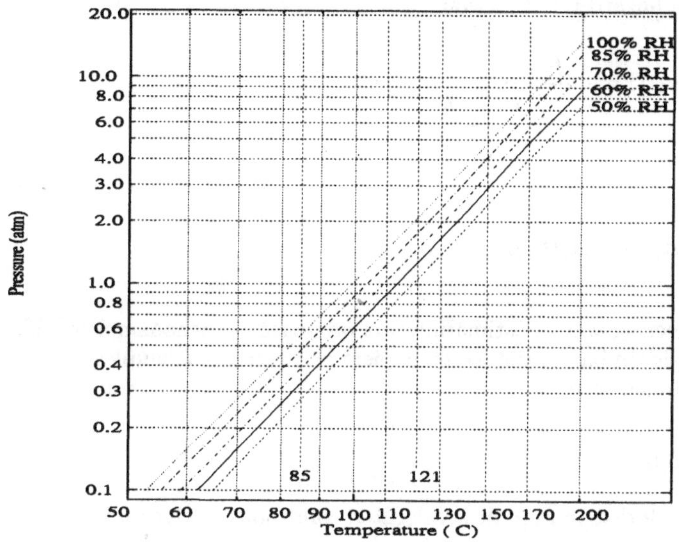

Figure 5.1: Pressures under different combinations of temperature and humidity.

5.1.1. Time Transformation Models

ALT results have to be adjusted according to some time transformation models (or the acceleration models) to provide predictions on the performance of the component in its normal use condition. The time transformation model is chosen so that the relationship between the parameters of the failure distribution and the stressed condition, which can be derived from an analysis of the physical mechanisms of failure of the component, is known [324].

Let the subscripts 1 and 2 refer to normal conditions and accelerated conditions, respectively, and η be the time transformation factor (or acceleration factor). Tobias and Trindada [421] summarize the several functional relationships between the accelerated and normal conditions :

1. The time to failure

$$t_1 = \eta t_2.$$

2. The cumulative distribution

$$F_1(t) = F_2(t/\eta). \tag{5.1}$$

3. The probability density function

$$f_1(t) = 1/\eta f_2(t/\eta).$$

4. The hazard rate function

$$h_1(t) = \frac{f_1(t)}{1 - F_1(t)} = \frac{1}{\eta} h_2(t/\eta). \tag{5.2}$$

5. The reliability

$$R_1(t) = R_2(t/\eta).$$

An engineering way of estimating η is proposed by Schoonmaker [368], whose approach is similar to the Xerox copiers and printers examples (both with $\eta = 30$) mentioned above.

□ Example 5.1

The Weibull distribution at accelerated conditions is given by

$$F_2(t) = 1 - \exp\left[- \left(\frac{t}{\theta_2} \right)^{\beta_2} \right], \quad \beta_2 \geq 1, \quad \theta_2 > 0.$$

From Eq. (5.1) and (5.2), the Weibull distribution at normal conditions is

$$F_1(t) = 1 - \exp\left[- \left(\frac{t}{\eta\theta_2} \right)^{\beta_2} \right]$$

and the hazard rate function is

$$h_1(t) = \frac{\beta_2}{\eta\theta_2} \left(\frac{t}{\eta\theta_2} \right)^{\beta_2 - 1}.$$

Assuming the same shape factors at both accelerated conditions and normal conditions ($\beta_1 = \beta_2$) and $\theta_1 = \eta\theta_2$,

$$F_1(t) = 1 - \exp\left[- \left(\frac{t}{\theta_1} \right)^{\beta_1} \right], \quad h_1(t) = \frac{1}{\eta^{\beta_2}} h_2(t).$$

If the shape parameters at different stress levels are significantly different, then the linear acceleration is invalid or the Weibull distribution becomes inappropriate [121]. □

In the next sections, some proposed models are introduced for the decision of η, which is usually unknown in ALTs. The time transformation models introduced later in this section are basically for the temperature stressing; the transformation models for other stressed tests will be described when the corresponding stress is introduced.

5.1.2. The Arrhenius Equation

The Arrhenius reaction rate model was determined empirically by Svante Arrhenius in 1889 to describe the effect of temperature on the rate of inversion of sucrose and has been used to calculate the speed of reaction at a specified temperature. The Arrhenius equation provides the reaction rate (RR) below

$$RR(T) = A\exp(-E_a/kT) \tag{5.3}$$

where

- A constant that can be a function of temperature
- E_a electronic activation energy; in eV (electron voltage)
- k Boltzmann's constant, $8.617 \times 10^{-5}\text{eV}/^o\text{K}$
- T temperature (^{o}K).

Eq. (5.3) shows that the rate a transition will occur due to the thermal energy is proportional to $\exp(-E_a/kT)$ and the time for the transition to be completed is given by

$$t = C\exp(E_a/kT) \tag{5.4}$$

where C is a temperature-independent constant [48]. The Arrhenius equation is the most frequently used time transformation model for temperature acceleration and can adequately describe the rate of many processes responsible for the degradation and failure of electronic components, including ion drift, impurity diffusion, intermetallic compound formation, molecular changes in insulating materials, creep, crystallographic changes, and microscopic structural material rearrangement [48]. The Arrhenius equation enables us to obtain a model of device reliability at normal operating temperature using data obtained from life tests at elevated temperatures.

Modifying the Arrhenius equation for burn-in, the equation for acceleration factor η is obtained [13]:

$$\eta = t_{T_1}/t_{T_2} = \exp\left[\frac{E_a}{k}\left(\frac{1}{T_1} - \frac{1}{T_2}\right)\right] \tag{5.5}$$

where

- T_1 ambient temperature (^oK)
- T_2 burn-in temperature (^oK)
- t_{T_1} time at ambient temperature (T_1)
- t_{T_2} time at burn-in temperature (T_2).

From Eq. (5.5),

$$t_{T_2} = t_{T_1}\exp\left[-\frac{E_a}{k}\left(\frac{1}{T_1} - \frac{1}{T_2}\right)\right]. \tag{5.6}$$

The ambient temperature T_1 is set to 30°C to 50°C, and the burn-in temperature T_2 is usually set to 125°C or 175°C for components and 70°C to 90°C for subsystems or systems depending on the components used in the burn-in subsystems and systems; we have to suitably choose the subsystem or system burn-in temperature so that the components which can not sustain high temperature will not be damaged. Figure 5.2 illustrates the idea: we apply a greater

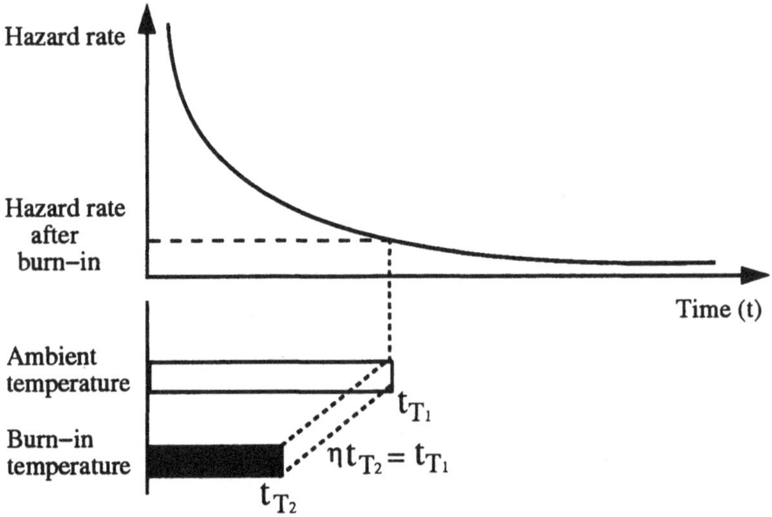

Figure 5.2: Implement burn-in at different temperatures to achieve the same effect.

stress ($T_2 > T_1$) to weed out the defective (weaker) items in a shorter time ($t_{T_2} < t_{T_1}$). In other words, the effect of putting the component in the stressed condition for time t is equivalent to using it under the normal condition for ηt.

□ Example 5.2

For a given activation energy of $0.4eV$, if a device is to be burned-in with a temperature of 125^oC for 7 days, the acceleration factor η for the ambient temperature of 40^oC and the equivalent time t_{T_1} at this ambient temperature are calculated as:

$$
\begin{aligned}
T_1 &= 40 + 273 = 313^oK \\
T_2 &= 125 + 273 = 398^oK \\
t_{T_2} &= 24 \times 7 = 168 \text{ hrs} \\
\eta &= \exp[\tfrac{0.4}{8.617\times10^{-5}}(1/313 - 1/398)] = 23.744 \\
t_{T_1} &= 168 \times 23.744 = 3989 \text{ hrs.}
\end{aligned}
$$

A larger η can be expected for higher T_2. Figure 5.3 shows the ηs for different E_a. □

The most difficult part of using the Arrhenius equation is the determination of the activation energy E_a because it is dependent on temperature.

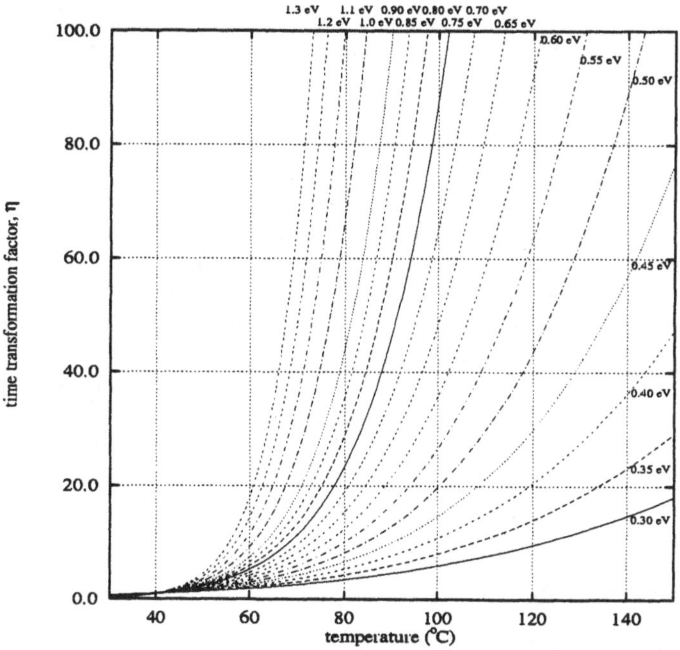

Figure 5.3: The time transformation factor for different activation energy.

Hallberg [173] assumes a lognormal-like hazard rate function and shows that the E_a and temperature are incorporated in a hazard rate function. Both Hallberg [173] and Zimmerman [462] use some E_as with practical examples. The range of $0.3eV$ to $1.5eV$ is typically used for the principal failure mechanisms; however, they do not give any explanation on their choices for E_as. McPherson [276] discusses the reasons why $0.4eV$ is not always correct. For hybrid microelectronic devices, Blanks [48] suggests activation energies between $0.35eV$ and $0.45eV$ up to 150°C. Bellcore [37] uses $0.4eV$ for device burn-in. The HCI failure is accelerated at low temperatures (e.g., −20°C) and the $E_a = -0.2eV$ may be used. It is estimated that for a certain product, the HCI lifetime at -80°C is about two times worse than at 25°C, which gives $E_a = -0.033eV$ by the Arrhenius model. A brief overview and some related references for the Arrhenius equation can be found in [430] by Vaccaro.

In the next example, a practical method to estimate E_a under the Arrhenius model is illustrated. Since most burn-in systems provide both temperature and voltage accelerations, a commonly used voltage acceleration model is also introduced in the example.

□ Example 5.3

There are two independent accelerations for the HTB test (which is also called the high temperature operating life, HTOL, test if it's done at 125°C and 7.0V for the 5V ICs): the temperature and the voltage acceleration. Because these two acceleration effects are independent, the overall acceleration factor can be obtained by multiplying the temperature acceleration factor by the voltage acceleration factor.

Let β (in 1/volt) be a voltage acceleration multiplier. The samples used for estimating the parameters (i.e., E_a and β) must be screened by the FT and the QC tests. To be more specific, the samples must go through the test flow used in mass production because the estimated E_a and β are mainly used to calculate the HTB failure rates in terms of FIT; most customers specify the failure rate after the 1,000-hr HTOL test to be less than 100 FIT at certain confidence level. An E-test program, which is usually similar to the QC E-test program, is used as the post-HTB test.

Suppose the Arrhenius model is used for the temperature acceleration (this model is the dominant method in calculating the temperature acceleration factor for most IC makers till 1997). The E_a can be estimated by conducting HTB tests at 125°C and 150°C). The HTB test beyond 150°C is not recommended because, for this particular memory IC with a plastic package type at the temperature higher than 150°C,

1. over-stress is found,
2. the self-heating effect becomes severe, and
3. the lifetimes of the capacitors on the burn-in boards rapidly decrease.

To verify if the voltage acceleration effect is indeed independent of the temperature acceleration, the internal voltages of 5 good ICs are measured on the MosAid tester (which is an engineering-type memory tester) at different temperatures (25°C, 40°C, 55°C, 70°C, 80°C, and 125°C). If the curves of the internal voltage at these six temperatures are almost the same, then, the voltage acceleration effects of these two HTB tests can be treated the same because the applied biases of the HTB-tests are both 7.0V.

To have a reliable estimate, large sample sizes are needed (preferably, more than 10,000 samples for each temperature; refer to Example 6.5 for details). Suppose 10,000 ICs (whose external input voltage is 5.0V at normal using condition) are put at both a 125°C and a 150°C 7.0V-HTB test. The samples are tested every 12 hrs by the E-test program at room temperatures (i.e., 25°C). The samples are considered good if they pass the E-test. The HTB tests continue for 168-hr. Two and four failures are found from the HTB test at 125°C and at 150°C, respectively. Let the acceleration effect at T_i be $RR(T_i)$. Thus, T_1=125+273=398(°K), T_2=150+273=423(°K), and $RR(T_1)$=2, $RR(T_2)$=4. Based on the Arrhenius model, the E_a can be estimated by

$$E_a = k \ln \left[\frac{RR(T_2)}{RR(T_1)} \right] / (1/T_1 - 1/T_2) = 0.4 \text{ (eV)}$$

where k is the Boltzmann's constant ($8.617 \times 10^{-5} eV/^oK$). The regression analysis may be applied to estimate E_a if more than two temperatures are used (as shown in Example 6.5).

Similarly, at least two different voltages are needed to estimate β. Let A_V be the voltage acceleration factor. The simplest and the most widely used voltage acceleration model is

$$A_V = \exp[\beta(V_{si} - V_0)] = \exp(\beta \Delta V) \tag{5.7}$$

where V_0 and V_{si} is the internal voltage at the normal using and at the stressed condition i, respectively. From Eq. (5.7), the β can be estimated by

$$\beta = \ln[RR(V_{s2})/RR(V_{s1})]/(V_{s2} - V_{s1}) \tag{5.8}$$

where $RR(V_{si})$ is the acceleration effect at V_{si}. The β is dependent on the thickness of the gate oxide, (T_{ox}), when it is less than 80Å. That is, the smaller the T_{ox}, the larger the β can be expected if $T_{ox} < 80$Å. Suppose 10,000 samples are subjected to 7.0V and 7.5V 125°C HTB tests. It is known that $(V_{s2} - V_{s1}) =$ 0.3V. After 168-hr HTB tests at 125°C, it is found that $RR(V_{s2}) = 5$ and $RR(V_{s1}) = 3$ after the E-test at 25°C. From Eq. (5.8), β is 1.7 (1/volt).

One important point is, the more frequently one tests the samples, the more reliable the results although this may not be practically feasible due to the cost consideration. The E-test is expensive and time-consuming. This notion is similar to the operating characteristic (OC) curve in the QC sampling plans: the larger the sample size, the lower the probability of making a wrong decision (e.g., the smaller the type-II error, will be at a fixed type-I error).

The HTB tests in this example can also be used to verify if the burn-in time is long enough to remove the potential defects by monitoring the failure percentage at the first 48-hr HTOL test. One possible passing criterion is the failure percentage before the first 48-hr HTOL test must be less than 0.2%. Similarly, we can also use the test results to see if the burn-in time can be reduced. Many IC makers set their own rules on reducing the burn-in time. Basically, the EFR test with sufficient samples must be conducted to see if the failure rate (in FIT) before the 48-hr HTOL test is below 50 FIT, for example.

The estimated method introduced in this example is a time censored test (type I or time censoring, see Section 6.4.1), which means that HTB tests are terminated at a pre-specified test time and $RR(T_i)$ ($RR(V_i)$) is defined as the number of failure after this fixed stress duration at T_i (V_i). If $RR(T_i)$ ($RR(V_i)$) is defined as the time to reach a pre-specified failure percentage (or a failure number, if the initial sample sizes for the tests are the same) at T_i (V_i), then we have a failure censored test (type II censoring or failure censoring) and E_a and β can be estimated by changing $\ln[RR(T_2)/RR(T_1)]$ ($\ln[RR(V_{s2})/RR(V_{s1})]$) in the equations in this example to $\ln[RR(T_1)/RR(T_2)]$ ($\ln[RR(V_{s1})/RR(V_{s2})]$). Furthermore, if resources (samples, tests, chambers, and person-power) permit, a more reliable estimators of E_a and β can be obtained (if more than two test conditions are used). The regression is used to estimate parameters. Practically, the failure censored tests are preferred by most analysts in the IC industries. A more detailed example on deciding the E_a and β is given in Example 6.5. □

5.1.3. The Eyring Reaction Rate Model

The Arrhenius model is useful in determining the effects of temperature stress to life. Temperature is not the only stress factor, however; other stresses have been considered to accelerate the lifetime. To describe the interaction of additional stresses, the Eyring modification to the Arrhenius reaction rate model is given by

$$\lambda = A\exp(\frac{E_{a1}}{kT})\exp(\frac{E_{a2}E}{kT})\exp(E_{a3}E) \tag{5.9}$$

where E_{a1}, E_{a2}, and E_{a3} are constants representing the activation energies at which various failure mechanisms occur, T is a temperature, E is the electric field, and k is the Boltzmann's constant. The first term of Eq. (5.9) shows an Arrhenius type relationship for temperature dependence. The second term refers to the interaction of the stress and temperature, and the final term comes from stress other than temperature.

5.1.4. Mountsinger's Law

Mountsinger's law is used to determine the life of insulating materials used in electric machines [48]:

$$\begin{aligned} \lambda = &\ A2^{T/\Delta T} \\ t = &\ C2^{-T/\Delta T} \end{aligned}$$

where λ is the failure rate, t is a failure time, ΔT is the temperature increment to double λ or halve t;

$$\Delta T = \frac{\ln\ 2}{B} = \frac{0.69}{B},$$

and T can be any unit of temperature, such as °C, and A, B, C are constants that depend on the materials. Provided that the temperature range is small enough to treat T^2 as approximately constant, the Arrhenius equation and the Mountsinger's law are equivalent with $B = E_a/kT^2$ [48]. Blanks [48] compares the Eyring, Arrhenius, and Mountsinger models and includes some reported activation energies for component burn-in.

5.1.5. The Power Law

The power law is first derived under the considerations of activation energy and kinetic theory for the ALTs of dielectric capacitors. It is also used by Rawicz [341] for the ALT for incandescent lamps where burn-in is proved to be effective. The model is given by

$$\text{Hazard rate} = e^{\gamma_o}V^{\gamma_1}$$

where γ_o and γ_1 are parameters dependent on material properties and may change according to applied stress and V is an applied voltage.

5.2. Environmental Stress Screening (ESS)

ESS is a process of eliminating defective parts from the production batch [358]. The concept of the screening process is to accelerate the lifetime of devices such that they begin operation with a failure rate beyond the infant mortality region. MIL-STD-883E, JEDEC #22, and BS9300/9400 specify the screening procedures and test conditions.

5.2.1. Characteristics of ESS

There are two types of defects: patent defects (or hard defects [409]) and latent defects. Patent defects are detectable by functional testing or inspection, but latent defects are not detectable until they are transformed into patent defects. Final tests separate functional products from malfunctional ones. However, products with latent defects may not be eliminated by the final tests and would have a very short lifetime in the field. The failure of these weak parts is the main reason for the high early failures commonly observed in the infant mortality period. A process of detection and elimination of such devices is called ESS. ESS can be defined as a process to accelerate the aging of latent defects by applying excessive stresses without shortening their useful life. Its goal is to cause latent defects resulting from manufacturing mistakes to become detectable failures without doing damage to good material. ESS is thus not a direct correction for circuit design or manufacturing problems but a temporary measurement that is frequently used to enable product shipments to be made without compromising or affecting final product reliability until corrective repair activities to fix the true root causes are initiated. The ESS for electronic equipment is well described in MIL-HDBK-344A.

Although ESS and burn-in are similar in purpose and method, some distinctions exist. Jarvis [200] believes that

- burn-in does not affect the steady-state hazard rate of the system, whereas ESS lowers the steady-state hazard rate,
- ESS exposes systems to environmental extremes outside specification limits, whereas burn-in tests are conducted under environmental conditions within specification limits, and
- burn-in is conducted at higher levels of assembly and for longer durations than ESS.

He concludes that removal of patent defects lessens the frequency of infant mortality failures, whereas removal of latent defects improves the steady-state hazard rate. ESS and burn-in are widely used in the electronic industry to eliminate early failures. When an electronics system requires highly reliable performance, it is subjected to both ESS and burn-in [118].

5.2.2. Optimization of ESS

ESS optimization is based on the study of the bimodal distribution, which can be described as a mixture distribution in Section 6.1.2. Perlstein *et al.* [325, 326] assume a bimodal distribution and find the optimal screen duration and screen effectiveness for a system of components by using superposition of the individual component renewal processes.

For the ESS analysis, Barlow *et al.* [26] use the mixture of exponential distributions which is given by

$$f(t \mid p, \lambda_d, \lambda_g) = p\lambda_d e^{-\lambda_d t} + (1-p)\lambda_g e^{-\lambda_g t} \tag{5.10}$$

where

- p the proportion of the defective parts in the population
- λ_d the hazard rate of a defective part
- λ_g the hazard rate of a good part, $\lambda_d \gg \lambda_g$.

They explicitly assume that all the defective parts' lives can be described by an exponential distribution with parameter λ_d, and the good parts' lives by an exponential distribution with parameter λ_g; that is, all the defective parts are homogeneous, as are the good ones.

The mixture of exponential model is also applied by Jensen and Petersen [203]. The hazard rates for the mixture of exponential model with $\lambda_d = 1$, $\lambda_g = 0.01$, and different ps are depicted in Figure 5.4. Two important implications

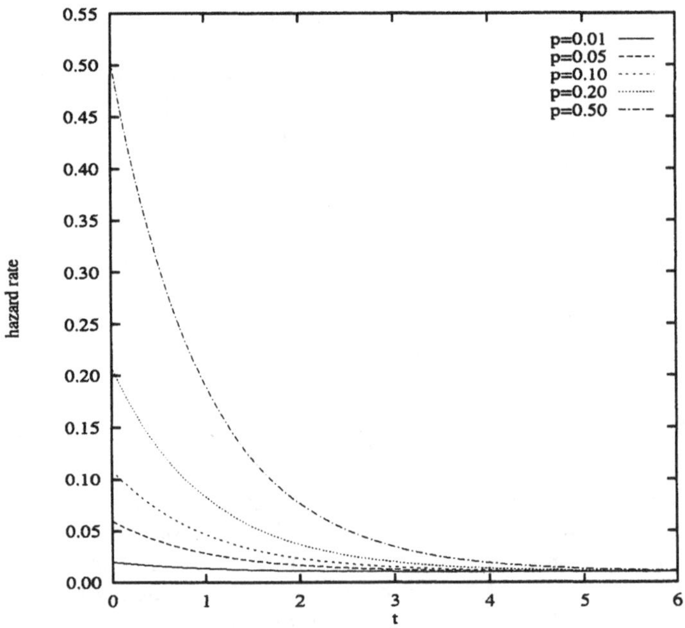

Figure 5.4: The hazard rate for mixture of exponential distribution.

can be obtained from Figure 5.4. First, the mixture of exponential model has DFR, which makes it a proper choice for describing the behavior of infant mortality, and the other is the larger the p, the larger the curvature. The model is also applied by Marcus and Blumenthal [263] for a sequential test study.

Reddy and Dietrich [344] also use the concept of mixed exponential distributions to study two level (component-level and unit-level) ESS . In this model, components are replaced upon failure and modeled using the renewal theory while unit level failures are minimally repaired. The optimal screen durations are found by minimizing the expected life-cycle cost, which includes cost of screening, in-house repair, and warranty. Pohl and Dietrich [330, 331] write that a good component subjected to stress at multiple levels will have the same failure characteristics as a good component that has not undergone screening by using a mixed Weibull distribution. The optimal screening strategy is defined by the selection of optimal stress levels and the duration at each level; the optimal screening durations at all levels are those which minimize the life-cycle costs. Their life-cycle cost model is separated into component level costs, unit (printed circuit board) level costs, and system level costs.

English *et al.* [118] discuss the truncated bathtub curves for latent failures and a mixed population. In the electronics industry, since the failure times are often truncated at some time T due to obsolescence, the probability that failures occur prior to time t, given that $t < T$, is significantly different from the situation where truncation is not considered. Consequently, if it is known that the obsolescence time is T, then the hazard rate function is given by [452]:

$$h(t) = \frac{f_0(t)}{F_0(T) - F_0(t)}, \quad 0 \leq t \leq T$$

where $f_0(t)$ and $F_0(t)$ are the pdf and the CDF of the original failure distribution, respectively.

For a good design of ESS, Hnatek [190] suggests some important requirements: knowledge of material performance under a wide range of operating and environmental stresses; quantitative relationships between defective sizes, magnitudes, and others and acceleration of degradation mechanisms; knowledge of threshold levels; knowledge of degradation mechanisms that cause device failure due to specific defects; detailed stress analysis with advanced methods; statistical design and mathematical modeling; and precise testing instrumentation to detect faulty circuit components.

5.2.3. Accelerating Stresses

An efficient ESS requires an acceleration of the mechanisms that give rise to the infant mortality. Voltage, temperature, humidity, radiation, chemicals, shock, and vibration are frequently used as stresses to accelerate failure since the reliability of ICs is affected by one or a combination of stresses in the operating environment. For example, when the temperature stresses are not effective for some mechanisms, over-voltage, over-current or some other environmental stresses may be used as accelerating stresses. The test conditions should be selected depending on the nature and degree of the failure mechanisms causing early failures, and each of the stresses has degrading effects that depend on the stress levels. It is known that failure mechanisms are typically chemical and physical processes, which are highly sensitive to the ambient temperature. Sabnis [358] discusses the effects of some accelerating stresses.

Voltage Stress

TDDB, EM, and HCI damage are the well-known effects of the electrical stress on ICs. Due to randomly distributed defects, an oxide stored at a fixed electric field wears out and, eventually, breaks down. EM in metals occurs as a result of a flow of electron wind, resulting in open or short circuits of interconnects. The time to failure due to oxide breakdown and hot-carrier damage decreases exponentially as the voltage stress is increased and the lifetimes due to EM are almost inversely proportional to the square of the current density [358] (which is usually described by the Black model). The instability problems do aggravate as the voltage is increased, but the effects are not very quantifiable. A simple, and the most widely used transformation model for voltage stress, is shown in Eq. (5.7) with an example.

Temperature Stress

The effects of temperature stress are almost always modeled by the famous Arrhenius equation. Oxide breakdown, EM, and many packaging-related failure mechanisms are aggravated by the increase of the ambient temperature. On the contrary, the HCI effect is known to be alleviated with increasing temperature. Section 5.1 describes several proposed time transformation models for temperature stress.

Humidity Stress

Moisture is present in nature and has detrimental effects on ICs. The packages in which the chips are housed and the materials that are used to secure the chips contain chemical impurities. The moisture, in combination with the chemicals and an electrical bias, can corrode the metals. The packages themselves can swell as a result of moisture absorption and create stresses that damage the ICs. The most commonly used humidity-related tests are HAST, PCT, and THB. Similar to HAST, the samples for THB are under voltage stress as used for HAST described earlier. However, there is no bias applied during PCT, whose test condition is 121°C and 100%RH (about 2 atm by Figure 5.1). It is obvious that HAST is the most severe of these three tests.

There is no universally recognized humidity acceleration model thus far. Two models are introduced here. Let X be the humidity activation coefficient. Sinnadurai [389] proposes the following temperature-humidity acceleration model

$$A = \frac{t_{amb}}{t_s} = \exp\left[X(RH_s^n - RH_{amb}^n) + \frac{E_a}{k}\left(\frac{1}{T_{amb}} - \frac{1}{T_s}\right)\right] \tag{5.11}$$

where RH is the relative humidity (in %) and T_{amb} and T_s are in absolute temperature. Recall that the Boltzmann's constant, k, is 8.617×10^{-5} $eV/^oK$. The n in Eq. (5.11) denotes the function of physics-of-failure related to component type [389, p 190]. Sinnadurai uses $X = 0.00044$ and $n = 2$ for the HAST test conditions. The acceleration factor, A_f, is equal to [415, p 10]

$$A_f = \left(\frac{RH_1}{RH_2}\right)^n \exp\left[\frac{E_a}{k}\left(\frac{1}{T_1} - \frac{1}{T_2}\right)\right].$$

Tam [415] uses $E_a = 0.9$ eV and $n = -3.0$ concerning the quality of plastic packaging devices whereas Sinnadurai [389] uses $E_a = 0.79$ eV and $n = -2.66$. In Tam's research, a table for the THB acceleration effect also contains the $E_a = 0.90$ (as well as 0.95 and 0.79).

Radiation Stress

There is a direct impact of radiation in the environment on the electrical properties of ICs. Even the materials that are used to fabricate and package the ICs can be contaminated with radioactive elements that emit γ or β rays or α particles. The radiation creates electron-hole pairs in oxides and silicon. These generated carriers are either trapped at the Si/SiO_2 interface or are collected in charge storage areas, upsetting the ICs either permanently or temporarily. An increase in the radiation level causes increased degrading effects on the ICs and is achieved by using radioactive materials like Am-241. The radiation stress test for DRAMs is called the accelerated soft error rate (ASER) test because the electron-hole pairs generated by the α particles may alter the states of the capacitors in DRAM memory cells either from 0 to 1 or 1 to 0. This usually won't do permanent damages to the capacitors. Therefore, a "soft-error" is used to distinguish from the permanent failure (the "solid failure") caused, for example, by ESD.

One test condition is to expose the samples to the Am-241 α-particle source with the intensity of 7×10^7 α per (cm^2-hr) for 10 minutes. The number of failed bits in each DRAM sample must be less than 10 after the E-test to pass the ASER test. It is estimated that the failure rate under the above test condition is equivalent to 1,000 FITs. That is, the ASER failure rate must be kept under 1,000 FITs to meet most customers' requirements. It is reported [335] that the cell capacitance at about 40 to 50fF (1fF $= 10^{-15}$ Faraday) can avoid the $\alpha-$radiation related soft error problems; a capacitor of 50fF charged to 5V holds 250fC (1fF $= 10^{-15}$ Coulomb) of charge, which corresponds to about 1.5 million electrons and is about the minimum level to withstand an $\alpha-$particle hit. However, 40 to 50fF cell capacitance was used for the 64Kb DRAM, which has become obsolete, and 26fF cell capacitance is what is commonly set for the 16Mb DRAM, the main-stream product in 1996. This means that the soft error phenomena has become very severe for high density DRAM's due to the smaller charge capacity of the scaled capacitors. Special designs must be applied to protect the capacitors from α-particle induced soft errors such as the error detection and correction circuit (ECC) and the implanted capacitor cell used by Motorola and the AT&T [335].

ESD Stress

ESD damage is perhaps the most insidious failure mechanism of electronic devices. The static electricity generated by the well-known triboelectric effect can destroy ICs when it discharges through them. The magnitude of the damage depends on the potential and the rate of discharge; the peak values of the natural ESD pulses can be in the tens of kilovolts. Although protective networks are designed and implemented on the chip to be tested for their ability and effectiveness against ESD, they are incapable of withstanding such levels. Therefore, extreme precautions must be taken to eliminate or minimize the generation of triboelectricity in the environment around the electronic equipment. In the IC industry, the most commonly used ESD models are the human body model (HBM), the machine model (MM), and the charges device model (CDM). The test method can be seen in MIL-STD-883E method 3015.7, EIAJ ED-4701 method c-111, and the ESD association standard ESD-S5.1~S5.3. The basic requirement for the ESD protection under these three models are: 2KV for HBM, 200V for MM, and 1KV for CDM. Comprehensive studies in ESD can be seen in [8].

Latch-up

As device scaling continues, the latch-up sensitivity of a CMOS circuit increases. Over-voltage stress, voltage transients, radiation induced photocurrent, or the input protection circuit being over-driven may trigger latch-up. Latch-up can be avoided early at the design stage. For example, the distance between the NMOS and the PMOS is relatively large in circuits carrying large current, such as I/O circuits. Many manufacturing preventive actions for latch-up have been proposed and implemented, such as epitaxy on a heavy doped substrate, guard rings, well controlled design rules, epitaxy/ buried layer CMOS structure, physical barriers to lateral currents, Schottky clamps, trench isolation, total dielectric isolation, and the use of silicon-on-insulator (SOI). Recently, the epi-wafer (which grows single-crystal semiconductor layers upon a single-crystal semiconductor substrate) has been employed for complicated ICs, like the microprocessors. The JEDEC Standard #17 titled *Latch-Up in CMOS Integrated Circuits* by the Electronic Industries Association (EIA) has become the industrial standard on the latch-up test.

Other Stresses

Chemicals in the air attack materials and cause corrosion. The salt that is present in nature, along with moisture, is damaging to ICs. In some areas, sulfur dioxide is a very common atmospheric pollutant. Increasing levels of chemicals accelerate degradation.

Physical movements of ICs and the forces exerted in the process of mounting and removing the packages at various stages of manufacture can cause breakage of leads, packages, wire bonds, and even chips. These shocks and vibrations can reduce the life of ICs.

5.2.4. Screening Tests

Amerasekera and Campbell [7] list some principal screening tests for semiconductor devices. They recommend combining high-temperature tests and mechanical stress tests to obtain an effective screening. Some of screening tests are also used as an accelerating test.

High Temperature Storage (HTS) Test

HTS is essentially a bake at temperatures much higher than burn-in. No bias is applied, and the device is not electrically activated. Therefore, the time transformation models in Section 5.1 can be applied for HTS. Temperatures are normally 250^oC for hermetic devices, or 150^oC for PEDs (plastic encapsulated devices). The purpose of HTS is to detect the quality of molding and wiring material used in the assembly processes and to monitor the EM and stress migration (SM) lifetimes. It can also be used as a screen against contamination, bulk Si defects, metallization and moisture problems, and is of particular value when testing MOS devices, especially the memory ICs. Usually, the samples must pass the E-tests after 1,000 hr HTS.

High Temperature with Bias (HTB) Test

HTB is sometimes called HTOL (under certain test conditions, see Example 5.3), which is performed to accelerate thermally activated failure mechanisms. The typical stress ambient is 125^oC with bias applied equal to or greater than the data sheet nominal value. Testing can be performed either with dynamic signals applied to the device or in static bias configuration for a typical test duration of 1,000 hrs. Elevated voltages are combined with high temperatures to make an effective screen against many failure mechanisms. HTB is particularly useful when screening thin films, transistors, metals, and capacitors defects in the MOS devices. Contaminations from particles and chemicals can also be found by HTB. A detailed description on HTOL appears in Chapter 6.

Electrical Over-stress (EOS)

Oxide defects are not very responsive to temperature stresses; however, voltage stressing forces defective oxides to fail prematurely. Stressing may take the form of a continuous electrical over-stress combined with an elevated ambient temperature as in the HTB. Electrical over-stress techniques are of value in MOS devices where the integrity of the gate oxide can be very effectively determined.

Temperature Cycling (T/C) and Thermal Shock (T/S)

T/C and T/S are performed by alternatively stressing devices at hot and cold temperature extremes. The terms T/C and T/S are frequently used interchangeably, but there is an important difference. T/C is a less severe screen, which uses an air-to-air conditioning medium and may require several minutes of transfer time when transferring between temperature extremes. T/S uses a liquid-to-liquid medium to provide the severe temperature shock environment and does not require a dwell time at room temperature when transferring between temperature extremes. Thus, T/S is a much more severe screen than T/C. T/C can be used to simulate the temperature change during operating and transportation; the test condition is 10 cycles, with a 10-min dwell time at the specified temperature extremes and a 5-min maximum transfer time between temperatures. T/C is an important test to monitor the quality of assembly, such as the material of the molding compound and the wire bond operation. It is also used to monitor the reliability of metal and the passivation. When the die size is large, T/C becomes more important because the dice may easily crack during assembly and under dramatic temperature changes when used in the field. The T/C test conditions for the PED's and the ceramic ICs are different. The most common T/C test condition is defined in MIL-STD-883E method 1011, condition C, where the temperature extremes are -65°C to +150°C with at least 10 minutes dwell time at each temperature; the samples must pass the E-test after 500-cycle T/C to be qualified.

Similar to humidity, there is no widely recognized time transformation model for T/C. AT&T proposes the following model to derive the acceleration factor for T/C ($A_{T/C}$, in days):

$$A_{T/C} = N(\Delta T_t/\Delta T_u)^n \tag{5.12}$$

where ΔT_t and ΔT_u are the temperature differences at test and at the using condition, respectively. The N is the number of the T/C test and the n in Eq. (5.12) is a constant; AT&T uses n =4. One obvious drawback for the model in Eq. (5.12) is that it does not consider the dwelling time and the transition rate between the two extreme temperatures.

□ Example 5.4

An IC will be used in luxury sedans, which may experience temperatures between -30°C and 120°C (i.e., $\Delta T_u = 150$°C). The IC supplier is asked to follow MIL-STD-883E, method 1011, condition E (i.e., between 300°C and −65°C, $\Delta T_t = 365^o$), to conduct T/C for 2,000-cycle. The dwelling time at each 300°C and −65°C is 10 minutes. The car maker believes that the 2,000-cycle T/C is equivalent to use of the IC for 10 years. From Eq. (5.12), the n can be obtained by ln(87,600/2,000)/ln(365/150) = 4.25. This T/C test may take as long as three weeks if the times for E-tests, temperature transitions, and defrosting (the T/C chamber will automatically defrost if needed) are included. □

Mechanical Shock Tests

Mechanical shock tests are used to test package integrity. The use of mechanical shock and vibration may detect devices with the potential for particulate contamination and material failure mechanisms. Typical test conditions are: acceleration = 1500G, orientation = Y1 plane, t=0.5 msec, and number of pulse = 5.

Centrifugal Spinning

Centrifugal spinning is a form of mechanical test, also used to determine package integrity. A component could be tested up to a stress level of 30,000Gs with the exact level being determined by the type of device and its required application.

Package Leakage Tests

Package leakage tests may check for either gross leaks or fine leaks in hermetically sealed devices. A fine leak test requires that a device be immersed in a radioactive gas (e.g., Kr^{85}) with helium mass spectrometry used to detect radioactive leakage from the package. Gross leak tests require that the device be immersed in hot fluorocarbon for about 30 seconds; if a hole exists in the seal, bubbles will be observed emerging from the package.

Humidity Testing

Humidity tests are usually carried out in an elevated temperature ambient. Testing at 85°C/85% RH with bias (THB) or 120°C/2 atmospheres pressure at 100% RH (PCT) indicates the resistance of the devices to the ingress of moisture and associated failure mechanisms, like mobile ions, metal corrosion, and delamination between die/ die pad & die/ molding compound. However, these tests could be destructive and the use of sample screening is recommended to evaluate the reliability of a batch. The acceleration models can be seen in Section 5.2.3.

Particle Impact Noise Detection (PIND) Test

The PIND test is capable of detecting loose particles down to 25μm in diameter in a packaged device. Sensitive acoustic monitors are attached to the package, and the device is shaken. The movement of very small particles within the package can therefore be detected.

Constant Acceleration

Constant acceleration screening is performed by placing units in a specially designed centrifuge fixture and submitting them to a centrifuge speed that results in a predetermined force being applied. Constant acceleration is not effective on solid non-cavity-type devices but is still effective in locating die lifting and package defects. Typical test conditions are: stress level = 30Kg, orientation = Y1 plane, and t=1 min.

Burn-in

Burn-in is an effective screening method used in predicting, achieving, and enhancing field reliability of ICs. It is the most effective screen in detecting die-related faults through time, bias, current, and temperature accelerating factors to activate the time-temperature-dependent failure mechanisms to the point of detection in a relatively short period of time. A typical burn-in would require devices to be subjected to a temperature stress of 125^oC for a minimum of 48 hrs. Infant mortalities are greatly influenced by the burn-in test, and a significant improvement in failure rates has been reported after burn-in. Reducing burn-in periods may well result in transferring infant mortalities from the manufacturing stage to the customer. There are several different types of burn-in in use, each of which presents a variation of the burn-in stresses created. Among them, steady-state burn-in, static burn-in, dynamic burn-in, and TDBI are widely used for semiconductor devices. A detailed description of each method is presented in Chapter 6.

As described above, each ESS test is used to screen certain defects. One other important issue is to decide the sample size for the test. The next example shows the most common method to set the sample size.

□ Example 5.5

Most reliability tests must be done during production qualification (see Example 2.1). One important issue during production qualification is deciding the sample size for each test item. The sample sizes for almost all tests are usually set higher than those determined for reliability monitoring because certain levels of maximum percent defective (MPD) must be met. The MPD is derived by assuming the β, the customer's risk or the type-II error, to be 0.1. From the binomial distribution, the sample size, n, satisfies

$$\beta = \sum_{i=0}^{x} \binom{n}{i} (\text{MPD})^i (1 - \text{MPD})^{n-i} \qquad (5.13)$$

where x is the number of rejects allowed. For MPD $\leq 1\%$, the Poisson distribution is used

$$\beta = \sum_{i=0}^{x} \frac{e^{-\lambda}\lambda^i}{i!} \qquad (5.14)$$

where $\lambda = n(\text{MPD})$. Therefore, n is equal to $\ln\beta/\ln(1\text{-MPD})$ for $x = 0$ when MPD $> 1\%$. If a minimum of 76 samples is required for HAST and no failure is allowed (e.g., n=76 and x=0), then, from Eq. (5.13), it is equivalent to 3% MPD. If MPD is 0.5%, from Eq. (5.14), 461 samples are needed if $x = 0$. The MPD table can be seen in, take one for example, Appendix D in MIL-PRF-38535D. For a given x, the smaller the MPD, the more samples are needed. The most commonly used scenario is $x = 0$.

The notion of the β is smaller than that of the lot tolerance percent defective (LTPD), which is usually seen in the QC fields. The LTPD is especially important to guarantee the quality of a group (batch, lot, and others) of products. The MPD is usually between 1% and 10% for most IC makers. For example, the sample size for the package-related reliability tests is 45 and 76 when MPD is 5% and 3%, respectively. The sample size for the die qualification tests, such as HTOL, LTOL, and HTS, is usually larger than that for the package-related tests. The usual MPD for the die qualification tests under β=0.1 is between 1% and 2%, which corresponds to 116~ 231 samples. The MPD levels for most

IC makers are between 3%∼ 10% and 2.5%∼5% for reliability monitoring and for production qualification, respectively.

To avoid possible quality variation between lots, samples are drawn from at least 3 qualified lots. The sample size for the HTOL is usually much larger than those for the package-related tests, since the main purpose of HTOL is to calculate the early (1 year) and the long-term (10 years) failure rate. A minimal sample size of 600 is recommended for the HTOL. That is, if the samples are from 3 lots, the minimum sample from each lot is 200. Package-related tests are usually used to monitor the assembly quality and stability of the assembly houses. They are done on a monthly or quarterly basis. Should the quality of a subcontractor become stable (failure percentages are below a specified level for 6 months) and reliable (all samples pass the electrical tests), a smaller sample size may be used to save costs because all the samples for the package-related tests are scrapped. □

5.3. Failures and Reliability Prediction

5.3.1. Failures in Semiconductor Devices

Generally, the mechanisms of semiconductor failures are classified into three main areas [7]: electrical stress failures, intrinsic failures, and extrinsic failures.

Electrical Stress Failures

Electrical stress failures are user-related mechanisms, and the cause of such failures is generally misuse. EOS and ESD, due to poor design of equipment or careless handling of components, are major causes of electrical stress failures, which could enhance the aging of components and the possibility of intrinsic or extrinsic failures. Since ESD is an event-related failure, it is not possible to do a screening test against it. A major problem of ESD damage is the formation of latent defects which are extremely difficult to detect.

Intrinsic Failures

Failures inherent to the semiconductor die itself are defined as intrinsic. Intrinsic failure mechanisms tend to be the result of the wafer fabrication (or front-end) of the manufacturing process and include crystal defects, dislocations and processing defects. Gate oxide breakdown, ionic contamination, surface charge spreading, charge effects, piping, and dislocations are important examples of intrinsic failure mechanisms. Time-dependent oxide breakdown occurs at weaknesses in the oxide layer due to poor processing or an uneven oxide growth. Failures of MOS devices due to oxide breakdown during device operational life are very low because it is possible to screen most defective devices before they reach the market. It is important that any defective gate oxides are detected at the final testing stage. Contamination is introduced by the environment, human contact, processing materials, and packaging. Since the mobility of ions is highly temperature dependent as well as influenced by an electric field, typical screens for ionic contamination are an HTS test between 150°C and 250°C and an HTB test.

Extrinsic Failures

Extrinsic failures result from the device packaging, metallization, bonding, die attachment failures, particulate contamination, and radiation of semiconductor manufacturing. That is, extrinsic conditions affecting the reliability of components vary according to the packaging and interconnection processes of semiconductor manufacturing. As technologies mature and problems in the manufacturers' fabrication lines are ironed out, intrinsic failures are reduced, thereby making extrinsic failures all the more important to device reliability.

The two main types of packaging used are ceramic and plastic. In the early microelectronic devices, plastic packages were commonly used. The failure rate of a plastic package has decreased to less than 50 failures per billion device hours [84, 438], and the encapsulation of IC chips with advanced molding techniques makes a plastic package relatively cost-effective. The failure rate of a ceramic package is usually higher than that of a plastic package because of the moisture inside a ceramic package [84]. Other advantages of a plastic package are light weight and small size compared with a ceramic package.

There are several studies on the reliability of packages. Hughes [195] shows that, due to the continued improvement in the reliability of plastic encapsulated devices, a plastic package is a good cost-effective alternative to a ceramic encapsulated package. Fayette [139] studies on burn-in for GaAs ICs. His conclusion is that correct materials and process choices are very important in order to eliminate infant mortality and the need for burn-in because burn-in is no longer a cost-effective screen for highly integrated devices. In the evaluation of requirements for plastic encapsulated devices for space applications, Leonard [247] concludes that burn-in is not necessary as a reliability assurance technique for plastic encapsulated devices. Hughes [195] and Olsson [308] evaluate the effects of 100% burn-in and T/C on the corrosion resistance of plastic encapsulated TTL ICs and conclude that burn-in and T/C are not always effective as screens for moisture-related corrosion of aluminum metallization.

Aluminum is a widely used element for metallization and interconnections in semiconductor devices. By using aluminum alloys, it is possible to form better contacts and to assure higher reliability in the metallization. Mechanisms affecting the reliability of the metallization are [7]: corrosion, electromigration, contact migration, and mechanical stress relaxation.

□ Example 5.6

Routine reliability tests (RRT) are important for IC manufacturers. The RRT can be done at both wafer and package level, and the wafer level test is usually done at higher stresses. The package level RRT is used mainly for monitoring subcontractors, whereas the wafer level RRT is used to assess the quality of the final products and to permit for quick response to deterioration in the production line.

If a gate operates at 3.3V as a normal operating condition, the wafer level test may be done at 16V or even higher. A major reason for this is the capacity of the test equipment, which usually can only handle one wafer at a time, and only one sample can be tested each time. One important drawback is that the correlation between the normal and the stressed condition can not be clearly observed. We do not know how well the extrapolation works. To compensate for this, samples from short-loop tests or from scribe-lines are assembled in side-braze and put in stress tests. The stress levels are usually lower than the wafer level measurement. Take the 3.3V-gate above for example; the side-brazed samples may be stressed at 10V∼12V with more samples and longer test/stress time (e.g., 2 weeks). The wafer level measurement can first give analysts ideas about the samples so that an earlier feedback and alarm is possible. To clarify the marginal cases, package level tests can be issued, although they are more expensive. It may take more than $2 to assemble a single side-braze. For normal package level tests, the sample size is at least 60, and some may exceed 2,000. On the other hand, the only cost for the wafer level measurement is in the wearout of the probe-pins, which can be probed for at least 100 times, and each pin costs about $10.

For package level RRT, the tests related to package quality are HAST, PCT, T/C, THB, and T/S. The HTOL, HTS, and LTOL are the most important and common die qualification tests for the full-device package level RRT which are used to monitor the die quality. Devices for the package level RRT are sampled directly from the warehouse storing the final products so that they best describe the quality of the outgoing goods. Suppose a HAST test is conducted on 45 samples (the corresponding MPD is 5%). The passing criterion is no failure at and before 96 test-hrs.

It must be mentioned here that the samples for the package-related RRT must be "pre-conditioned." The pre-conditioned test is used to categorize the ICs into five ranks:

- **I** No need for moisture proof bag.
- **II** Need moisture proof bag and should be used within 1 year.
- **III** Need moisture proof bag and should be used within 168 hrs.
- **IV** Need moisture proof bag and should be used within 72 hrs.
- **V** Need moisture proof bag and should be used within 24 hrs.

The higher the ranking a product achieves, the better the quality. The ICs are categorized into these five ranks based on the results of the tests (in that order): baking (125°C for 12∼24 hrs), IR (infrared)-reflow or vapor phase reflow (VPR) for at least two times (according to customers' spec.), soaking, and sonic acoustic tomograph (SAT) or cross section inspection. Baking is used to remove humidity remaining in the ICs. The soaking conditions, which simulate storage conditions, for each rank are in Table 5.1. The IR-reflow simulates the operation when the ICs are mounted on the PC board. Five to twenty cycles T/C may be required by some customers in pre-condition. Products are qualified for a certain rank if the samples do not have serious delamination (as defined in MIL-STD-883E method 2030) under SAT and pass the E-tests. Presently, level

Table 5.1: Pre-condition of the 5 ranks.

Rank	Temperature (°C)	Humidity (%RH)	Time (hrs)
I	85	85	168
II	85	60	168
III	30	60	168
IV	30	60	72
V	30	60	24

III or above (i.e., level I or II) is required by almost all major customers. The JEDEC standard JESD22-A113 *Predictioning of Plastic Surface Mount Devices Prior to Reliability Testing* gives detailed descriptions on the procedures and test conditions for precondition.

The stress from HAST is several times more severe than THB. This can be explained by the usual passing criterion for HAST and THB: the minimum requirement for HAST and THB is 96 hrs and 1,000 hrs, respectively. Thus, it's usually used as a substitute for THB to save time (although some engineers claim that there are some failure modes which cannot be seen from HAST). The most commonly seen failure mode from HAST is the popcorn failure. The samples with this failure mode show severe delamination after SAT examination. Sometimes, package cracks can be observed. The samples with package cracks usually fail the DC electrical tests. The HAST failures reflect the quality of the assembly lines. Physical analyses reveals that poor clean room conditions and poor raw material handling are the two main causes leading to HAST failures. □

5.3.2. Reliability Modeling

As can be seen from a review of the shortcomings of current reliability models, it is difficult to develop a viable reliability model. Rickers [345] of the Rome Air Development Center (RADC), who developed the MIL-HDBK-217 model, has said that the model was founded on a substantial database consisting of over 1010 device hrs. Such a database must be collected from accelerated life tests, screening, burn-in, reliability demonstration and field experience in addition to device characterization and malfunction data.

A number of reliability models have been developed which are considered valid for a specific application. Amerasekera and Campbell [7] describe two reliability models. The reliability model, which comes closest to universal usage, is the reliability model in the MIL-HDBK-217 published by the USA DOD. However, the validity and usefulness of this model is much debated. Other widely used models are those presented by French Telecom (CNET) [304] and the BT [52].

5.3.3. Strength-Stress Model

The strength-stress model considers uncertainties about the actual environmental stress to be encountered as well as the properties of a component, which is described in Section 3.3. One way to incorporate the stress-strength model into generic hazard rates of electronic components is the use of stress factors, which are multiplied with the generic hazard rate to obtain the estimated hazard rate based on the environment, the quality of the device, and expected stress level. More information is available in electronic handbooks and standards (like MIL-STD-883E and Bellcore manual [37]).

5.3.4. Reliability Models in MIL-HDBK-217

The MIL-HDBK-217 is the most commonly used standard database of failure rates for electronic components. This provides constant hazard rate models for all types of electronic components, taking into account factors that are considered likely to affect reliability. Since the method is mainly applied to repairable systems, the models relate to the component failure rate contribution to the system, i.e. the failure rate of the 'socket' into which that component type fits. MIL-HDBK-217 assumes independent, identically exponentially distributed times to failure for all components.

Microelectronic Monolithic ICs

For silicon ICs, the total device failure rate, λ_p, is given by

$$\lambda_p = \pi_Q[C_1\pi_T\pi_V + (C_2 + C_3)\pi_E]\pi_L \quad 10^6\text{hr}^{-1} \tag{5.15}$$

where

π_Q	quality factor,
π_T	temperature weighting factor,
π_V	voltage derating stress factor,
π_E	application environment factor,
π_L	learning factor,
C_1, C_2, C_3	cost coefficients.

Value of π_Q is determined by the level of screening applied to the component, and ranges from 0.5 for an S-level high reliability screen through eight stages to 35.0 for a commercial, non-military specified component. The π_T is related to the junction temperature and the package thermal resistance. The π_V is dependent on V_{CC}, with V_{DD}=5V having a π_V=1. Twenty levels of π_E have been defined, ranging from Ground Benign (G_B) with a factor of 0.38 to Missile Launch (M_L) with a factor of 13.0. The values of π_L are either 10 or 1, depending on the stability of conditions and controls of the manufacturing process. A period of six months of continuous production is recommended before π_L is reduced from 10 to 1.

Discrete Semiconductors

The general failure rate model for discrete transistors is

$$\lambda_p = \lambda_b(\pi_A \pi_R \pi_C \pi_{S2} \pi_Q \pi_E) \;\; 10^6 \text{ hr}^{-1} \tag{5.16}$$

where

λ_b	base failure rate,
π_A	application factor,
π_R	transistor power rating,
π_C	complexity factor,
π_{S2}	voltage stress factor,
π_Q	factors for Si and Ge transistors,
π_E	values for Si and Ge transistors.

The π_A accounts for the effect of application in terms of circuit function. The π_R accounts for the effect of maximum power or current rating. Values of π_R range from 1.0 to 5.0 according to the power rating. The π_C accounts for the effect of multiple devices in a single package. The π_C values vary between 0.7 and 1.2 depending on the device complexity. The π_{S2} adjusts the model for a second electrical stress (application voltage) in addition to the wattage included in λ_b. The π_Q factors vary between 0.12 and 12.0 depending on the screening level and the encapsulation. The λ_b is related to the influence of electrical and temperature stresses on the device.

The base failure rates for discrete semiconductors are given in MIL-HDBK-217 [429], as functions of temperature and electrical stress:

$$\lambda_b = A \exp\left[\frac{N_T}{T + (\delta T)S}\right] \exp\left[\frac{T + (\delta T)S}{T_M}\right]^p \tag{5.17}$$

where

A	a failure rate scaling factor
N_T, T_M, P	shaping parameters
T	operating temperature in °K, ambient or case, as applicable
δT	difference between typical maximum allowable temperature with no junction current or power and the typical maximum allowable temperature with full rated junction current or power
S	stress ratio of operated electrical stress to rated electrical stress.

One of the most common objections to the MIL-HDBK-217 model is the value attributed to the factors. O'Connor [304] claims that the military handbook models do not account for system level reliability and quality control activities such as burn-in. O'Connor argues against the high values given to the π_Q factors, and he suggests they should be reduced by a factor of 103. Blanks

[48] discusses the disparity between the temperature dependence of failures of observed and calculated data. Calculated data use a much lower temperature dependence than observed data would imply.

For the other electronic components, many stress factors are introduced by Bellcore [37], Fuqua [157], and Blanks [48]. Blanks [48] discusses the temperature dependence of failure rate and suggests different weighting scenarios for each electronic component.

Resistors

The failure rate temperature dependence of resistors is quite complicated, but for some types it follows the Mountsinger's law.

Capacitors

According to the MIL-HDBK-217 data, mica, ceramic and glass fixed capacitors follow Mountsinger's law.

Inductive devices

The failure rate obeys neither Arrhenius nor Mountsinger but is proportional to

$$\exp(a \cdot T^G)$$

where

- a a constant
- T hot-spot temperature in °K
- G depending on the class of insulation and varies from 3.8 to 15.6.

Connectors

The failure rate of this kind of component can be expressed as

$$\lambda_P = \lambda_b \pi_E \pi_P + N\lambda_C$$

where

- λ_b a temperature dependent parameter, and can be reasonably well approximated by the Mountsinger relation within a specific temperature range
- π_E an environmental stress factor
- π_P a factor for active number of pins (from MIL-HDBK-217)
- λ_C a function of the frequency of connect/disconnect actions, and the $N\lambda_C$ can be neglected for frequencies below 40 per 1,000 hrs.

Hybrid Microelectronic Devices

The stress factors approximately follow the Arrhenius relation with activation energies between 0.35 eV and 0.45 eV up to 150°C [48].

5.3.5. Reliability Model of British Telecommunications (BT)

BT has confidence in its own model, which is the all multiplier expression:

$$\lambda_p = \lambda_b \pi_T \pi_Q \pi_E \tag{5.18}$$

where λ_b, π_T, π_Q, and π_E are defined in the Eqs. (5.15) and (5.16). The base failure rate and the factors in this model have all been estimated from BT data; hence BT can claim that reliability predictions are less than a factor of two out, when compared with their own field failure data. This is no guarantee, however, that the same model will be applicable to other organizations. BT uses three levels of π_Q compared to the eight levels used in the military handbook. The levels are dependent on levels of screening specified or approved by BT and are dependent on the package type. Values of Q range between 4 and 0.5. Only three levels of π_E are used by BT, compared with 20 levels given in the military handbook. This is because BT components are either benign, ground fixed, or ground mobile with values of 1.0, 1.5 and 8.0, respectively. All these values have been estimated from an analysis of BT failure data [7].

For simplicity, the following stress factors are used in this book: temperature (π_T), electric (π_S), environmental (π_E), and quality (π_Q). Let $\lambda_{G_{ij}}$ be the generic failure rate of the ith component in subsystem j. The component generic failure rate λ_G can be adjusted by

$$\begin{aligned} \lambda'_{G_{ij}} &= \lambda_{G_{ij}} \pi_{T_i} \pi_{S_i} \pi_{Q_i} \\ \lambda_{ij} &= \pi_E \lambda'_{G_{ij}}, \end{aligned} \tag{5.19}$$

and the subsystem failure rate λ_j can be computed by

$$\lambda_j = \pi_E \sum_{i=1}^{n} N_{ij} \lambda'_{G_{ij}}$$

where N_{ij} is the number of ith component in subsystem j and n is the number of different components.

□ Example 5.7

There are basically two kinds of failure analysis (FA): electrical FA and physical FA. The electrical FA helps to clarify the failure modes, failed address(es), and solid or soft errors. It is usually done on a memory/logic tester, a curve tracer, I-V/C-V meters, or a parametric analyzer. From the failure modes, we can distinguish between DC or functional failures. The DC failures usually result from assembly processes, shipping, and handling (e.g., the ESD events). Consider the following DC test characteristics:

- After pre-condition and HAST, ICs may result in severe delamination, with popcorn failures as a worse situation. Most of these samples may fail DC tests.
- The bonded wires may twist together and show large I_{CC}.

- DC failures are common for ICs with ESD issues.
- Samples with too much solder at pins may form bridges between pins after IR-reflow and cause pin-to-pin shorts.
- A cracked die can be easily found by the DC tests.

As to functional tests, many test patterns are designed to help identify failure modes and addresses. Some testers are able to show the failed bit map graphically. After the failed addresses are located, the sample may be de-capped by chemicals. The focused ion beam (FIB) can precisely cut the chip at the failed location. Then scanning electron microscopy (SEM) and energy dispersive X-ray spectroscopy (EDX) can be used to find out the cause of the failure and to analyze the property of the foreign objects if there are any. Samples with delamination may still pass the DC tests but are likely to fail the functional tests. The delamination may be caused by the contamination of the chip during back grinding or molding. If the top or back surface of the die is not clean, the adhesion between the die and the molding compound may become worse and lead to delamination after pre-condition, HAST, PCT, or THB. At least three kinds of failure can be expected with this small gap between die and molding compound:

1. The difference in expansion coefficient of temperature (ECT) of air, die, and molding compound may cause die or package crack,
2. The Na, K, F ions may cause DC failures, and
3. The heat dissipation of the chip may become inefficient and accelerate the wearout processes.

It is found that a foreign object remaining at the back-side of the die will cause popcorn and package crack. From SEM and EDX, or electron spectroscopy for chemical analysis (ESCA) analyses, the UV tape residue is detected on the back-side of a popcorn sample. Before reaching this conclusion, one has taken several actions: (1) checked the status and maintenance records of the die bonder and molding machines, (2) checked the manufacturing processes and records of die bond and molding, (3) checked the expiration dates of the lead frame, tape, and molding compound used in the defective dice. The usable date of the UV tape is later found to be expired when it is used in manufacturing. This also illustrates the importance of management of the raw material in an assembly house. A rule of thumb is to have a first-in first-out (FIFO) control system, which requires well-trained engineers, and a well-managed database and computer system. Many equipments are important for physical FA, such as (the primary purposes are listed after each equipment)

- Secondary Ion Mass Spectrometry (SIMS): contamination (surface, heavy metal, gate oxide, SiO_2, SiO_2/Si interface, BP-TEOS, B, P, and Na) and junction depth (doping level, implant profile and dosage, W_xSi_y, F&O contents)
- Auger Electron Spectrometry (AES): particle, small area junction depth, defect and bond pad analysis

- Rutherford Back-scattering Spectrometry (RBS): W_xSi_y, TiSi dopant level, Cu content in Al films, H_2 content in SiN, and plasma exposure induced chemical change
- Transmission Electron Microscopy (TEM): gate oxide/TiN/Ti/TiSi thickness and thin film structure
- Fourier Transform Infrared (FTIR) spectroscopy: H_2 content in SiN, water content in SOG
- X-ray Diffraction (XRD): Al rocking curve, grain size, and organic materials
- Total Reflectance X-ray Fluorescence (TXRF): surface contamination
- X-ray Fluorescence (XRF): quantitative and qualitative information of concentration and stiochiometry of materials
- Atomic Force Microscopy (AFM): surface roughness, pin-hole, grain size, and particle analysis
- Induced Couple Plasma (ICP)-mass-spectrometry: trace analysis for cations
- Ion Chromatography (IC)-mass-spectrometry: trace analysis for anions
- Gas Chromatography (GC)-mass-spectrometry: trace analysis for organic compounds
- UV-spectrometry: trace analysis for the SiO_2 contents in water
- Atomic Analysis (AA)-mass-spectrometry: trace analysis for cations
- SAT: package void, delamination, and die crack
- X-ray: wire sweep, wire bond condition, and die crack
- Liquid Crystal (LC): heat sources in a die
- Emission Microscope (EMMI): hot spot examinations

Other methods like current induced voltage alteration (CIVA), light induced voltage alteration (LIVA), optical beam induced current (OBIC), electric beam induced current (EBIC), and laser scanning microscope (LSM), can all be used to find the defective position in bad dice. Costs for each of them vary greatly, and the selection of equipment depends on the FA purposes. The hitting source defines the resolution. Generally, equipment using electrons, ions, light, and X-rays as hitting sources have the highest to the lowest resolution. Hence, SEM and TEM usually give more recognizable pictures on thin film structures.

It is worth mentioning that automatic test equipment (ATE) is becoming more and more important because some failures are difficult or impossible to see when the chip is not functioning. A special handling unit is required to hold the sample, and test patterns must be input to observe the chip performance during dynamic operations. The hot spots can be identified by the aforementioned equipment so that this information can feedback to the TD or R&D department for improvements. The hot spots may be removed by re-designing the masks or the circuit layout. Without real-time ATE tests, it is difficult to find the root causes of failures, and one may thus spend as long as 3 months on physical FA without reaching any solid conclusion. Actually, it is estimated that, on average, the causes of over one third of the failures are unknown, which demonstrates the importance of built-in-reliability. □

5.4. Conclusions

In this chapter, ESS, ALT, and reliability prediction models are described. Fuqua [157] describes ESS as a series of tests conducted under environmental stresses to disclose latent defects. Thus, the goal of ESS is to cause latent defects resulting from manufacturing mistakes to become detectable failures without doing damage to otherwise good material, and an efficient ESS requires acceleration of the mechanisms that give rise to infant mortality.

By using accelerated life tests, manufacturers can assure the reliability of semiconductor products and the quality of ongoing processes and designs. Generally, qualification is needed whenever fundamental changes occur in the semiconductor manufacturing systems. Such changes include, for example, the introduction of new technology or processes, modification of current processes, and design revision.

6. Burn-in Performance, Cost, and Statistical Analysis

Today, customers expect high reliability, low cost, and versatile goods. Some practices related to successful production systems have been discussed in Chapter 1. Statistical analyses can further enhance reliability analysis. In this chapter, we review some widely used techniques for reliability analysis, especially those for burn-in. A product, which will also be called a system or an item, may contain subsystems or modules. For simple products, like a flashlight, we may use components to indicate the modules (that is, the bulb of a flashlight is treated as a component as well as a module). Formal definitions for component, subsystem, and system will be given in Section 6.1.3. This chapter begins by listing basic definitions and considerations related to burn-in which will be used hereafter. A summary of analytic techniques and previous work on burn-in follows. The last section of the chapter describes some cost parameters that are usually taken into account in optimization models.

Depending on component properties and operations efficiency, we can apply different levels of burn-in and screening procedures to weed out infant mortalities in the manufacturing shop [230]. But, in most applications, the reliability of the resulting systems after burn-in is worse than what was forecasted due to incompatibility, as described in Section 6.1.5, not only among components but also among different subsystems and at the system level. As mentioned earlier, we concentrate on the discussion of temperature as a stress factor; however, it should be noted that most techniques proposed in the later parts of this book can be applied to other stress factors, such as vibration, humidity, and voltage, as long as a proper time transformation model is provided.

6.1. Design of Burn-in

6.1.1. Purpose and Definition

Burn-in can be defined or described in many different ways. Bergman [40] defines burn-in in a general way as a pre-usage operation of components performed in order to screen out the substandard components, often in a severe environment. Jensen and Petersen [203] have almost the same definition as Bergman. One important contribution of their book is to popularize the idea that components and systems to which burn-in has been applied have lifetimes that can be modeled as mixtures of statistical distributions [49]. In the AT&T manual [13], burn-in is described as an effective means for screening out defects contributing to infant mortality. This screening typically combines electrical stresses with temperature over a period of time in order to activate the temperature-dependent and voltage-dependent failure mechanisms in a relatively short time. Nelson [298] describes burn-in as running units under design or accelerated conditions for a suitable length of time. Tobias and Trindade [421], however, restrict burn-in to high stress only and require that it should be done prior to shipment. However, almost all IC makers still perform 100% burn-in at various durations to screen defective products.

We define burn-in as a technique used to weed-out infant mortality by applying higher than usual levels of stress to speed up the deterioration of electronic devices. Burn-in is complete when we are reasonably sure that all the weak devices have failed, thus leaving the remaining devices in a reliable state. One of the major problems associated with burn-in is the determination of exactly how long the burn-in process should continue, balancing appropriately the needs of reliability and the total costs [203]. Some common concerns for burn-in include:

- How far should we go to reduce infant mortality by burn-in?
- What are the savings from burn-in?
- Under what environmental conditions should burn-in be performed?
- Should burn-in be accomplished at the system, subsystem, or component level?
- Who should be in charge of burn-in—the vendor, buyer, or a third party?
- What would be the expected life after burn-in? How does it differ from the expected life without burn-in?
- Is burn-in always necessary and economical?
- Are there any side effects of burn-in?
- How will the industry benefit from burn-in data?

Generally, burn-in is a screening technique used to improve product reliability. It is one of the most widely-used engineering methods for removing weak devices from a standard population, which usually consists of various engineering systems composed of devices, parts, or components [49]. Burn-in is very useful for highly integrated circuit systems [497, 516, 230].

6.1.2. Mixture of Distributions

Devices which become more prone to failure throughout their life will not benefit from burn-in because burn-in stochastically weakens the residual lifetime. For burn-in to be effective, lifetimes should have high failure rates initially and then improve [49]. The type of bathtub-shaped hazard rates of Figure 3.1 would seem to be appropriate for burn-in, since burn-in removes high infant mortality. Jensen and Petersen [203] explain that there are reasons why many systems and components have bathtub-shaped hazard rates. Many industrial populations are heterogeneous, and there are only a small number of different subpopulations. Although members of these subpopulations do not, strictly speaking, have bathtub-shaped hazard rates, sampling from them produces a mixture of these subpopulations, and these mixtures often have bathtub-shaped hazard rates. Initially, the higher failure rate of the weaker subpopulation dominates until this subpopulation dies out. However, after that, the lower rate of the stronger subpopulation takes over so that the failure rate decreases from the higher to the lower level [49].

Notice that if $F_i(t)$ for $i_1, \ldots, i_k$ is a DFR distribution in t, $a_i \geq 0$ for $i_1, \ldots, i_k$, and $\sum_{i=1}^{k} a_i = 1$, then $F(t) = \sum_{i=1}^{k} a_i F_i(t)$ is also a DFR distribution. However, a mixture of IFR distributions is not always IFR [32]. Therefore it is valid to use a mixture of distributions to model the infant mortality period if each subpopulation has a DFR distribution. See Table 3.2 for the preservation of the mixture of two DFR distributions.

□ Example 6.1

Sampling from two exponential distributions–one with a small mean and one with a large mean–results in a distribution with DFR, which is a special case of the bathtub shaped hazard rate where no aging is expected. The mixture of more complex distributions produces a distribution with more typical bathtub shaped hazard rates. Tobias and Trindade [421] present the following distribution as a mixture of three distributions.

$$\begin{aligned} F(t) &= 0.002\Phi\left(\frac{\ln(t/2700)}{0.8}\right) + 0.001\left(1 - e^{-(t/400)^{0.5}}\right) \\ &\quad + 0.97\left[1 - e^{-10^{-7}t}\left(1 - \Phi\left(\frac{\ln(t/97500)}{0.8}\right)\right)\right] \end{aligned} \tag{6.1}$$

where Φ is the standard normal CDF. The left tail of the distribution in Eq. (6.1) is very steep and represents the infant mortality period. Many studies provide models for bathtub shaped hazard rates [13, 630, 668, 738]. One simple way of obtaining them is by mixing standard distributions such as exponential, gamma, and Weibull [49]. □

The AT&T reliability model [13] for the hazard rate of an electronics component is a mixture of a Weibull and an exponential distribution. The early part of the hazard rate is modeled by a Weibull distribution with DFR, and the latter part is modeled by an exponential distribution. Therefore, the exponential distribution is appropriate for the steady-state region of the model. The hazard rate of the AT&T model is given by

$$h(t) = \begin{cases} h_1 t^{-\alpha}, & 0 < t < t_c \\ h_L, & t_c \leq t \end{cases} \tag{6.2}$$

where $0 \leq \alpha < 1$, $h_1 = h(1)$, and t_c, the crossover time, is taken to be equal to 10^4 hrs. The h_L is the long-term hazard rate after the crossover time. Eq. (6.2) does not have an aging period since AT&T claims that AT&T electronic equipment tends not to age before it is replaced. It is assumed that aging needs not be considered, provided that

- devices are protected by appropriate specifications for life testing and qualifications,
- devices are not stored or used in conditions beyond the allowable ratings, and
- devices are not used in products whose field life is longer than the device design life.

Chien and Kuo [76] present optimal burn-in for IC systems based on the AT&T model.

□ **Example 6.2**

Assume the following for an IC: the long-term hazard rate, h_L=10FITs and α=0.6. What percentage might be expected to fail in the first month of operation and then in 2 years of operation ?
The first month of operation is within the infant mortality period. From Eq. (6.2), at $t = 10^4$ hrs

$$h_1(10,000)^{-\alpha} = h_L.$$

Therefore,

$$h_1 = (10)(10,000)^{0.6} = 2512 \text{ FITs} \simeq 0.251 \times 10^{-5} \text{ /h}.$$

The number of hours in a month is 365×24/12=730. Since the CDF of a Weibull distribution in the AT&T model has the form:

$$F(t) = 1 - \exp[\frac{-h_1 t^{1-\alpha}}{1-\alpha}],$$

the expected percentage failing in the first month of operation is

$$F(730) = 1 - e^{-0.0003} = 0.03\%.$$

In 2 years (17520 hrs) of operation, since the first 10,000 hrs is infant mortality, the expected percentage failing in the first 10,000 hrs of operation is

$$F(10,000) = 1 - e^{-0.0005} = 0.05\%$$

Beyond 10,000 hrs, since the hazard rate is constant, $h(t) = h_L = 10$ FITs $= 10^{-8}$/h,

$$F(t) = 1 - \exp[-\int_{10,000}^{17520} 10^{-8} dt] = 1 - \exp[-0.0000752] = 0.0075\%.$$

The total expected failures in 2 years is $0.05 + 0.0075 = 0.0575\%$ of the original population. □

6.1.3. Levels of Burn-in

Three levels (from the lowest to the highest: component, subsystem, and system) are defined in this book as shown in Figure 6.1. Assume 5 components in subsystem 1, and 5 subsystems in the system. We can apply burn-in to different levels as depicted in Figure 6.2, where h_c, h_u, and h_s denote the hazard rate of a component after component, subsystem, and system burn-in, respectively. In Figure 6.2, the component, subsystem, and system burn-in times are t_1, $(t_2 - t_1)$, $(t_3 - t_2)$, respectively. Since different burn-in temperatures may be used for different levels, the efficiencies of removing the defects may not be the same.

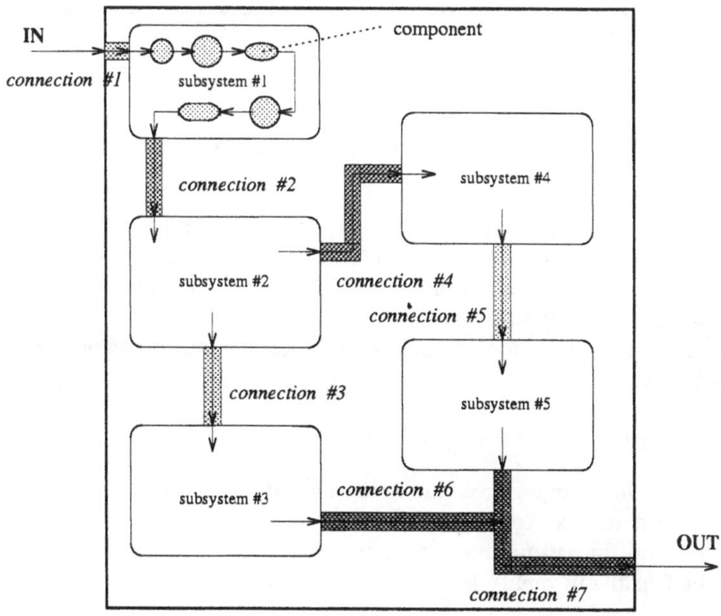

Figure 6.1: A three-level system.

Eight burn-in policies in Table 6.1 can be chosen: no burn-in, component only, subsystem only, system only, component and subsystem, component and system, subsystem and system, and all levels burn-in. These will be probed in depth in later chapters. The machine-level burn-in (system burn-in) is reported in the study of Elkings and Sweetland [119] where they list some disadvantages of this burn-in scenario:

- It is too expensive.
- It has limited use in screening "design" problems; the design problems may be found during burn-in but are not removed since the repairs are done on the original design.

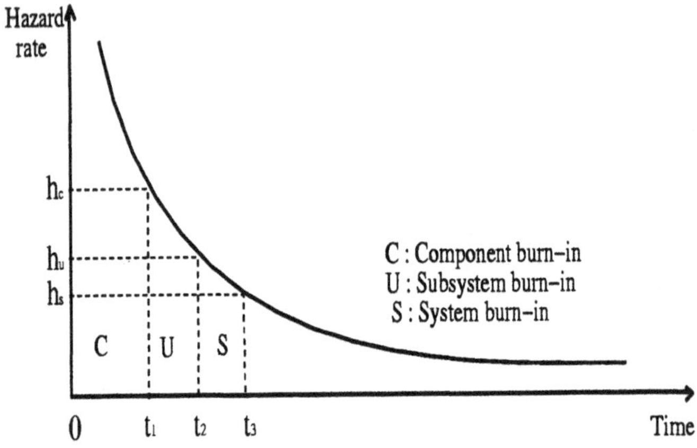

Figure 6.2: Burn-in applied at different levels.

These two shortcomings of system burn-in are quite common in practice; however, as reported by Chien and Kuo [76, 81], system burn-in is still favored because it can remove more "incompatibility" (which will be introduced in Section 6.1.5) than lower-level burn-ins.

A burn-in strategy for complex and expensive ICs is introduced in the example below.

□ Example 6.3

In general, smaller sample sizes can be used at later burn-in stages to save cost. However, for a new and complex IC, we need large samples for engineering evaluations and for estimating parameters (e.g., E_a and β) for burn-in. This will not be feasible in reality if

- the cost per chip is high,
- the process is not fixed,

Table 6.1: The eight burn-in policies.

Policy	Description
O	no burn-in
I	component burn-in only
II	subsystem burn-in only
III	system burn-in only
IV	both component and subsystem burn-in
V	both component and system burn-in
VI	both subsystem and system burn-in
VII	all levels burn-in

- the lot-to-lot conformity varies (e.g., out-of the 3σ control chart), and
- the yield is low (e.g., less than 30%).

In addition to chip-related issues, we must also consider test equipment-related issues:

- the number of samples in a burn-in chamber may be small,
- burn-in board (BIB) for the product may be expensive, and
- BIB can probably hold limited samples per board.

One possible action is to compare this new IC with similar ones or its prior generations to learn about its characteristics and the associated parameters. For example, to estimate parameters for a P7 chip, Intel may compare the complexity among 386, 486, Pentium, and Pentium-Pro to narrow down the ranges of the possible values for parameters so that the sample sizes of the HTB tests may be decreased. However, even though a smaller sample size is acceptable, the time needed for FA may be too long because of the complexity of this microprocessor and, thus, may delay its introduction to the market. We need to find another alternative, which is the use of the "test-chip."

The selected test chip may be a new design specifically produced for testing complex and expensive ICs. It may also be an existing product that resembles the new ICs. Whichever it is, there is one strict rule: the test chips and the new ICs must be made on the same production line because most parameters are process dependent. In other words, the equipment used to make the test chips must be a subset of those producing the new ICs; the raw materials (wafers, gas, chemicals, and others) must also be the same or from the same suppliers. The reasons for all this care is to increase the correlation of the test results using the test chip. For example, the 256Kb static random access memory (SRAM) can be used as the test chip for the microprocessors made by the $0.8\mu m$ technology because

1. these two ICs may be designed by the same design rule,
2. the cache memory in the microprocessors is exactly the SRAM structure, and
3. the production line is able to manufacture these two kinds of ICs with the same equipment and the raw materials.

For the microprocessors made by the $0.45\mu m$, $0.3\mu m$, and $0.18\mu m$ technology, the 1Mb, 4Mb, and 8Mb SRAM can be considered as the test chip, respectively.

Suppose a test chip is chosen for a microprocessor. The HTB tests on the microprocessor (at smaller sample sizes) and the test chip (at much larger sample sizes) are done simultaneously. After performing the HTB tests at different temperatures and stressed voltages on the test chip, parameters can be estimated. The HTB tests can be terminated either at a certain time (e.g., 1,000 hrs) or when a certain failure percentage is reached (e.g., 0.5%). The latter method requires a proper test during burn-in (TDBI) or in-situ test equipment; the TDBI for microprocessors will be extremely expensive (as high as \$1.5 million for each chamber). Detailed electrical and physical FAs on the failed microprocessors and on the failed test chips are performed if any failure is detected. The correlation between the microprocessor and the test chip can be verified from the failure modes obtained from the FA and the Poisson yield model: $Y=\exp(dA)$, where Y is the HTB test yield, d is the defect density (in $1/mm^2$), and A is the chip area (in mm^2). (Generally, for process technologies between 0.7~0.9μm, d is between $0 \sim 0.05$ and $0.25 \sim 0.30$ for memory and logic products, respectively.) Good correlation is derived if the estimated HTB test yields of the microprocessor obtained by the d from the test chips fall within the 95% confidence interval. Re-designing or re-selecting the test chip may be necessary if large deviations are identified. □

6.1.4. Burn-in Procedures

To correctly analyze burn-in results, it is necessary to distinguish between burn-in procedures. As described in Chapter 2, burn-in occurs between two FT stages, which are called the pre- and post-burn-in tests . Certain test items will be used for calculating burn-in failure rates by lots. (The number of wafers in a lot varies; most IC makers set 25 wafers in a lot.)

A typical burn-in chamber consists of a temperature oven and a control system, which is either a PC or a workstation. The temperature oven contains power suppliers, compressors, and an interface with the control system. Usually, a driver board goes with each BIB to send and receive signals to the DUTs. The accuracy, channel number, and precision of the driver boards greatly affect the price of the burn-in chambers. For static and simple dynamic burn-in systems (which will be introduced below), the driver boards may only send DC bias or simple signals to the DUTs (it is called the "write" operation) and may not be able to get the feedback of the DUTs (the "read" operation). The modern TDBI system is able to perform "write" and "read" operations with short time delays, which makes it capable of detecting any failure during burn-in. The control system is composed of a database to store programs and test results and the software tools for designing test programs, setting biases and temperatures, and writing the test patterns during stressing and functional tests.

Some burn-in chambers need a board-checker to see if the DUTs are mounted properly on the BIBs. Advanced burn-in chambers provide in-situ tests to check the contacts of the DUTs. Usually, an initial test is used to perform functional tests to make sure the DUTs are electrically good samples at room temperature. Bad DUTs will be dismounted. Take a TDBI chamber for example. Before temperature and voltage stresses, another functional test at high temperature (e.g., 70°C or 80°C) will be performed to test the samples' characteristics. Then the oven will ramp-up to the stressed temperature (e.g., 125°C) and stay at that temperature for a specified duration with the stressed test pattern (for example, the read-modify-write, RMW, pattern) applying on the DUTs. After burn-in, functional tests will be on the DUTs to estimate the burn-in failure rate. Of course, simple dynamic burn-in chambers cannot perform the complicated in-situ pre- and post-burn-in tests mentioned above. Instead, the DUTs should be tested by a FT program on a memory or a logic tester, whose test cost is much higher than that done on the TDBI system. This phenomenon will become more significant when the ICs are more complicated.

Two distinct types of burn-in, static burn-in and dynamic burn-in, are discussed in the AT&T manual [13] according to the scenario of applying stresses. Jensen and Petersen [203] classify burn-in procedures by repair. In the sequential burn-in procedure, items are not repaired during burn-in. However, Markov burn-in is a method of handling the burn-in of repairable systems. Among the many burn-in methods, static burn-in, steady-state burn-in, dynamic burn-in (DBI), and TDBI are frequently used for semiconductor devices [352]. To determine which type of burn-in is appropriate for a given device, we need to know the types of possible defects and the extent to which these defects are activated by the various techniques [190].

Static Burn-in

Stresses are applied to the samples at a fixed level or on an elevated pattern. Static burn-in can be performed only at the component level. Static burn-in is more effective than dynamic burn-in for defects resulting from corrosion or contamination and for typical SSI or MSI commercially made ICs where the majority of defects are caused by contamination and intermetallic formation [13]. Static burn-in is similar to steady-state burn-in, but in static burn-in, samples are powered, and the outputs are loaded for maximum power dissipation.

Steady-state Burn-in

A forward or reverse bias voltage is applied while burn-in is being performed. As many junctions as possible must be biased. Steady-state reverse bias burn-in is used for digital and linear circuits, mainly with NPN inputs and outputs [190]. Steady-state burn-in is most effective in detecting failures caused by contamination, corrosion, and intermetallic formation. Since there is a possibility that steady-state burn-in causes IC parameter shift, after burn-in samples may have high leakage currents and be slower than they were before the burn-in.

Dynamic Burn-in (DBI)

Stresses are exercised on the samples to simulate the real operating environments. For DBI, there are two options: component level burn-in and subsystem and system level burn-in. If the stresses are at the subsystem or system level, lower temperature should be used so that unexpected damages can be avoided. Consequently, subsystem and system burn-in is generally limited to lower temperatures than component burn-in [13]. DBI is less effective than static burn-in at screening out surface problems but is reported to be a more effective technique for LSI and complex VLSI integrated circuits [13]. Wurnik and Pelloth [447] list the practical performance of DBI and also consider the setting of the E_a in the Arrhenius equation for this burn-in scenario.

For some complex devices, steady-state burn-in and static burn-in may not be adequate. Since the external biases or loads do not effectively stress internal nodes, DBI is used to stress them. DBI places live signals on the clock, address, and data lines of ICs, and these signals stress internal nodes. That is, stresses are applied to the samples to simulate real operating environments.

Two types of the most common dynamic burn-in are parallel excitation, where all devices are connected in parallel, and ring counterexcitation, where the devices are connected in series [190]. Thus, the output of one device drives the input of the next in the ring counterexcitation. However, all devices of parallel excitation are driven by the same source.

Test During Burn-in (TDBI)

Electrical tests after burn-in usually take a long time and thus increase cost. TDBI is similar to the dynamic burn-in method, but it has a difference in that the DUTs are also cycled with a functional test pattern [189]. The functional test patterns are usually used as test signals, and the tester monitors the outputs of the samples under test, checking for functional failures. TDBI is frequently used for DRAMs and SRAMs due to relatively long test time but are not cost effective for most VLSI circuits.

Sequential Burn-in

Burn-in is continued until a certain time between failures has been achieved. Let W_i be the ith waiting time between failures after the transformation. The burn-in is terminated when

$$W_i \geq W^*, \tag{6.3}$$

where W^* is the normalized waiting time [203]. The tabulated values for W^* is in the Appendix 2 of Jensen and Pertersen [203]. We now assume that the weak subpopulation may follow a Weibull distribution of Eq. (3.17). Then

$$Y_i = \ln(1 - F(t_i)) = \left(\frac{t_i}{\eta}\right)^\beta .$$

Eq. (6.3) of the condition to terminate the burn-in is

$$W_i = Y_i - Y_{i-1} \geq W^*.$$

That is,

$$\left(\frac{t_i}{\eta}\right)^\beta - \left(\frac{t_{i-1}}{\eta}\right)^\beta \geq W^*.$$

Instead of computing waiting time beforehand, it is convenient to compute the increment Δ,

$$\Delta \equiv t_i^\beta - t_{i-1}^\beta \geq W^*(\eta)^\beta,$$

whenever a failure occurs [203].

□ **Example 6.4**

A manufacturer puts a batch of 50 devices in a sequential burn-in test. From previous experience, it is known that the early life of devices follows a Weibull distribution with the parameters: β=0.5, η=2(hrs). The average proportion of weak devices in a batch is 10%. After the test the manufacturer wants to guarantee, with 95% confidence, that no weak devices are left in the batch. To determine when to stop the burn-in, we obtain W^*=3.0 and $W^*(\eta)^\beta$=4.24 from the table A2.1 in the Appendix 2 of Jensen and Pertersen [203].

Thus, $\Delta \geq 4.24$ is the stopping criterion for this burn-in test. That is, burn-in can be stopped when the t_i is such that $t_i^{0.5} - t_{i-1}^{0.5} \geq 4.24$. □

Markov Burn-in

Markov burn-in is used for repairable burn-in models where the failed samples are fixed after some exponential repair time. The system is put on burn-in for a predetermined period. During the burn-in test, the system is continuously monitored for failures, and when a failure occurs, the failed sample is taken out of the test to be repaired. When the failed component has been found and the repair completed, the system is put on the test again, and the burn-in time restarted from zero. If the system completes the test period without a failure, then it is taken out of burn-in and after additional inspections, is ready to leave the factory. This is called Markov burn-in because the theory of Markov chains is used in the evaluation of the results of the burn-in procedure [203].

Stress Burn-in

In the IC industry, DBI and TDBI are commonly exercised. Most burn-ins are conducted under a very stressful environment; hence we recognize them as stress burn-in.

Implementation

In addition to these burn-in methods, sometimes high voltage is applied during burn-in, because some failures, which are not temperature-dependent, such as gate oxide breakdown, are not well detected by normal burn-in. Thus, high-voltage stress tests can be defined as burn-in screens. Many IC memory manufacturers use high-voltage stress tests instead of dynamic burn-in to detect oxide defects in MOS ICs [189].

The Arrhenius equation is also used for DBI and TDBI although the model can only be applied if the degradation occurs at constant stress. Basically, we do not consider the repairable-type burn-in models because of the complexity of the repair mechanism. The repair time distribution is assumed to be exponential in the Markov burn-in procedure. However, this assumption, though it can result in technically tractable outcomes, is usually too strong in practice. Choosing a reasonable type of burn-in and specific stress conditions depends on the device technologies and the reliability requirements.

Example 5.3 briefly explains procedures to estimate E_a and β. Here, a more detailed example depicts a complete and practical technique to estimate these two parameters.

□ Example 6.5

In an IC life-cycle, re-designs, evaluations, and FAs repeat many times to find the "optimal" process recipe. The optimal process is derived based on the trade-offs among

- customers' requirements,
- the chip performance and characteristics,
- available equipments, raw materials, and technologies,
- time to market, and
- cost.

To fix the manufacturing processes, except the front-end operations, we have to make sure the programs and conditions (temperatures, test flows, durations, and others) of all back-end operations such as the WAT, WP, WBI (if it is built-in), assembly, FT, burn-in, and the QC-gate E-test are all decided and formally released so that the product is ready for the production (or full) Qual. (refer to Example 2.1). The production Qual. is to make sure the product meets industrial standards and customers' requirements. We must keep in mind that the activation energy for the Arrhenius model, E_a, and the voltage acceleration multiplier, β, cannot be estimated until the processes and the test conditions are all fixed. These two parameters are required to represent the long-term failure rate in terms of FIT. We must bear in mind that what customers really care about is the E_a and β after

- the product is in its mass production stage, and
- they represent the product characteristics after the ICs are used in the field.

Therefore, the ICs sampled for estimating E_a and β must be the ones randomly selected from the final products, which are ready to ship to the customers. Although changes may still be made for yield enhancement after the product is fully released for mass production, the E_a and β derived earlier for this product can still be used as references. The changes categorized to be minor (not affecting product's quality and reliability) may not need to be qualified by the time- and cost-consuming reliability tests. Most IC makers have internal criteria for major & minor changes; the corresponding reliability tests are usually defined for the corresponding major change. For example, it is obvious that the thickness change of the gate polysilicon is a major change and the 1,000-hr HTOL, 48-hr EFR, 1,000-hr LTB, and HCI, gate-oxide TDDBs (both package and wafer level measurement) must be performed to decide if the temporary engineering change notice (TECN) is qualified to become an official ECN. From the above explanations, readers may understand that semiconductor industries are highly risky (although the return of investment may also be high when the products are at good prices) because any major change needs a long time and a great deal of money (labor, handling, stressing, testing, and analyzing) to verify and to be qualified; any mis-operation and mis-judgement will induce large losses.

The sample size for estimating E_a and β may be small (e.g., less than 1,000 per condition) for new or expensive ICs. Should this happen, a longer test time may be used as a remedy. However, it should be kept in mind that failure modes may be different if the HTB test duration is longer than a certain time (e.g., 5,000-hr for most modern ICs). If we know from previous experience that the overall acceleration factor (including both the temperature and the voltage acceleration) is 1,000, then 1,000-hr may be the maximum HTB test duration for estimating E_a and β because this long test time may accelerate the samples into the aging region, where failure modes are generally different from those in the infant mortality and in the random failure region. The most common failure mechanisms in the aging region are HCI, gate oxide and capacitor-film TDDB, and EM. In other words, the large sample size is particularly important for the experiments to estimate E_a and β.

The minimum sample size, n, needed for each test condition can be decided by (for details, see Appendix 5D in Grosh [168]):

$$n = \chi^2_{1-\alpha}(2x+2)/(2p) \tag{6.4}$$

where $\chi^2_{1-\alpha}(2r)$ is the $100(1-\alpha)^{th}$ percentile of the χ^2 distribution with $2r$ degrees of freedom, x is the expected number of failure in the 125°C/7.0V HTB test (which is the common HTOL test condition for the 5.0V ICs), and p is the average of the accumulated failure percentage in the z-hr HTOL test. The z value ranges from 168 to 1,000 depends on how long the estimating experiment will continue. Critical values of χ^2 distribution for given $1-\alpha$ and degree of freedom are shown in Figure 6.3. Figure 6.4 shows the sample size at different xs and ps at 95% confidence level (C.L.). By Eq. (6.4) and let $x = 8$, the sample sizes under different confidence levels are plotted in Figure 6.5. Some IC makers use as many as 10,000 samples per condition to estimate E_a and β; from Figure 6.4, we may expect 4 failures at 95% confidence level if p, from previous experience, is 0.09%.

To estimate E_a, we must fix the applied bias (e.g., 7.0V) and use at least two different temperatures. Similarly, we need at least two HTB tests with different voltages at the same temperature (e.g., 125°C) to estimate β. The HTOL test is usually one of the chosen HTB tests so that the EFR and the long-term failure rate of the product can also be derived from the estimating experiment; this is an extra benefit gained from the experiment. That is, the simplest combination to estimate E_a and β is to issue the following three HTB tests: 150°C/7.0V, 125°C/7.0V, and 125°C/7.5V; the first two are for E_a, and the last two, for β. This simplest combination explains why we use the p obtained from the previous HTOL test(s) to decide n: there should be more

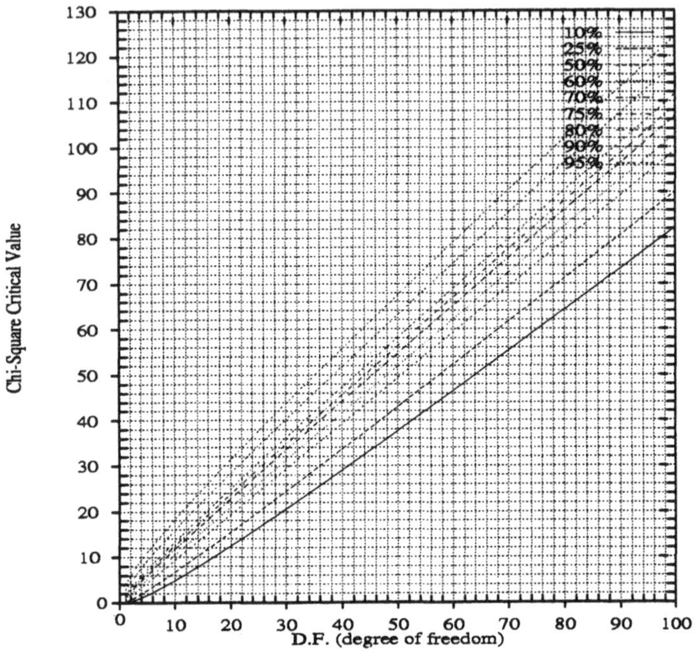

Figure 6.3: χ^2 critical values.

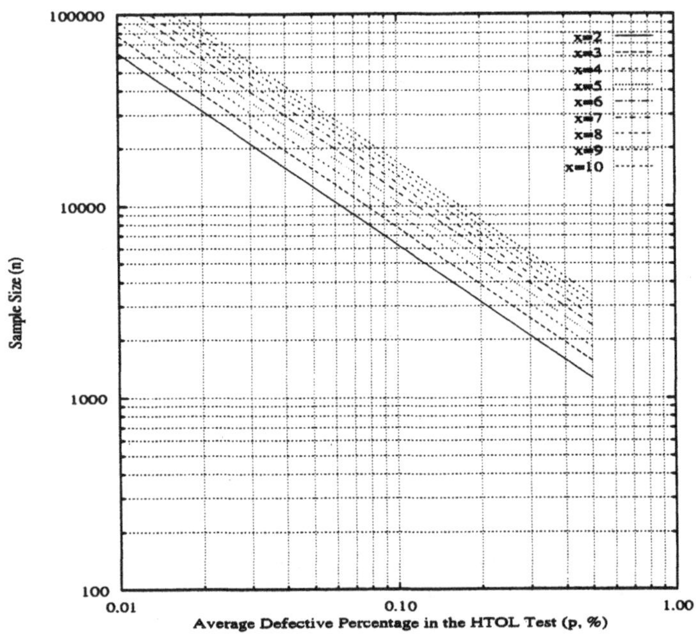

Figure 6.4: The sample size needed for the experiment at 95% C.L.

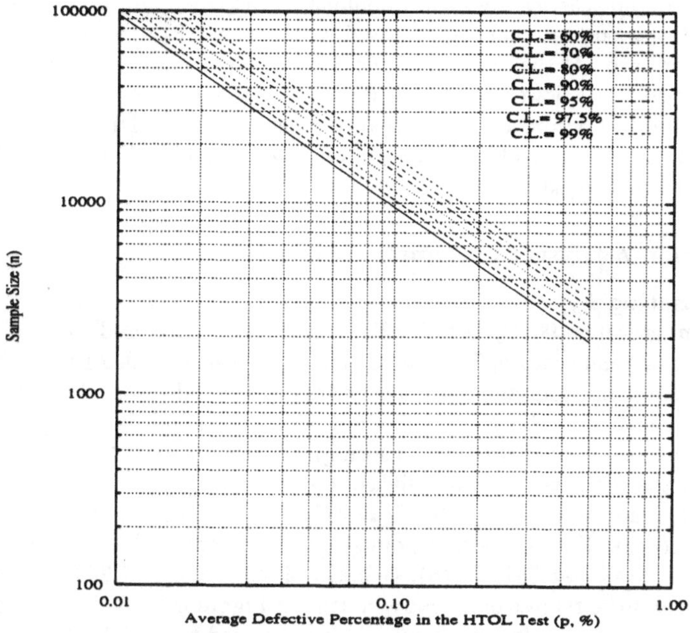

Figure 6.5: The sample size needed for the experiment when $x = 8$.

failures from the other two HTB tests because of the higher temperature and voltage.

Eqs. (5.5) and (5.7) can be used to obtain E_a and β for the two-condition case (see Example 5.3). When there are three or more conditions for estimating E_a (e.g., 90°C, 110°C, 125°C, 150°C), a regression line can be drawn from the Arrhenius plot whose Y-axis and X-axis is $\ln(RR(T_s))$ and $(1/T_s - 1/T_j)/k$, respectively. The value of the slope is the estimated E_a. As before, T_j and T_s is the junction and the stressing temperature in °K, respectively, and k is the Boltzmann's constant. The $RR(T_s)$ is either the time to reach a failure percentage or the failure percentage at a test point at T_s. Similarly, if more than three voltages (e.g., 6.5V, 7.0V, 7.5V, and 7.8V for an IC with external input voltage of 5.0V at normal using condition) are used in the experiments for β, β is equal to the value of the slope of the plot with Y-axis and X-axis being $\ln(RR(V_s))$ and $(V_{int,s} - V_{int}) = \Delta V_{int}$, respectively, where $V_{int,s}$ and V_{int} is the internal voltage at the stressing and at the normal using condition, respectively. To decrease the gate electrical field, most DRAMs have internal voltage regulators to reduce the external voltage (e.g., 5V) to lower levels (e.g., 3.3V and 2.8V). The external voltage of DRAMs, for example, is designed to be equal to the bias used on the motherboard. However, as the gate channel length decreases from 0.36μm to 0.32μm, 0.28μm, 0.22μm, and 0.18μm, the internal operating voltage has to be reduced accordingly from 3.3V to 2.8V, 2.5V, 2.2V, and 1.8V, respectively, in order to alleviate the short channel effect and to increase the hot carrier injection immunity. This trend will make the devices operate faster so that they can be used with newer CPUs, which evolve in an even faster path. At the same time, the size shrinking trend forces motherboard designers to use a lower operating bias in the board circuits (including the chip-set and the peripheral interfaces) to meet CPUs and DRAMs specs. The MosAid tester can provide information on $V_{int,s}$ and V_{int}. Finally, the $RR(V_s)$ is defined similarly to $RR(T_s)$.

Regression Analysis to Estimate E_a and β

(1)**Estimating E_a**
Four temperatures (90°C, 110°C, 125°C, and 150°C) are used to estimate E_a. The junction temperature, T_j, is 55°C. The temperature higher than 150°C will damage the capacitors and the fuses on the BIBs and, thus, the highest test temperature is set to be 150°C. The four HTB tests will be terminated when 0.1% of the samples fail (type II censoring). Suitable TDBI systems are used in the tests; the time between two adjacent test points is 8-hr. The time needed to reach 0.1% failures at 90°C, 110°C, 125°C, 150°C is 168-hr, 112-hr, 56-hr, 32-hr, respectively. Hence, the four points in the Arrhenius plot have coordinates (−3.411, 5.124), (−5.081, 4.718), (−6.223, 4.025), and (−7.946, 3.466). The test data and the regression line are plotted in Figure 6.6, where the coefficient of determination (R^2) is 0.9753 and E_a is estimated to be 0.38 (the regression line is $\ln(RR(T_s)) = 6.4869 + 0.3801\, x$, where x is $(1/T_s - 1/T_j)/k$).

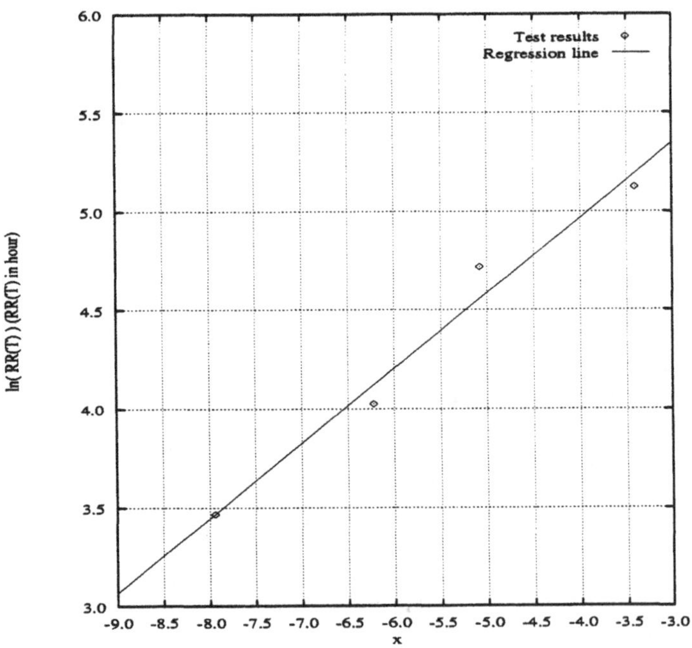

Figure 6.6: Regression to estimate E_a.

(2)Estimating β

Suppose four voltages (6.5V, 7.0V, 7.5V, and 7.8V) are used to estimate β; the corresponding internal voltage is 4.4V, 4.7V, 5.0V, and 5.2V, respectively, and we know V_{int} = 3.3V. The input voltage higher than 8.2V is believed to greatly enhance the self-heating effect at 125°C and, thus, the highest stress voltage is set to be 7.8V so that we still maintain a 0.4V guard-band. After 332-hr (two weeks) of stress tests (type I censoring), the failure percentages are 0.5%, 0.9%, 2.1%, and 3.8%, respectively. Hence, the values for the Y-axis are −5.298, −4.711, −3.860, and −3.270 when the internal voltage differences are 1.1V, 1.4V, 1.7V, and 1.9V, respectively. The test data and the regression line are plotted in Figure 6.7, where R^2 is 0.9917 and β is estimated to be 2.56 (the regression line is $\ln(RR(V_s)) = -\ 8.1844 + 2.5568\ \Delta V_{int}$).

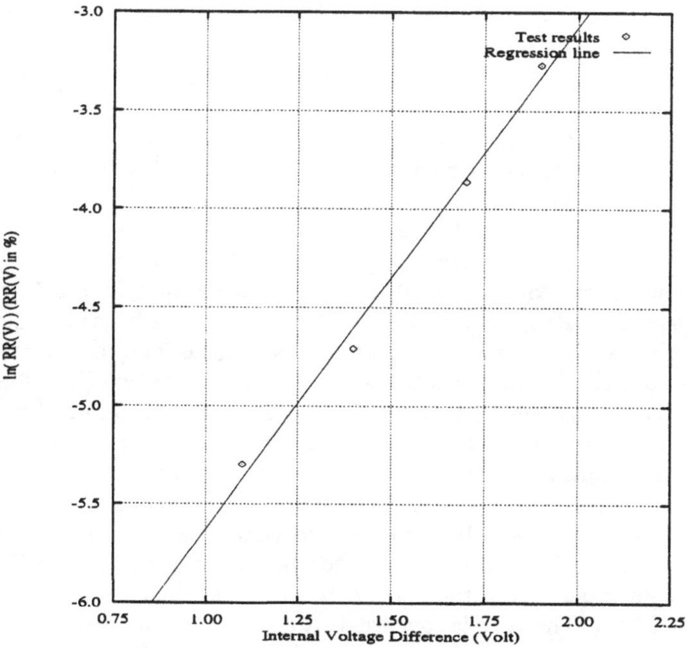

Figure 6.7: Regression to estimate β.

Two Kinds of Estimating Methods

Two kinds of estimating methods will be introduced in detail based on available resources, meaning the stressing and the testing equipment. The first approach (method #1) best suits the needs when enough TDBI or in-situ testing systems are available. However, for most IC makers, the second method (method #2) seems to be more proper due to lack of the required systems to apply method #1. Both of them can provide information on the EFR and the long-term failure rate.

(1) Method #1 (Type II Censoring Test)

To derive reliable E_a and β, the time between two adjacent test points must be as short as possible. That is one of the reasons that IC makers are increasingly using TDBI or equivalent in-situ testing chambers which are able to increase temperatures to at least 150°C during stressing. Some advanced burn-in chambers provide Weibull, lognormal, and normal plots for model fitting and are able to estimate the parameters (such as the shape, the scale, and the location parameter of the Weibull distribution) based on the test records stored in their database. Users can see if a certain model is the best from the regression line and R^2. Some TDBI systems can record the failed addresses and display failed-bit maps after incorporating the address/data scramble formula. Some can even mask the failed bit(s) to gather more data on the failed samples.

Even though TDBI systems are available, we still need to take the following two issues into account.

First, full correlation tests between the tester and the TDBI system must be completed. The test item used to identify burn-in failures is usually the first functional test item in the post-burn-in test program, which is originally designed to be executed on a memory or a logic tester. This special test item is arranged in this manner to save precious testing time because the FT tests will be aborted once the sample fails a test item (this setting can be de-activated in the debugging mode so that the sample will be tested by all test items). The memory or the logic tester offers much higher timing accuracy than the most advanced TDBI system thus far which accounts for a huge price difference between the two. A common memory tester can cost as much as 2.5 million US dollars; an advanced TDBI chamber costs only one-half million US dollars. The correlation test must ensure that the TDBI system is able to correctly identify the burn-in failures using the test item originally designed for the tester. A core control program incorporating the timing, the bias, and the test pattern has to be designed and stored in the TDBI computers. Thousands of samples (both electrically failed and pass ICs) are used to verify that all of the failures can be detected by the TDBI system at the specified temperature (e.g., 70°C). The TDBI system is qualified if the burn-in escape is less than a pre-specified level (e.g., 100 PPM). The correlation test is a time-consuming task; it may take more than two months to fully correlate a TDBI system (some fine-tuneing of the driver boards and BIB re-designs may be required).

Second, suppose a TDBI system is qualified as a replacement of the tester. Unless specified, only one temperature is allowed in a TDBI chamber, which implies we need at least two TDBI chambers for estimating E_a. Besides, the capacity of one TDBI chamber must be large enough to contain 10,000 samples (as mentioned above), which is not possible for most TDBI systems. The capacity of a burn-in chamber depends on the package type of the sample and the burn-in board (BIB) dimension. The larger the BIB, the more samples it can hold. Similarly, more samples can be on a BIB if they are assembled by a smaller package type. One BIB of a certain TDBI chamber can hold 168 26/24 small outline package J-lead (SOJ) 300 mili-inch (mil) ICs and only 120 50-pin thin small outline package type II (TSOP-II) 400-mil ICs. If the TDBI system has 12 driver boards, the full capacity is only 2,016 26/24 SOJ 300-mil and 1,440 50-pin TSOP-II 400-mil ICs. This means, if the ICs are packaged as 26/24 300-mil SOJ, we need 5 TDBI chambers. Furthermore, the chambers have to ramp-up and down during the experiment because the stressing (e.g., 125°C) and the testing temperatures (e.g., 80°C) are different. The samples cannot be tested at the stressing temperature because it is beyond the specifications of the sample, which is between 0 and 70°C for most DRAMs, for example. To be more specific, the samples are burned-in (stressed) at 125°C (or 150°C) and 7.0V for a certain period of time with the stress pattern (read-modify-write, RMW) and the chamber must ramp-down to the test temperature defined for the product (e.g., 80°C, if 10°C guard-band is required); the chamber will stay at 80°C for some time (e.g., 3 minutes) to assure the junction temperature of the samples is about 80°C. It may take more than 30 minutes to reduce and to stabilize the temperature inside the chamber from the stressing to the testing temperature. Besides, BIBs which can sustain a 150°C test, for example, are much more expensive than those used only below 130°C. The useful life of a BIB is usually around 10,000-hr only. In summary, although many TDBI systems are now available, not many IC makers use of them for the E_a and β experiments due to the limitations mentioned above.

Suppose enough TDBI chambers are available for the E_a experiments and consider the example in Table 6.2 for DRAMs. The initial sample size at each temperature (125°C and 150°C) is 10,000. These 20,000 ICs are randomly sampled from 20 lots (1,000 ICs per lot) in the warehouse storing the final products (ready for shipping out to customers) and the fab-out dates cover the last 3 months. Hence, we believe the samples best represent the quality of the out-going products. The experiment will continue for 500-hr (about 3 weeks). From Table 6.2, where E_a is estimated by

Table 6.2: Experiment 2 for estimating E_a.

cum. failures	0.10%	0.15%	0.20%	0.25%	0.30%
cum. # of failures	10	15	20	25	30
150°C cum. burn-in hrs (t_1)	80	135	210	300	401
125°C cum. burn-in hrs (t_2)	146	247	382	N/A	N/A
E_a (eV)	0.349	0.351	0.347	N/A	N/A

$$K \ln(t_2/t_1)/(1/398 - 1/423) = 0.58 \ln(t_2/t_1). \tag{6.5}$$

It is obvious that the E_a is about 0.35eV.

The test records of the above example can be analyzed to see if the Weibull distribution can be used to describe the failure behavior of the product by using the Weibull plotting paper. Based on the data in Table 6.2, two lines can be drawn on the Weibull plotting paper, as shown in Figure 6.8. The Weibull distribution seems to be a good choice as the points of both the 125°C and the 150°C HTB tests fall on two separate parallel lines with shape and scale parameters of 0.78 and 3660 and 0.78 and 1770, respectively. (We use $F(t) = 1 - \exp(-(t/\alpha)^m)$, where m and α is the Weibull shape and scale parameter, respectively.) We can check to see if the failure mechanisms of the two test conditions are different from the slopes of the lines. Special attention

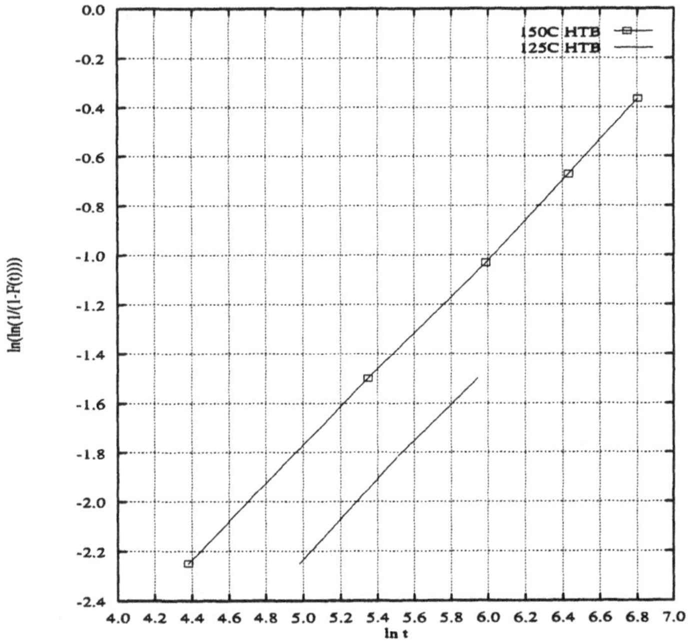

Figure 6.8: Test results on a Weibull plot.

must be paid if the slopes are significantly different (e.g., 0.4 vs. 0.8), which implies they may be over-stressed at the higher temperature. From Figure 6.8, the failure behaviors for the two HTB tests are similar because the two lines are almost parallel. From Figure 6.8, when $y = -1.6$ (for example), the x-value of the 150°C and the 125°C line is 5.2 and 5.8, respectively. Thus, the E_a can be estimated by

$$E_a = (5.8 - 5.2)K/(1/398 - 1/423) = (0.6)(0.58) = 0.348(eV) \tag{6.6}$$

which is very close to the one estimated by Eq. (6.5).

Let us see how we can derive the long-term failure rate provided the data in Table 6.2. If the voltage acceleration factor is known to be 10.8, then the overall acceleration factor (after incorporating the temperature acceleration with $E_a = 0.35$eV) is 95.36 if the junction temperature at the normal use condition and the stressed temperature is 55°C and 125°C, respectively. This is a relatively small acceleration factor; certain IC makers use an overall acceleration factor of over 750. The long-term failure rate based on 382-hr 125°C/7.0V HTB test (which is used by most IC makers as the HTOL test condition for the 5V ICs) is $(0.2\%)/382/95.36\times10^9 = 54.9$ (FIT). The failure rate under 60% UCL (upper confidence level) is $54.9\times\ \chi^2_{0.60}(42)/(2\times20)$=59.1 (FIT). The multiplier under the $100(1-\alpha)\%$ UCL, $\omega_{1-\alpha}$, can be determined by

$$\omega_{1-\alpha} = \begin{cases} \chi^2_{1-\alpha}(2)/2, & \text{if } x = 0 \\ \chi^2_{1-\alpha}(2x+2)/(2x), & \text{if } x \geq 1 \end{cases} \tag{6.7}$$

where x is the number of failure in the HTOL test. Readers can find detailed derivations for Eq. (6.7) in Grosh [168, Appendix 5D].

(2) Method #2 (Type I Censoring Test)

DBI chambers are used by most fabs and testing houses. Compared with a TDBI chamber, the price for a DBI chamber is comparably lower; in fact it may be only one tenth that of a TDBI chamber. The main advantage of a DBI chamber is its large capacity; usually, a DBI chamber can hold more than 12K (some can contain more than 20K) 26/24 SOJ 300-mil ICs (compared with 2K ICs per TDBI chamber described in method #1). The huge price difference of the DBI and the TDBI chamber primarily results from

- the number of I/O channels,
- the timing accuracy and precision of the I/O signals,
- the data management system,
- data analysis tools, and
- the capability to generate test patterns: for example, march, checker-board, read modify write, cross, fuse, pause, butterfly pattern and other N-, N^2- and $N^{3/2}$-degree test patterns (where N is the number of cell to read/ write; e.g., for a 4M×4 DRAM, N =4M)

A DRAM maker is planning to conduct an experiment to estimate E_a using DBI chambers and memory testers. This estimation approach will test the samples at chosen points and the experiment will terminate at a pre-specified time. Usually, the experiment duration varies from one to two weeks.

Several things must be mentioned before introducing the example.

1. Because the samples have to be tested by the memory testers, they must be loaded and unloaded from the stress chambers and testers at the test points. This is a time-consuming task due to the large sample size. To make things worse, some unexpected damage may occur on the samples, like lead bending, ESD, and lost sample(s).

2. Stress may not accumulate to the level needed to illustrate latent defects. Two-hr burn-in can be set as the shortest burn-in duration between two adjacent test points to ensure the samples are actually stressed.

3. From this estimating scenario, we can verify if the burn-in time is long enough by observing the failure rates of the first few test points. The common requirement is, after the first 48-hr 125°C/7.0V HTOL test, the failure rate should be less than 100FIT (if E_a and β are known) or the

defective percentage less than 0.2% (if E_a and β are unknown); this is exactly the way burn-in time is determined: use a pre-specified burn-in time (based on previous experiences or some other rules); put all samples in a 48-hr EFR test (whose test conditions are the same as the HTOL test); and see if the EFR is below 0.2%. If a higher EFR is obtained, a longer burn-in time is used and lots are selected to go for the test flow with this longer burn-in time; the EFR is calculated after the 48-hr EFR test. It is suggested that the longest burn-in time be 96-hr to save on cost and to reduce cycle-time. If a 96-hr burn-in time must be used, it implies that the processes require lots of improvement and that the burn-in time can only be reduced when the processes are improved and the EFR is below the acceptable level. Burn-in time reduction is a never-ending mission for product engineers.

Most companies have internal guidelines to set the initial burn-in times. To mention some of them:

- if a new IC which is directly shrunk from its prior generation, the burn-in time can be set to be half of the initial burn-in time of its prior generation,
- for two DRAMs with different memory capacities from the same process technology, the one with smaller memory capacity must have a shorter burn-in time, and
- the DRAMs with the same memory capacity may have the same initial burn-in time even though they are made by different process technologies (e.g., 0.40μm, 0.36μm, and 0.32μm).

4. To save on cost, some samples after certain test duration (e.g., 48-hr) can be taken out from the DBI chambers. The samples taken out earlier (e.g., at or before 48-hr HTB test) can be sold at normal price if they pass the QC E-test. The samples that experience longer HTB stress may be sold at lower prices, to special customers, or even be scrapped when the test time is longer than the specified workable duration (e.g., 3,000-hr).

Consider the data in Table 6.3, where some samples are removed during tests and the tests continue for 168-hrs (one week). The experiment may actually take as long as 10 to 12 days because the samples have to be loaded and unloaded from the DBI chambers and be tested on memory testers. Similar to Eq. (6.6), the E_a can be estimated by

$$E_a = k \ln(t_{p_{150}}/t_{p_{125}})/(1/398 - 1/423) = 0.58 \ln(t_{p_{150}}/t_{p_{125}}) \qquad (6.8)$$

where $t_{p_{150}}$ and $t_{p_{125}}$ are the time to reach the same cumulative (from the beginning of the HTB tests) failure percentages in PPM at 150°C and 125°C HTB test, respectively. From Table 6.3, we observe

1. To save cost, some samples are removed from the tests.
2. The failure percentage of the HTOL test (125°C) in the first 48-hrs is 550 PPM, which is under the acceptable level (i.e., 0.2%). Therefore, the burn-in time for the present process is sufficient to remove defective parts.

Table 6.3: Experiment 1 for estimating E_a.

Burn-in (hrs)	150°C Sample size	# of failures	p_{150}	125°C sample size	# of failures	p_{125}	E_a (eV)
0 - 24	20,000	14	700	20,000	10	500	N/A
24 - 48	10,000	2	900	10,000	2	700	0.402
48 - 72	9,998	1	1,000	9,998	1	800	0.402
72 - 96	9,997	1	1,100	9,997	1	900	0.402
96 - 168	5,000	1	1,300	5,000	1	1,100	0.324

3. The estimated E_as fall between 0.324 and 0.402eV, which is the usual range based on our prior knowledge for this kind of DRAM and process.

4. The E_a after 72-hr HTB tests is estimated by linear interpolation.

5. After incorporating the voltage acceleration factor, if the overall acceleration factor is 300, the 168-hr long-term failure rate from the HTOL test result is 21.8 FIT.

6. We can also prepare the Weibull plot as seen in Figure 6.8 from the data in Table 6.3. However, because we test the samples every 24-hrs, the fit may not be good. We must also be aware that the Weibull distribution is not the only choice to model the failure behavior in the infant mortality region. Some companies use the lognormal distribution.

It has to be emphasized that the E_as in Tables 6.2 and 6.3 are "overall" activation energies because we do not allocate the failures according to their failure modes. However, in practical applications, thorough electrical and physical FAs are performed on the failed parts to find process problems and to estimate the E_as of different failure modes. From our experience, E_a of the gate oxide/capacitor film breakdown, of the interlayer defect, of silicon defect, and of the refresh degradations are 0.3eV, 0.6eV, 0.5eV, and 0.5eV, respectively.

The last reminder is: both E_a and β are material, process, and design dependent; the E_a and β derived in other fabs cannot be used directly. □

6.1.5. Incompatibility and Compatibility Factors

Suppose a system contains two statistically independent components in series; the system reliability should be the product of each component's reliability. Unfortunately, one is likely to observe a system reliability with a value less than the product. How much less depends on the reliability of the connecting element that ties the two components together, the number of components, and the degree of care in assembly, among other variables [228]. This incompatibility factor, which exists not only at the subsystem but also at the system level, is used to incorporate the reliability loss due to poor manufacturability, workmanship, and design strategy. In other words, human factors are also important in system reliability analysis. With new technologies, many modern products are more complicated, containing both electronic and mechanical components in a system. In addition to this, the effect of incompatibility on reliability is becoming more important.

Examples of incompatibility causes include bad welding joints, inconsistent components when specific components are used at the same time, bad product design or manufacturing processes, and human handling of the components. As shown in Figure 6.1, for example, bad welding joints may occur in connections 1 to 7. Software is another issue which has to be taken into account in system reliability. A robot, for example, is a combination of three major elements: electronic components, mechanical parts, and software. Therefore, to model the reliability of a robot, one has to consider the incompatibility among these three elements.

The existence of incompatibility is widely accepted by most field engineers; however, no good model has been reported. Compatibility factors may be used to incorporate reliability loss due to poor manufacturability.

□ Example 6.6

Except for the central processing unit (CPU), DRAMs are the most important semiconductor components in a PC. The modern PC mother boards provide slots for the DRAMs, which are assembled in a small board called a single in-line memory module (SIMM) or dual in-line memory module (DIMM). Presently, most PCs still use SIMMs as the major memory units.

A 72-pin SIMM is the main-stream product in the market in mid 1997. Customers can choose the SIMM with memory from 1MB (mega-byte) to 32MB for printers, from 1MB to 128MB for notebook and desktop PCs, and as large as 256MB for workstations. One type of the 16MB SIMM, for example, needs eight 16Mb DRAMs. The SIMM can be tested by an E-test program; most SIMM makers use the Eureka CST SIMM tester, which is able to check the AC and DC parameters, the refresh time, the input/output (I/O) voltage and leakage, and to run simple functional patterns (e.g., march and checkerboard). Like CST, most SIMM test programs are able to identify the failed IC(s) on the tested SIMMs. The failed SIMMs can be repaired by replacing the failed IC(s). Of course, failure of the SIMMs on the CST test, for example, may result from the defective board.

After the initial E-test, the SIMMs are inserted on the slots on the PC mother board and are further tested by simulation software like QAplus, WinStone, WFWG, WINCheckit Pro, and BAPCo for several cycles to ensure they work well with the CPU, the chip-set, and the mother board. It may take as long as 30 minutes to complete a single-cycle test depending on the CPU. This is another type of burn-in: running several applications in a short time to expedite the operations which may be equivalent to using the PC for several weeks. The simulation software will execute database, spread-sheet, word processing, graphics, and mathematical calculation applications (e.g., the Microsoft Word/ Excel/ Works/ Access/ Powerpoint, Lotus 1-2-3, WordPerfect, Borland dBASE/ Paradox, Adobe PageMaker, and Corel Draw) for several times. Different kinds of mother boards with several chip-sets must be tested with the SIMMs to ensure the compatibility. Some basic requirements for the mother board used to test the SIMMs are

1. it is the main-stream product in the present market,
2. the mother board can only be used to test the SIMMs for three months because, first, the connections of the SIMM slots may wear-out and, second, this mother board may not retain its main stream position under the fast PC evolution,
3. the SIMM makers must have more than one identical mother board (with the same CPU and chip set) so that the possibility of issuing false alarms can be reduced which may also greatly save debugging time when failure is encountered (because identical mother boards can be used to narrow down the root causes of the failure).

It is frequently seen that SIMMs using ICs from certain makers cannot pass the simulation software tests on selected mother boards and chip-sets. Let us suppose a customer wants to expand his PC memory. The new SIMMs he purchases may not work well with the SIMMs originally on his PC. This kind of incompatibility is encountered very often even though the customer buys the SIMMs with the ICs from the same maker. Therefore, burn-in at different levels has become an important issue. In this example, the IC on the SIMM, the SIMM, and the PC mother board (with the CPU and the chip-set) are the component, subsystem, and system, respectively. □

6.2. Performance and Cost Modeling

Block and Savits [49] consider some basic criteria for determining the optimal burn-in time. Recall that the bathtub shaped hazard rate of Figure 3.1 has two changing points, t_1 and t_2. Generally, burn-in duration ends at around the first changing point t_1, and the optimality criteria give us such a burn-in time.

6.2.1. Performance Criteria

The following performance based criteria should be incorporated into a general cost structure [49]. For the four criteria mentioned below, the optimal burn-in time t_b^* occurs before the first changing point t_1 [236, 279, 322]. This result is very important since it provides an upper bound for burn-in time [49, 225].

Maximum Mission Probability

Let τ be a fixed mission time and $R(t) = 1 - F(t)$. Find burn-in time t_b which maximizes

$$R_b(\tau) = \frac{R(t_b + \tau)}{R(t_b)}, \tag{6.9}$$

i.e., find t_b such that given reliability to time t_b, the probability of completing the mission, is as large as possible.

Maximum Mean Residual Life

Let X be a lifetime. Find the burn-in time t_b that maximizes

$$E[X - t_b | X > t_b], \tag{6.10}$$

i.e., find the burn-in time which gives the maximum mean residual life.

Minimum Burned-in Components

Let $\{N_b(t), t \geq 0\}$ be a renewal process of lifetimes which are bounded in for t_b time units. For fixed mission time τ, find burn-in time t_b which minimizes

$$E[N_b(t)] \tag{6.11}$$

i.e., the mean number of burned-in components which fail during the mission time τ.

Maximum Warranty Period

For a fixed α, $0 < \alpha < 1$, find the burn-in time t_b which maximizes

$$\tau = q_\alpha(b) = F_b^{-1}(\alpha) = inf\{x \geq 0 : R_b(x) \leq 1 - \alpha\}, \tag{6.12}$$

i.e., find the burn-in time which gives the maximal warranty period τ for which at most α-percent of items will fail.

6.2.2. Cost Functions

Discussions of several cost functions are given by Kuo and Kuo [230], Leemis and Beneke [245] and Nguyen and Murthy [300]. They are used to find the burn-in time which minimizes cost. As suggested by Kuo [225, 230], the following cost factors must be considered:

Burn-in Fixed Cost (B)

To perform burn-in tests, suitable furnaces have to be prepared, installed, and maintained. Extra costs include the training of the operators and the set-up of the furnaces. All of these costs are used to estimate burn-in fixed cost.

Burn-in Variable Cost (c_v)

Applying burn-in over time means that certain costs are incurred, such as the wages of the operators, the cost of electric power (because some burn-in furnaces consume a lot of energy), and the depreciation of the furnace. The burn-in variable cost is \$ per unit time and, thus, is time-dependent.

Shop Repair Cost (c_s)

The manufacturing costs of ICs increase as the number of layers increases. It is economical to repair the failed samples if they are repairable. Shop repair occurs before the items are shipped to the retailers or the customers.

Field Repair Cost (c_f)

Once the items are used by customers, it costs the vendors much more to fix a failure. This results from losses from warranty, the expenses of repairing crew(s), and the shipping and handling of possible back-log for the necessary subsystem(s) or component(s). In general, field repair cost is several times higher than shop repair cost.

Total Allowable Cost (TC)

Although it is true that with longer burn-in times, more dormant defects can be removed and hence higher reliability attained, burn-in cost may increase unacceptably. In other words, the trade-off resides in total allowable cost and the length of burn-in time; the choice is a managerial one which must be determined by available resources and budget.

Factors Used to Indicate Loss of Credibility (l)

Users tend not to buy products that have a bad reputation as determined either by their own experience or that of associates. We define this phenomenon as the loss of credibility. Every manufacturers' ultimate objective is to stay competitive and profitable which depends heavily on product credibility. Kuo [225] introduces the use of penalty term l to take this into consideration. He uses $l = 0$, $l = 1$, and $l = 3$ for negligible (low penalty), average (medium penalty), and severe loss (high penalty) of credibility, respectively.

Expected Total Cost Function

The expected total cost of the system to be used in the book is given by [225]

$$\begin{aligned} C_s &= \text{fixed burn-in cost} + \text{variable burn-in cost} \\ &\quad + \text{field failure cost} + \text{penalty cost} \\ &= \sum_{j=1}^{n} \sum_{i=1}^{n_j+1} [c_{ij} + c_{ij,b} + e_{ij,b} c_{ij,s} + (1+l) e_{ij,t} c_{ij,f}] \end{aligned} \quad (6.13)$$

where

- n_j number of component in unit j
- c_{ij} cost of component i in unit j
- $c_{ij,b}$ burn-in cost for component i in unit j
- $c_{ij,f}$ field repair cost for component i in unit j
- $c_{ij,s}$ shop repair cost for component i in unit j
- $t_{ij,b}$ burn-in time for component i in unit j
- $e_{ij,b}$ expected fraction of failure during burn-in for component i in unit j
- $e_{ij,t}$ expected fraction of failure at time t after $t_{ij,b}$ hrs of burn-in
- l ratio used to calculate the cost of the loss of credibility.

All aforementioned cost factors are used hereafter along with the superscripts, c, u, or s, to represent component, subsystem, or system, respectively. The cost optimization model is the most commonly used formulation to determine the optimal burn-in time; this can be found in many reports and papers, such as [65, 74, 76, 77, 78, 81, 83, 225, 264].

6.2.3. Burn-in Cost Function Overview

Watson and Wells [439] study the problem of burn-in as a method for increasing reliability of devices with hazard rates that decrease at least part of the time. They assume that an infinite number of devices are available and that unlimited burn-in periods are applicable. Their purpose for using burn-in is to increase the mean residual life. Cost is not considered as a factor related to burn-in problems.

Whenever infant mortality is not specified by a particular distribution but rather by a DFR, Lawrence [238] defines the clear upper and lower limits on burn-in time to achieve a specific mean residual life. The limits depend upon the DFR assumption, a knowledge of the first moment, and a percentile of the failure distribution.

Barlow *et al.* [28] present an upper limit on the MLE of the decreasing hazard rate function. They show that the upper limit for the hazard rate function of an exponential life distribution can be a conservative upper limit for that of any distribution with DFR. If $X_1 \leq X_2 \leq \cdots \leq X_r$ are censored observations with DFR, then the conservative $100(1-\alpha)$ percent upper limit on the MLE of the hazard rate function at time X_r is given by

$$\frac{\chi^2_{1-\alpha}(2r)}{2(\sum_1^r X_i + (n-r)X_r)}$$

where $\chi^2_{1-\alpha}(2r)$ is the $100(1-\alpha)^{th}$ percentile of the chi-square distribution with $2r$ degrees of freedom and n is the sample size.

Ninomiya and Harada [301] introduce the basic concept of bimodality by describing the multilayer debugging process. This procedure addresses questions about the reliability of systems that survive burn-in, but it is effective only when the failure rate of the weak subpopulation is much greater than that of the strong subpopulation.

Marcus and Blumenthal [263] introduce a method to screen early failures based on the concept of failure-free intervals. In their method, burn-in continues until a predetermined failure-free time between failures is observed.

In a report by Fox [154], the total annual cost consists of the manufacturing, installation, and design costs of a system, prorated over the useful system life, the annual maintenance cost, and the annual operating cost. All of the above costs have been expressed as functions of the system failure rate. A minimum annual cost measured against the failure rate is obtained through a simple mathematical model. Fox imposes two assumptions: no infant mortality and no corrective maintenance. The second assumption indicates the replacement of a failed unit by a new one. Whether or not this is the optimal maintenance policy is not considered. Presumably, a minimum annual cost is obtained with an acceptable failure rate without requiring thorough burn-in.

Cozzolino [93] develops a cost model which can be solved by dynamic programming. Washburn [437] introduces a model based on the three parameter generalized gamma distribution which is presented by Stacy [397]. Washburn's objective function to be minimized is given by

$$c(t) = c_v t + c_s n(1 - R(t)) + \frac{sk}{P_E(t)} \tag{6.14}$$

where

- s sale price of the units that survive the burn-in process
- n number of units placed on a burn-in test
- k minimum number of units needed after burn-in
- $P_E(t)$ Relative efficiency of the units.

It may be seen that $k/P_E(t)$ in the third term of Eq. (6.14) represents the required number of units to satisfy mission requirements. The optimal burn-in time t_b^* satisfies $c^{'}(t_b^*) = 0$.

Weiss and Dishon [442] consider two cases for replacement of burn-in failures in the determination of a burn-in policy:

- Burn-in failures are replaced at the end of the program by units that have not been burned-in.

- More than the required number of units should be burned-in.

Failures are replaced with new burned-in units either at the end of the burn-in period or at intermediate times during burn-in.

Stewart and Johnson [407] present a Bayesian method to find optimal burn-in and replacement times that minimize the expected cost per unit time. Their cost model is given by

$$c(t) = \begin{cases} c_0 + B + c_v t_b, & 0 \le t \le t_b \\ c_0 + B + c_v t_b + c_p, & t_b < t < t_R \\ c_0 + B + c_v t_b + c_r, & t_R \le t \end{cases} \tag{6.15}$$

where

- c_0 cost per unit
- c_p cost of a replacement per unit that fails prior to the scheduled replacement time
- c_r cost of a scheduled replacement per unit that is in service
- t_b burn-in time for each unit
- t_R scheduled replacement time for each unit.

For the given CDF $F(t)$ and its pdf $f(t)$, the expected cost per unit service time is

$$EC = c_0 + \gamma B + c_v t_b + c_p[F(t_R) - F(t_b)] + c_r[1 - F(t_R)]$$

where γ=0 for $t_b = 0$ and γ=1 for $t_b > 0$, and the expected service time is

$$ES = \int_{t_b}^{t_R} (t - t_b)f(t)dt + (t_R - t_b)[1 - F(t_R)].$$

In order to find the optimal pair of burn-in time and service time under the posterior distribution of the hazard rate function, Stewart and Johnson [407] compute the values of EC/ES for a wide range of burn-in and replacement times and search for the minimum value. Their method is based on the multi-parameter Bayesian analysis, which can be applied to a function whose hazard rate is not bathtub shaped. When the reliability function is specified, Canfield [64] presents a model to determine the optimal burn-in and replacement times for units. Although his approach is much simpler than Stewart and Johnson's method, he does not address the effect of uncertainties.

Plesser and Field [329] present a cost-optimized burn-in model to minimize the mean total cost. They assume that

- the shape of the infant mortality curve does not change within the burn-in period,
- there is a clear dividing line between the infant mortality rate and the steady-state failure rate, and
- there are no constraints on the solution.

Practically, not all of these assumptions are necessary. In addition, the cost function which is the sum of costs due to the mean failures during burn-in and development is greatly simplified. The expected cost of operating a unit with N_b failures during burn-in and N_d failures during deployment is given by

$$c(t_b) = B + c_v t_b + Q_0 E[N_b] + Q_1 E[N_d] \tag{6.16}$$

where

Q_0	shop repair cost for a unit which fails during burn-in
Q_1	field repair cost for a unit which fails during deployment.

Since the expected number of failures in an interval is equal to the integral of the hazard rate function $h(t)$ over that interval, Eq. (6.16) is written as

$$c(t_b) = B + c_v t_b + Q_0 \int_0^{t_b} h(t)dt + Q_1 \int_{t_b}^{t_{IM}} h(t)dt + Q_1 h(t)(t_r - t_{IM}) \tag{6.17}$$

where t_{IM} is the time at which infant mortality ceases and t_r the time for retirement. From Eq. (6.17), $c'(t_b) = 0$ gives

$$h(t_b^*) = \frac{c_v + Q_1 h(t_{IM})}{Q_1 - Q_0}.$$

One requirement for using this method is that the hazard rate function must be known.

Nguyen and Murthy [300] develop a model to determine the optimal burn-in time to minimize the cost function, which includes manufacturing costs and warranty costs. They minimize the expected total cost per system under various warranty scenarios. Let $\nu(t_b)$ be the expected manufacturing cost per unit with burn-in time t_b. The expected manufacturing cost is given by

$$\nu(t_b) = \begin{cases} c_0 + B + c_v t_b + c_s \int_0^{t_b} h(t)dt & \text{for repairable products} \\ \frac{1}{R(t)}[c_0 + B + c_v \int_0^{t_b} R(t)dt] & \text{for nonrepairable products.} \end{cases}$$

For the warranty period T, the expected warranty cost, $w(T, t_b)$, depends on the types of warranty policy: the failure-free policy (for both repairable and nonrepairable products) and rebate policy (for nonrepairable products). Considering the failure-free policy, the expected warranty cost is given by

$$w(T, t_b) = \begin{cases} (c_s + c_f) \int_0^T h(t_b + t)dt & \text{for repairable products} \\ (\nu(t_b) + c_f)M(T) & \text{for nonrepairable products} \end{cases}$$

where c_f is the field repair cost that arises when a failure occurs during the warranty period and $M(T)$ the expected number of replacements in $[0,T]$, which is described by the renewal equation:

$$M(T) = F(T) + \int_0^T M(T - t)dF(t).$$

Thus, the expected total cost per unit is

$$c(T, t_b) = \nu(t_b) + w(T, t_b). \tag{6.18}$$

The optimal burn-in time t_b^* minimizes Eq. (6.18).

In a military airborne system, Reich [343] analyzes the reliability- associated life-cycle costs to include reliability and maintainability. With burn-in, the investment costs are increased to balance against the decrease in support costs. The support costs associated with reliability and maintainability can be controlled through changes in the inherent reliability and maintainability or by changes in maintenance practices. The central theme is to set up a strategy to explore a good investment return ratio and to minimize the life-cycle cost. Reich's approach to the determination of optimal burn-in savings is frequently practiced.

Studies by Cheng [72, 73] of the renewal process of lognormal devices, such as semiconductor lasers, show that burn-in can increase mean lifetime and reduce costly early field replacement. Cheng determines the optimal burn-in time by minimizing the system life-cycle cost. Again, constraints do not exist in the optimization process.

Kuo and Lingraji [231] report a study of a medical facility in the Topeka VA hospital, in which a machine selection guide is based upon a machine's total annual cost. Reliability is the only decision variable necessary to reach an optimal selection decision given the company's desired maintenance policy. If the use of different burn-ins were included as a criterion, a different selection guide would be obtained. Kuo [225] presents a cost optimization model with system and component reliability constraints. Chi and Kuo [75] add the capacity constraint to the previous cost optimization model and present a methodology to solve that problem.

Life-cycle cost is defined by Marko and Schoonmaker [264] as burn-in cost plus field failure cost. They use exhaustive search techniques to optimize spare module burn-in and thereby minimize a part of the life-cycle cost. Increased costs associated with added burn-in are compared to the field savings from reduced failures until an optimal solution is identified.

□ **Example 6.7**

One way to reduce burn-in time is to develop the wafer level burn-in. However, there are technical difficulties to solve, such as the wafer ECT, wafer handling, and a method for applying bias and test patterns. The component level burn-in time with temperature and voltage acceleration for most ICs is no longer than 96 hrs. At least two issues have to be checked when considering reduction of burn-in time. First, the stability of the burn-in failure rates must be established. The monitoring duration may be as long as three months to ensure that the trend of the burn-in failure rates can be verified. Usually, the alarm flag of burn-in failure rate can be determined if the highest allowed burn-in failure rate is given. Assume the acceleration factor is 750. Further suppose that the Weibull distribution with shape parameter 0.1 is believed to best describe the early failure behavior of the product. If the highest allowed burn-in failure rate after the product is released for one year is 100FITs, the corresponding obtainable highest failure rate in the 48-hr burn-in is 3.7%. This value can be used as the alarm flag. Similarly, the estimated shape parameter of the Weibull distribution can also shed light on the failure mechanism. Attention has to be paid to the lots with a Weibull shape parameter significantly deviating from 0.1.

Second, one or more experiments simulating the effects of reduced burn-in time must be conducted. For example, at least 3000 ICs may be sampled from 5 or more lots for the production flow containing the reduced burn-in time. That is, after the completed production processes, these ICs are put on burn-in for 48 hrs to simulate the failure rate equivalent to field use for one year. This test is called the EFR test by most IC makers. Maximum failure rates for different burn-in times are specified mainly by customers after incorporating the product characteristics. For example, 48-hr, 24-hr, and 12-hr burn-in times may be considered if the failure rate after the EFR test is less than 30FITs, 20FITs, and 10FITs, respectively.

Notice that all the aforementioned rules are guidelines and not strict standards. Proper requirements have to be designed based on the product characteristics. □

6.3. Burn-in Optimization

Kuo [225] introduces a constrained optimization model to find an optimal burn-in time for a repairable system. Although Kuo [225] and Plesser and Field [329] use similar cost functions, Kuo's model is different from the Plesser and Field's model in several ways:

- Kuo adds reliability constraints.
- Kuo models the burn-in process at the system and the component level.
- Kuo restricts the initial hazard rate corresponding to the Weibull distribution with a shape parameter less than 1.

Based on the expected total cost function of Eq. (6.13), the problem to be solved is

$$\begin{array}{ll} \text{Minimize} & C_s = f(t_{ij,b}) \\ \text{s.t.} & R_s(t|t_{ij,b}) \geq R_{s,min}(t) \\ & R_{ij}(t|t_{ij,b}) \geq R_{ij,min}(t) \\ & \text{(Other constraints)} \\ & 0 < t_{ij,b} < t_{ij,L} \quad \forall\, i,\ j. \end{array} \tag{6.19}$$

Kuo [225] suggests solving this problem of Eq. (6.19) as if there were no constraint and then checking the validity of the reliability constraint afterward. For the Weibull distribution:

$$f_{ij}(t) = \alpha_{ij} t^{-\beta_{ij}} \exp[-(\frac{\alpha_{ij}}{1-\beta_{ij}})t^{1-\beta_{ij}}],$$

the solution of the unconstrained optimization problem, given that C_s is Eq. (6.13), is obtained by calculating

$$t^*_{ij,b} = \left[\frac{\alpha_{ij}[(1+l)c_{ij,f} - c_{ij,s}]}{(1+l)c_{ij,f}\lambda_{ij,L} + B_0 r}\right]^{1/\beta_{ij}}, \tag{6.20}$$

where

$\lambda_{ij,L}$	Steady-state hazard rate for component i in unit j
B_0	Burn-in cost coefficient
r	Time coefficient of burn-in cost
α_{ij}	Scale parameter of the Weibull distribution $(0 < \alpha_{ij})$
β_{ij}	Shape parameter of the Weibull distribution $(0 < \beta_{ij} < 1)$.

If $t^*_{ij,b}$ obtained from Eq. (6.20) does not violate system and component reliability constraints, then the optimum solution has been reached; otherwise, nonlinear optimization techniques can be used to find an optimal solution.

□ **Example 6.8**

Another constrained optimization under reliability and capacity restrictions is given in Chi and Kuo [75]. If there is no minimum reliability constraint and if there is infinite capacity, then the burn-in optimization problem becomes quite simple. Namely, it finds the burn-in time optimizing the unconstrained life-cycle cost. Unfortunately, in most cases, physical constraints exist, and they cause difficulties in optimizing the problem. For example, when the reliability constraint is violated, we can increase the burn-in time to satisfy the reliability constraint up to the maximum system reliability obtainable by burn-in. However, the capacity constraint could be violated at some point because of the increased burn-in time.

Define $t_{ij,L}$ as the time at which the $\lambda_{ij,L}$ for component i of unit j is reached. The maximum system reliability obtainable by burn-in is the system reliability at a field operation time t after all components are burned in for up to $t_{ij,L}$. This is not always equal to the minimum reliability requirements. In contrast, the capacity constraint can be satisfied by decreasing burn-in time, but the reliability constraint might be violated.

When we consider a manufacturing system that produces various types of electronic systems, the capacity of burn-in facility available for a planning production period T_p is

$$F_t = V_t \times T_p$$

where V_t is the batch capacity of the burn-in facility in terms of columns or racks of ovens. The amount of capacity required for component i of unit j is F_{ij}. The capacity required for component i of unit j for the burn-in periods $t_{ij,b}$ is $F_{ij}t_{ij,b}$. Therefore, the capacity constraint becomes

$$\sum_i \sum_j F_{ij} t_{ij,b} \leq F_t. \tag{6.21}$$

In order to minimize the system cost while satisfying both the reliability and capacity constraints, the problem to solve is:

$$\begin{array}{ll} \text{Minimize} & C_s = f(t_{ij,b}) \\ \text{s.t.} & R_s(t|t_{ij,b}) \geq R_{s,min}(t) \\ & R_{ij}(t|t_{ij,b}) \geq R_{ij,min}(t) \\ & \sum_i \sum_j F_{ij} t_{ij,b} \leq F_t \\ & 0 < t_{ij,b} < t_{ij,L} \quad \forall\, i,\, j \end{array} \tag{6.22}$$

where $t_{ij,b}$ is the decision variable. The procedure to solve the problem of Eq. (6.22) follows these steps:

Step 1 Find the optimal burn-in period $t^*_{ij,b}$ for the unconstrained minimization using Eq (6.20).

Step 2 Check the reliability constraint. If it is satisfied, then go to Step 3; otherwise, go to Step 6.

Step 3 Check the capacity constraint. If it is satisfied, then we have an optimal burn-in period; otherwise, go to Step 4.

Step 4 Solve the problem with capacity constraint and go to Step 5.

Step 5 Check the reliability constraint. If it is satisfied, then we have an optimal burn-in period; otherwise, we can solve this problem by using goal programming [198]. The goal level depends on the strategy of the company.

Step 6 Solve the problem with the reliability constraint and go to Step 7.

Step 7 Check the capacity constraint. If it is satisfied, then we have an optimal burn-in period; otherwise, it is necessary to increase the capacity. In this case, since the system reliability obtained is the minimum reliability required, the optimal burn-in time can not be reduced without increasing the capacity.

If the burn-in temperature T_2 in the Arrhenius equation is less than the maximum allowable burn-in temperature, the capacity constraint can be partially or totally satisfied by increasing the burn-in temperature to the maximum temperature allowed for burn-in. If the capacity can not be increased, then a subcontract with independent, outside laboratories can be made for burning in some components. □

In addition to discussing the problems of minimizing life-cycle cost, Chandrasekaran [68] determines optimal burn-in time so as to maximize mean residual life. Mean-residual-life optimization is applied to situations of IFR, DFR, and combinations of IFR and DFR. With regard to the total cost optimization, Chandrasekaran determines optimal burn-in time using the following policies:

1. A burn-in cost proportional to the total burn-in time of all items,

2. A burn-in cost proportional to the number of items undergoing burn-in, including failures, and

3. A cost for replacing a failure, excluding the cost of the item itself.

A general guide is applied to a single unit and a whole system for unconstrained burn-in optimization.

Kuo and Kuo [230] address some key burn-in issues, and Leemis and Beneke [245] provide a literature review on burn-in. Block and Savits [49] summarize and compare some different descriptions of burn-in through the survey of recent research. Although most of the papers deal with electronic units, there are still some studies specifically addressing mechanical parts, such as Rawicz [342] and Vannoy [432]. Friedman's research [155] can serve as a helpful reference for hardware and software mixed systems. As mentioned in Chapter 1, many military standards and handbooks are used widely for burn-in analysis information. Among them, MIL-HDBK-217 and MIL-STD-883c are the most important ones (For an introduction to MIL-HDBK-217, see [259, 292, 304]).

6.4. Statistical Approaches for Burn-in Analysis

Considerations and statistical techniques for burn-in are introduced in this section. The maximum likelihood estimation (MLE) technique in the parametric approach is a very important and useful way of estimating parameters. In order to analyze burn-in data, the total time on test (TTT) and Weibull probability plots are frequently used. Through the TTT transform, we can graphically detect exponentiality, mean residual life, bathtub hazard rate, and others. The Bayesian approach is described in the last section.

6.4.1. The Maximum Likelihood Estimation Technique

When more complex production techniques are involved, models with more parameters play an increasingly important role. The more complicated a model becomes, the more parameters that can be expected. The maximum likelihood theory for estimation and statistical inference will be straightforward for increasingly advanced models if one has adequate statistical models and well-designed experiments.

It is necessary to have as many failures as possible for the MLE to exist because maximum likelihood estimates may not exist when there are not enough failures. After deriving the likelihood function, the MLEs can be obtained by solving maximum likelihood equations. After deriving the MLE of a sample, it will be helpful to have interval estimators. For some particular distributions like exponential and gamma distributions, we can use the relation between them and the Chi-square distribution. Let us consider the exponentially distributed random variable, $X \sim \mathrm{EXP}(\theta)$. Since

$$\frac{2\sum_{i=1}^{n} x_i}{\theta} \sim \chi^2(2n),$$

the two-sided, equal-tailed $100(1-\alpha)\%$ confidence interval for θ is

$$\left[\frac{2\sum_{i=1}^{n} x_i}{\chi^2_{1-\alpha/2}(2n)}, \frac{2\sum_{i=1}^{n} x_i}{\chi^2_{\alpha/2}(2n)}\right].$$

where $x_1, x_2, \ldots, x_n$ be the n observations.

We define $L_i(\vec{\theta})$, $F(t;\vec{\theta})$, $f(t;\vec{\theta})$ as the likelihood of parameter vector $\vec{\theta}$ of the ith observation, the cumulative distribution function at t with parameter vector $\vec{\theta}$, and the probability density function at t with parameter vector $\vec{\theta}$, repectively. Given different types of censored data, the likelihood function can be expressed accordingly as shown below.

Left-censoring

An observation is said to be left-censored at t^L if it is known only that the observation is less than or equal to t^L. The likelihood function of the left-censored observations is

$$L_i(\vec{\theta}; t_i^L) = F(t_i^L; \vec{\theta}) - F(-\infty; \vec{\theta}) = F(t_i^L; \vec{\theta})$$

$$L(\vec{\theta}) = \prod_{i=1}^{n} \left(f(t_i; \vec{\theta})\right)^{\delta_i} \left(F(t_i; \vec{\theta})\right)^{1-\delta_i}$$

where $\delta_i = 1$ if t_i is a failure time and $\delta_i = 0$ otherwise.

Right-censoring

An observation is said to be right-censored at t^U if the exact value of the observation is not known but only that it is greater than or equal to t^U. The likelihood function of the right-censored observations is

$$L_i(\vec{\theta}; t_i^U) = F(\infty; \vec{\theta}) - F(t_i^U; \vec{\theta}) = 1 - F(t_i^U; \vec{\theta})$$

$$L(\vec{\theta}) = \prod_{i=1}^{n} \Big(f(t_i; \vec{\theta})\Big)^{\delta_i} \Big(1 - F(t_i; \vec{\theta})\Big)^{1-\delta_i}$$

where $\delta_i = 1$ if t_i is a failure time and $\delta_i = 0$ otherwise .

Left- and Right-censoring

If we allow both right- and left-censoring and assume that observation begins after the $(\ell + 1)^{st}$ failure and stops at a time when only r units (out of n) are still alive (i.e., there will be ℓ left-censored observations and r right-censored observations), then

$$\begin{aligned} L(\vec{\theta}) &= \tfrac{n!}{r!\ell!} \Big(F(t_{\ell+1:n}; \vec{\theta})\Big)^{\ell} f(t_{\ell+1:n}; \vec{\theta}) f(t_{\ell+2:n}; \vec{\theta}) \cdots \\ &\quad \cdots f(t_{n-r-1:n}; \vec{\theta}) \Big(1 - F(t_{n-r-1:n}; \vec{\theta})\Big)^{r}. \end{aligned}$$

□ **Example 6.9**

In Figure 6.9, if we consider both left- and right-censoring and let all t_i^L and t_i^U be the same; i.e.,

$$\begin{aligned} t_i^L &= t^L = 1{,}500, \quad i = 1, 2, \ldots, 10 \\ t_i^U &= t^R = 5{,}500, \quad i = 1, 2, \ldots, 10, \end{aligned}$$

then the likelihood function will be

$$L(\vec{\theta}) = \Big(F(1{,}500)\Big)^3 f(2{,}000) f(2{,}700) f(3{,}500) f(4{,}600) \Big(1 - F(5{,}500)\Big)^3 . \quad \square$$

To generalize, allowing for random left and random right-censoring, we first define new notation for the censoring indicators. Let $\delta_i = 0$ if t_i is a right-censored observation, and $\delta_i = 1$ otherwise. Also, let $\gamma_i = 0$ if t_i is a left-censoring time, and $\gamma_i = 1$ otherwise. Then

$$L(\vec{\theta}) = \prod_{i=1}^{n} \Big(F(t_i; \vec{\theta})\Big)^{1-\gamma_i} \Big(f(t_i; \vec{\theta})\Big)^{(\delta_i)(\gamma_i)} \Big(1 - F(t_i; \vec{\theta})\Big)^{1-\delta_i} .$$

Left-truncation

Although left-truncation is similar to left-censoring, the concepts differ in important ways. Left-truncation arises in life-test applications when failures that occur before τ^L are not recorded and we have no information even on the number of units that failed before this time.

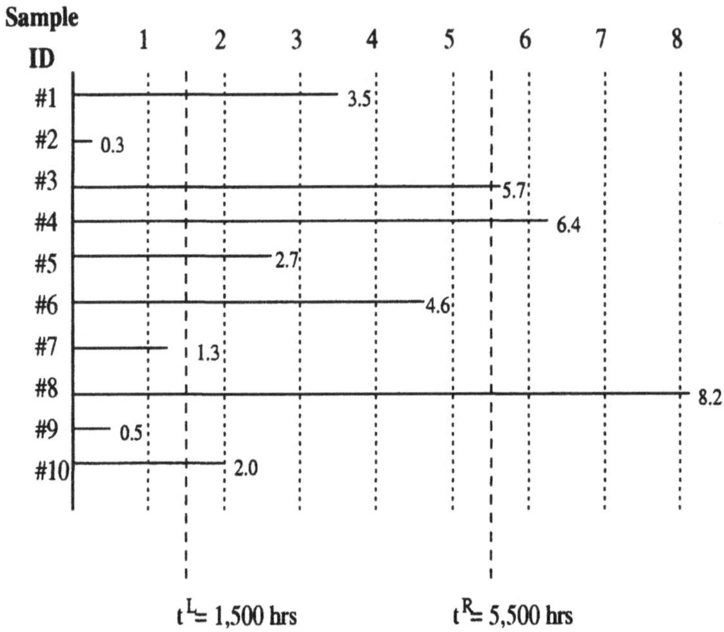

Figure 6.9: Failure time in 1,000 hrs for left- and right-censors.

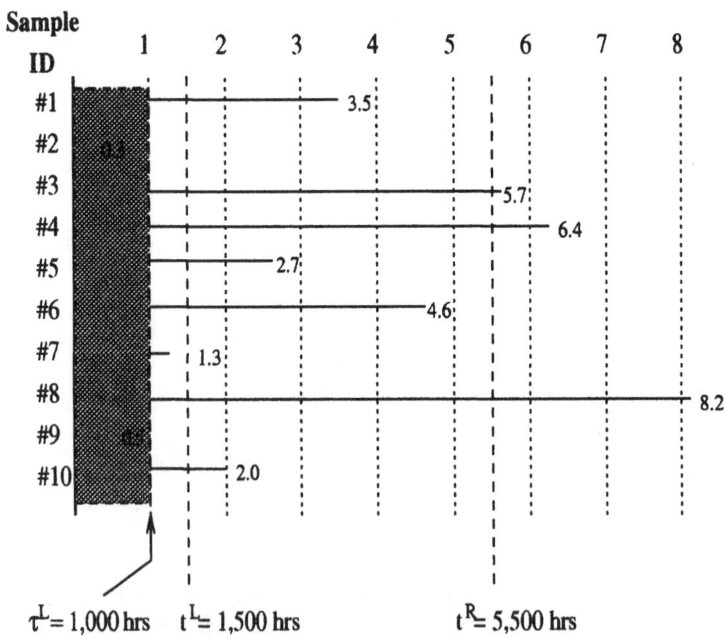

Figure 6.10: Failure time in 1,000 hrs for left-truncation.

If a random variable T_i is truncated when it falls below τ_i^L, then the probability of an observation is the conditional probability.

$$L_i(\vec{\theta}) = \Pr(t_i^L \le T_i < t_i^U | T > \tau_i^L) = \frac{F(t_i^U;\vec{\theta}) - F(t_i^L;\vec{\theta})}{1 - F(\tau_i^L;\vec{\theta})}.$$

Figure 6.10 clarifies the ideas of left-truncation.

Right-truncation

Right-truncation is similar to left-truncation and occurs when the upper tail of the values of the distribution are removed. If the random variable T_i is truncated when it lies above τ_i^U then the probability of an observation is

$$L_i(\vec{\theta}) = \Pr(t_i^L \le T_i < t_i^U | T \le \tau_i^U) = \frac{F(t_i^U;\vec{\theta}) - F(t_i^L;\vec{\theta})}{F(\tau_i^U;\vec{\theta})}. \tag{6.23}$$

The likelihood for the example in Figure 6.11, where the information after $\tau_i^U = 6,000$ is removed, can be easily derived from Eq. (6.23).

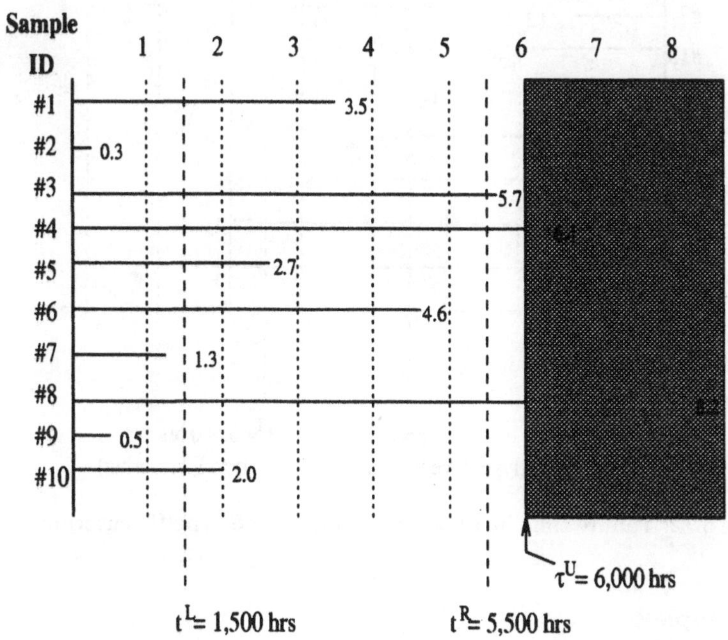

Figure 6.11: Failure time in 1,000 hrs for right-truncation.

Type I Censoring (Time Censoring)

The experiment is terminated at a pre-specified time, t^T. If n samples are tested and the failure times are $t_1, \ldots, t_n$, then the likelihood function is

$$L(\vec{\theta}) = \prod_{i=1}^{n} \Big(f(t_i)\Big)^{\delta_i} \Big(1 - F(t^T)\Big)^{1-\delta_i} \tag{6.24}$$

where δ_i is 1 if $t_i \leq t^T$ and 0 otherwise.

Type II Censoring (Failure Censoring)

The test is terminated once the rth failure is observed. If n samples are tested, then the likelihood is

$$L(\vec{\theta}) = f(t_1; \vec{\theta}) f(t_2; \vec{\theta}) \cdots f(t_r; \vec{\theta}) \Big(1 - F(t_r; \vec{\theta})\Big)^{n-r} \tag{6.25}$$

where t_i is the ith smallest failure time among the n observations.

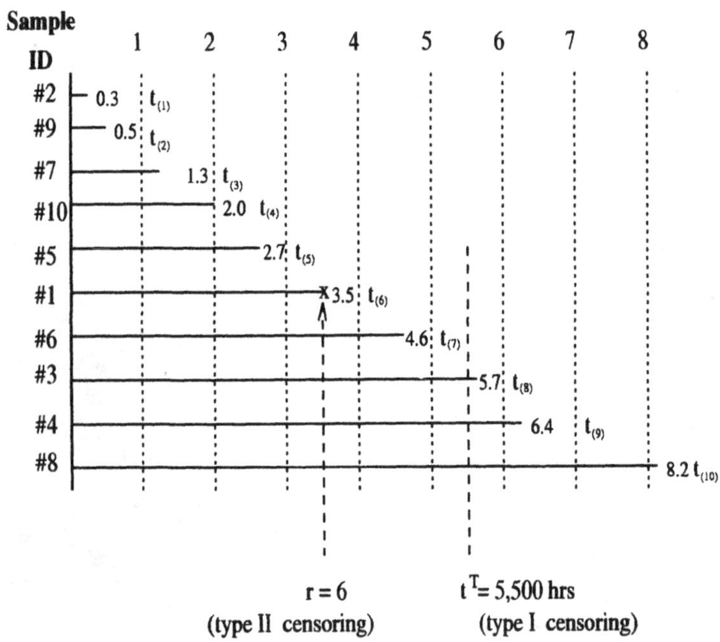

Figure 6.12: Failure time in 1,000 hrs for type I and type II censoring schemes.

□ Example 6.10

In Figure 6.12, the likelihood for type I censoring (t^T from Eq. (6.24)) is

$$L(\vec{\theta}) = f(300) f(500) f(1,300) f(2,000) f(2,700) f(3,500) f(4,600) \Big(1 - F(5,500)\Big)^3$$

and type II ($r = 6$ and from Eq. (6.25)) censoring is

$$L(\vec{\theta}) = f(300)f(500)f(1,300)f(2,000)f(2,700)f(3,500)\Big(1 - F(3,500)\Big)^4. \quad \square$$

6.4.2. Likelihood Ratio Test

The second approach is to use the corresponding score function or from the likelihood ratio test. The likelihood ratio test statistic for $\vec{\theta_1}$ is

$$R(\vec{x}) = \frac{L(\vec{\theta_1}, \hat{\vec{\theta_2}})}{L(\hat{\vec{\theta}})}$$

where $\hat{\vec{\theta}}$ is the MLE of $\vec{\theta}$, and, thus, $R(\vec{x}) \leq 1$. When the length of $\vec{\theta_2}$ is 0, this is a relative likelihood for $\vec{\theta} = \vec{\theta_1}$. Otherwise, we have a maximized relative likelihood for $\vec{\theta_1}$. When $\vec{\theta_1}$ is of length 1, $R(\vec{x})$ is a curve projected onto a plane and when $\vec{\theta_1}$ is of length 2 or more, $R(\vec{x})$ is a surface projected onto a hyperplane. In either case, the projection is in a direction perpendicular to the coordinate axes for $\vec{\theta_1}$. When $\vec{\theta_1}$ is of length 1 or 2, the $R(\vec{x})$ is usually displayed graphically for clear illustration. It can be shown that under certain regularity conditions, when n approaches infinity, we have the following approximation (see [191, pp 238–243], [192, pp 257–266], and [251, pp 307–309]) under $H_o : \vec{\theta} = \vec{\theta^o} = (\theta_1^o, \ldots, \theta_k^o)$:

$$\Lambda(\vec{x}) = -2\ln R(\vec{x}) = -2\ln\left(\frac{L(\theta_1^o, \ldots, \theta_k^o)}{L(\hat{\theta}_1, \ldots, \hat{\theta}_k)}\right) \sim \chi^2(k)$$

Again, consider n identically independent distribution exponential distributions $x_1, \ldots, x_n$ with parameter θ, then

$$\Lambda(\vec{x}) = -2n\ln(\frac{\hat{\theta}}{\theta}) + 2n(\frac{\hat{\theta}}{\theta} - 1)$$

where $\hat{\theta}$ is the restricted MLE (maximization over a subset of the parameter space) and θ is the unrestricted MLE (maximization over the entire parameter space). Suppose $\hat{\theta} = 100.0$, plot $\Lambda(\vec{x})$ over $0 \leq \theta \leq 250$, we have Figure 6.13, which is drawn for $n = 10$, 20, and 100. For the example shown in Figure 6.13, the degree of freedom is 1 (and, hence, $\chi^2_{0.05}(1) = 3.84$) and the 95% confidence interval for $n = 20$, for example, is approximately [60, 180]. Figure 6.13 also elucidates the fact that the interval (under the same confidence level) becomes smaller for larger n.

The MLE is very useful for censored data analysis where much research has been done. For exponentially distributed data, the study of Epstein [131] is among the earliest references for type I censor; the UMVUE of a two-parameter exponential distribution is presented by Basu [35]. Other good studies for type II censor are found in other references [34, 249, 262, 360].

Exact asymptotically efficient confidence bounds for the two-parameter Weibull model with type II censored data are described by Johns and Lieberman [205], where special tables are needed. D'Agostino [98] considers the same

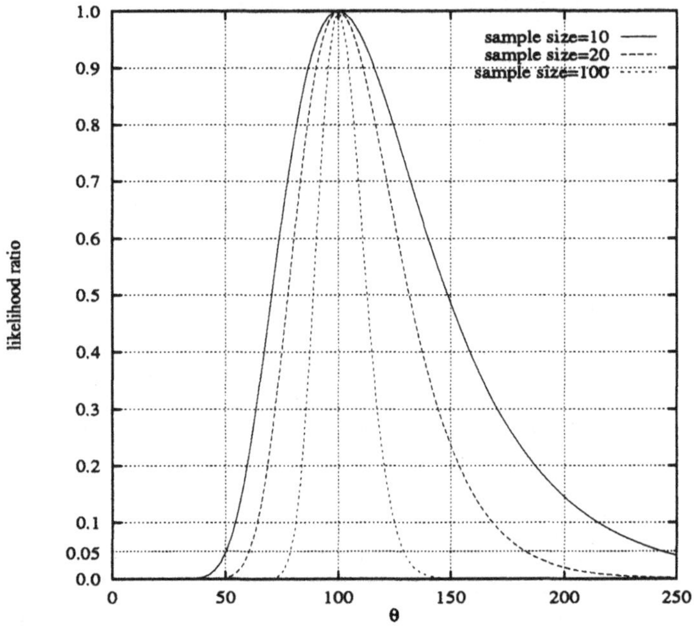

Figure 6.13: Use of the profile likelihood to determine confidence interval.

problems and he develops a technique which does not require special tables. The BLIE for the reliability in the Weibull model with type II censored data is derived by Mann [617]; he demonstrates that the BLIE has a smaller expected loss than the BLUE. The BLUE for the scale parameter of an extreme-value distribution using type II censored data is given by Bain [17], whose results are applicable to the Weibull model since the logarithms of Weibull deviates obey an extreme-value probability density law [381]. The MLE for the two-parameter Weibull model with censored data is also considered by Cohen [85]. The MLE for the three-parameter Weibull and gamma models with censored data is studied by Harter and Moore [180]. Research on normal, logistic, and extreme-value models appears in Harter and Moore [181, 182, 183], respectively.

6.4.3. Total Time on Test (TTT)

The concept of TTT was first introduced by Epstein and Sobel [131] and was later extended by Barlow and Campo [27]. The TTT is primarily a graphical technique. One of the important uses of the TTT has been to obtain approximate optimal solutions for age replacement and also to obtain approximate optimal burn-in times [49].

The TTT transform of a distribution function F with mean μ is defined as

$$\varphi_F(u) = \frac{1}{\mu}\int_0^{F^{-1}(u)} R(t)dt, \quad 0 \le u \le 1, \tag{6.26}$$

where $F^{-1}(u) = inf\{t : F(t) \ge u\}$ and $R(t)$=$1 - F(t)$. Barlow and Campo [27] show that

$$\frac{d}{du}\varphi_F(u)\mid_{u=F(t)} = \frac{1}{h(t)}. \tag{6.27}$$

Therefore, one can use Eq. (6.27) to determine the distribution function (also see [170]). If we prepare a TTT plot, which draws $\varphi_F(u)$ versus u, the CFR, IFR, and DFR functions have diagonal, convex, and concave curves, respectively, from Eq. (6.27). Some examples are shown in Figure 6.14, where the pdf, CDF, and hazard rate function of a Weibull distribution are derived from Eqs. (3.14), (3.15), and (3.16), by changing the parameter α=λ^β. Consider:

$$f(t) = \alpha\beta t^{\beta-1}\exp(-\alpha t^\beta),\ F(t) = 1 - \exp(-\alpha t^\beta),\ h(t) = \alpha\beta t^{\beta-1}. \tag{6.28}$$

From Eq. (6.28), we have

$$\varphi_F(u) = \frac{1}{\mu\beta\alpha^{1/\beta}} IG(\frac{1}{\beta}, -\ln(1-u)) \tag{6.29}$$

where IG is the incomplete gamma function:

$$IG(x;a) = \frac{1}{\Gamma(a)}\int_0^x e^{-t}t^{a-1}dt.$$

The exponential distribution can be derived by letting $\beta = 1$ in Eq. (6.28). Thus, from Eq. (6.29), the TTT transform of an exponential distribution with parameter $\alpha = \lambda$ (i.e., $\mu = \frac{1}{\lambda}$) is given by

$$\varphi_F(u) = \lambda\frac{1}{\lambda}IG(1, -\ln(1-u)) = u,$$

which is a straight line with a unit slope as shown in Figure 6.14. Generalized from Figure 6.14, the TTT plot for a bathtub hazard rate function in which a beta distribution of Eq. (3.11) is used, is shown in Figure 6.15. For the hazard rate functions in Figure 3.3, the corresponding TTT plot is given in Figure 6.15, which is derived by drawing the beta CDF function, $F(t)$, at the x-axis and $t - \int_0^t F(t)dt$ at the y-axis, $t \in [0,1]$.

Suppose n samples are put on test and m $(n \ge m)$ failure times, $t_{1:n},\ldots,t_{m:n}$, are recorded at the termination time t_T of the test; then, from Eq. (6.26), the TTT is computed by

$$TT_n(\frac{i}{n}) = \frac{1}{n}\left[\sum_{j=1}^{i} t_{j:n} + (n-i)t_{i:n}\right] \tag{6.30}$$

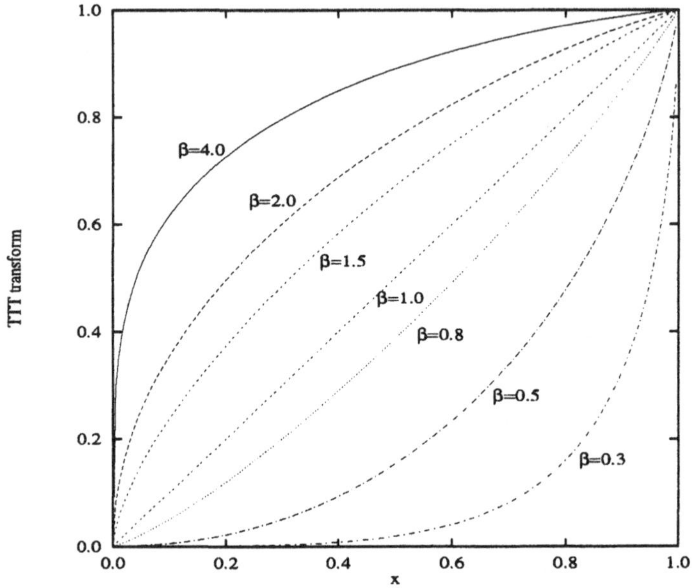

Figure 6.14: TTT plots for some Weibull distributions with the same scale parameter.

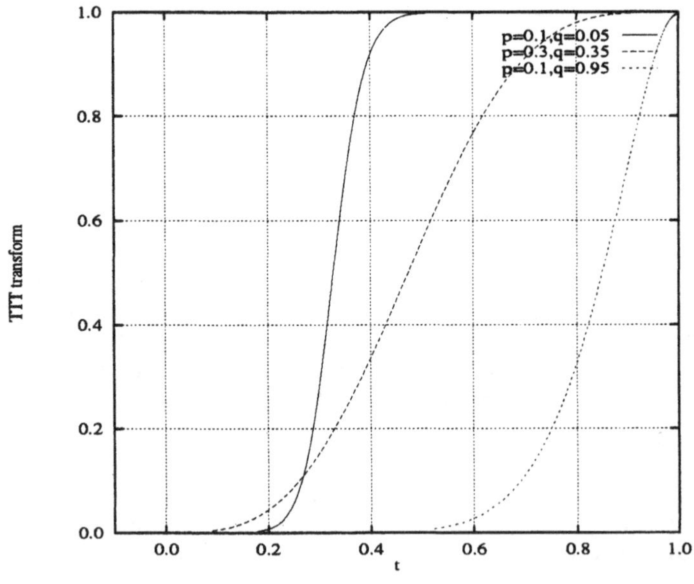

Figure 6.15: TTT plots for some beta distributions which have U-shaped hazard rates.

where $TT_n(0)=0$ and $t_{m+1:n}=\cdots=t_{n:n}=t_T$. The empirical (or the scaled) TTT plot is the plot of

$$\left(\frac{i}{n}, \frac{TT_n(\frac{i}{n})}{TT_n(1)}\right).$$

The IFR and DFR can be easily detected by the TTT transform as shown in Langberg *et al.* [233] and Wei [440]. Some other applications have been proposed based on the TTT transformation, such as testing exponentiality [441] and MRL [193], detecting a bathtub hazard rate [1, 222, 293], and setting an age replacement plan [39, 41, 216]. A burn-in model using the TTT transform is provided by Klefsjö [217]. The characterizations of distributions such as IFR, DFR, IFRA, DFRA, NBU, NBUE, NWU, NWUE (see Section 3.1.3) are studied in references [23, 193, 233]. Interested readers can refer to several references [23, 42] for extensive review of the TTT transform technique and its applications to reliability analysis.

6.4.4. Probability Plots

In burn-in analysis, Weibull probability analysis is frequently used due to the flexibility of the Weibull distribution in describing failures. From the different forms of Weibull distribution in Eq. (3.17),

$$R(t) = 1 - F(t) = \exp[-(\frac{t}{\eta})^{\beta}].$$

By taking double logarithms,

$$\ln(-\ln(1 - F(t))) = \beta \ln t - \beta \ln \eta. \tag{6.31}$$

Eq. (6.31) is a straight line of the form $y = ax + b$ and the slope of line is β. η is generally given by the 63% of life value ($F(\eta) = 63\%$). Therefore, the Weibull probability plot paper has a log log reciprocal ordinate scale and a log abscissa scale.

For the Weibull hazard rate function of Eq. (3.17):

$$h(t) = \frac{\beta}{\eta}\left(\frac{t}{\eta}\right)^{\beta-1},$$

the cumulative hazard rate function $H(t)$ is

$$H(t) = (t/\eta)^{\beta-1}. \tag{6.32}$$

By transforming Eq. (6.32), we have

$$\ln t = \frac{1}{\beta} \ln H + \ln \eta. \tag{6.33}$$

The Weibull hazard rate plot paper, therefore, has a log ordinate scale, and a log abscissa scale and the slope is $1/\beta$. O'connor [303] suggests the hazard plotting procedure which is given below

1. Tabulate the failure times in order.
2. For each failure, calculate the hazard interval ΔH_i:

$$\Delta H_i = \frac{1}{\text{number of items remaining after previous failures/censoring}}.$$

3. For each failure, calculate hazard function

$$H = \Delta H_1 + \Delta H_2 + \ldots + \Delta H_n.$$

4. Plot cumulative hazard against life value on the chosen hazard paper.

6.4.5. Bayesian Approach

For a pdf with parameter θ, if it is believed that θ comes from a pdf $g(\theta)$, which is called the prior distribution, then given sample observations $x_1, \ldots, x_n$, the posterior pdf is

$$f_{\theta|\vec{x}}(\theta) = \frac{f(x_1, \ldots, x_n \mid \theta)g(\theta)d\theta}{\int f(x_1, \ldots, x_n \mid \theta)g(\theta)d\theta}, \tag{6.34}$$

where $\vec{x} = (x_1, \ldots, x_n)$. The Bayes estimator, $\hat{\theta}$, is the one which minimizes the expected risk, which is the expectation of a loss function, $L(T; \theta)$; that is,

$$\hat{\theta} = \min_{\theta} E_\theta[R_{\hat{\theta}}(\theta)] = E_\theta[E_{\theta|\vec{x}}[L(\hat{\theta}; \theta)]]. \tag{6.35}$$

Under the square error loss,

$$L(\hat{\theta}; \theta) = (\hat{\theta} - \theta)^2 \tag{6.36}$$

it can be shown that the Bayes estimator is equal to the posterior mean, i.e.,

$$\hat{\theta} = E_{\theta|\vec{x}}[\theta] = \int \theta f_{\theta|\vec{x}}(\theta)d\theta. \tag{6.37}$$

The Bayesian approach for exponential censored data is provided by Bhattacharya [45]. Bayesian type II censoring data for a Weibull with unknown scale parameter is studied by Soland [393]. Extended work on both unknown scale and shape parameters is found in references [320, 394]. A summary of them is tabulated in Table 6.4 for references.

Jeffreys' vague prior (also called Jeffreys' prior or non-informative prior) in Table 6.4 is suggested by Jeffreys [201]:

$$g(\theta) \propto | I |^{\frac{1}{2}} = \frac{1}{\theta}$$

where I is the Fisher's information. Such a prior is invariant under parametric transformations [201]. That is, in general, an approximate noninformative prior

Table 6.4: Distributions used in the Bayes model.

Probability distributions	Prior distributions [References]
Binomial	Beta[110, 178, 179, 261, 318, 377, 395] Discrete[261], General-beta[327] m-beta[163, 269], Uniform[138, 318, 332]
Birnbaum-Saunders	Jeffreys' vague prior[310]
Burr	Gamma[315], Uniform[315]
Exponential	Beta[63, 156], Dirichlet process[221] Exponential[45, 275] Gamma[70, 156, 186, 272, 324, 349, 350, 361, 428, 443, 444, 455] Improper noninformative[45, 70, 169, 350] Inverted gamma[45, 184, 362], Jeffreys[384] Natural conjugate[58], Normal[120] Truncated Pareto[324] Uniform [45, 184, 169, 337, 349, 350]
Gamma	Exponential[124, 275], Gamma[253]
Inverse Gaussian	Jeffreys' vague prior [309]
Lognormal	Exponential[61], Inverted gamma[61, 370] Jeffreys[370], Normal[312] Uniform [61, 312, 370]
Mixed exponential	Beta[316], Diffuse Bayes[12] Inverted gamma[316], Jeffreys [316]
Normal	Jeffreys[388], Normal[67, 113, 285, 406]
Poisson	Gamma [185], Inverted uniform [185] Uniform [185]
Power-law	Beta [14], Inverted gamma [14] Jeffreys invariant [14], Log gamma [14] Gamma [14]
Weibull	Beta [425], Discrete [71] Exponential [62, 424], Extreme-value [425] Gamma [71, 177, 424] Inverted gamma [62, 319, 425] Inverse Weibull [132], Jeffreys [71] Log-normal [425], Normal [425] Poisson [425], Truncated-normal [425] Two-point [177], Uniform [62, 132, 177, 319, 424]

is taken proportional to the square root of Fisher's information, which is known as Jeffreys' Rule.

Among many Bayesian studies, Barlow [24] points out that, for the exponential life distribution and any prior distribution for the failure rate parameter, the posterior distribution has a DFR. For example, for a given gamma prior $g(\lambda \mid \kappa, \theta)$ for λ of the exponential life distribution, we will have

$$h(t \mid \kappa, \theta, n, T) = \frac{\kappa + n}{1/\theta + T + t}$$

where n is the number of observed failures, T is the TTT, and $h(t)$ is the hazard rate at time t. Therefore, instead of assuming DFR for a system at its infant mortality stage, we can suitably assign a prior for the parameter of the exponential life distribution, which has a CFR, to model the failure mechanism.

□ **Example 6.11**

Use simulation to generate data points from an exponential distribution with $\lambda = 0.000737035$.

$$t_1 = 231.1554, \quad t_2 = 731.4072$$
$$t_3 = 893.8713, \quad t_4 = 1873.3406$$

Recall that the TTT for type I censoring is given by

$$T = \sum_{i=1}^{m} t_{i:n} + (n-m)t_T$$

and the TTT for type II censoring

$$T = \sum_{i=1}^{m} t_{i:n} + (n-m)t_{m:n}$$

where t_T is the test termination time, m is the number of failures, and n is the total number on test. Take n=4, θ=0.5, κ=0.00147407, and use type II censoring, and the predictive hazard rate at t is

$$h(t \mid \kappa, \theta, n, T) = \frac{4.00147407}{5758.3249 + t},$$

and it is shown in Figure 6.16. □

A similar study is conducted by Rodriguez and Wechsler [351] for a discrete case: the sampling distribution is geometric instead of exponential. Rodrigues and Wechsler prove that, for any choice of the prior distribution, the posterior hazard rate is decreasing. They use the law of maturity to explain the result: given n successes and no failures thus far, the analyst becomes optimistic; that is, he expects less chance of a failure in the next run. Therefore,

$$h(n) = \Pr\{\text{the } (n+1)^{st} \text{ run is a failure} \mid \text{the first } n \text{ runs are successful}\}$$

is a DFR.

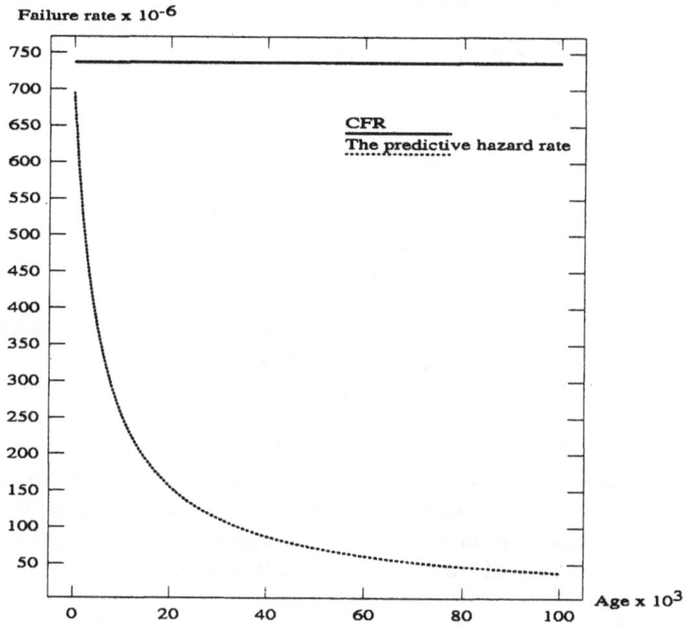

Figure 6.16: The predictive failure rate of a CFR system.

Barlow [24] suggests that the choice of the conjugate prior of the exponential distribution, the gamma distribution, is of special interest since the posterior will also be a gamma distribution. A conjugate prior is the one which remains unchanged in form after the arrival of sample information; that is, both prior and posterior pdfs belong to the same family with different parameters upon the arrival of new information. The conjugate priors are very popular because of their simplicity of interpretation and ease of mathematical manipulation.

Aside from the gamma conjugate prior for the exponential distribution, the beta conjugate prior can be assigned for a binomial trial with p_1 successes and q_1 failures. That is, consider the beta conjugate prior for a binomial distribution:

$$f(x \mid p_1, q_1) = \binom{p_1 + q_1}{p_1} x^{p_1} (1 - x)^{q_1}.$$

The posterior distribution is also a beta distribution with parameters $(p + p_1)$ and $(q + q_1)$. Likewise, if another binomial trial is taken with p_2 successes and q_2 failures, the parameters of the posterior distribution will be revised to be $(p + p_1 + p_3)$ and $(q + q_1 + q_2)$. Except for the exponential-gamma and the binomial-beta conjugate families, the normal-normal and Poisson-gamma conjugate families are also quite helpful in many Bayesian studies.

Understand that the consequences of selecting a certain prior distribution are very important in Bayesian analysis. To accomplish this, the following questions have to be answered:

- How is the prior to be identified?
- What is the prior to be represented?
- How does the prior behave in the presence of experimental (observed) data?
- What are the characteristics of the prior?

It is also the decision makers' (DM) role to explain what the prior distribution represents, whereas the experts must provide objective and quantitative statements of his or her degree of belief; i.e., prior identification is a cooperative effort between the expert and the DM [268]. The broad subject area involving quantification of subjectively held degree of belief information may be loosely subdivided into probability encoding and consensus. The DM must somehow combine or pool the distributions assessed by the experts to form a single consensus prior distribution for use in a Bayesian analysis [268]. This is another interesting topic in statistics: incorporating the experts'opinions. A formal development of conjugate prior distributions is presented by Raiffa and Schlaiffer [339].

As illustrated previously, the choice of a prior distribution for the parameters is usually subjective. One way to avoid specifying a prior is to apply the empirical Bayes (EB) procedures. There are two types of EB approaches. The parametric EB is the form when the prior is known but the prior parameter(s) is(are) unknown, and nonparametric EB is used when neither the prior distribution nor the parameter(s) is(are) known. The EB approach first proposed in Robbins [346, 347] is an extension of the standard Bayesian method. Rather than specifying an arbitrary density function [381], these studies use the former information, usually a set of previous estimates, to construct a prior distribution. Table 6.5 summarizes some sampling (life) distributions used in the EB analysis. The EB and the classical Bayesian approach deal with a known sam-

Table 6.5: Distributions used in the empirical Bayesian model.

Probability distributions	References
Binomial	[246, 266, 359, 378, 454]
Exponential	[160, 246, 248, 386]
Gamma	[55, 61]
Geometric	[359, 426]
Lognormal	[61, 311]
Negative binomial	[359]
Normal	[67, 114, 115, 246, 285, 291, 408]
Poisson	[38, 59, 246, 359]
Weibull	[38, 60, 62, 92, 424]

pling distribution. When the sampling distribution is unknown, the nonparametric method should be applied as shown in Chapter 7. The nonparametric Bayesian technique is another important way for analyzing data.

By applying the Bayesian method, we are able to make use of experts' opinions as our prior information and to reflect our degree of belief in previous experiences by assigning suitable parameters. This is especially important for the semiconductor industry since many new ICs are designed and manufactured under rapidly advancing technology, i.e., only a few, or even no, field reports are available for the analysts. In other words, many ICs are unique and, thus, it is very difficult to predict their performance. The Bayesian approach allows us to incorporate the previous knowledge and experts' opinions for better estimates, in the sense of a smaller mean square error (MSE) than other traditional methods like the MLE. The nonparametric Bayesian approach for reliability analysis will be introduced in Chapter 10

6.5. Conclusions

The importance of burn-in has been realized increasingly in recent years. It is a must to survive in today's highly competitive market. Although it is reported by Bailey [16] that burn-in for some TTL ICs does not have significant effect and by Pantic [314] that, for some components, burn-in procedures provide little, if any, benefit (an example of the molded linear IC from 1979 to 1983 is given, and Pantic uses the burn-in metrics factor to compare the burn-in effectiveness), burn-in is still recognized as an effective way of weeding out potential defects by many researchers and field practitioners [245].

The various factors including performance and cost related to burn-in procedures are reviewed in this chapter. The following chapters will propose different approaches, i.e., parametric, nonparametric, and Bayesian method, for burn-in analysis. We will incorporate the incompatibility introduced in Section 6.1.5 into our models because it is usually found in practice that the resultant reliability after burn-in is inferior to what was forecasted.

7. Nonparametric Reliability Analysis

Generally, there are two methods for making inferences from a set of failure records, that is, parametric and nonparametric. A parametric procedure should be used when all the required assumptions for the chosen parametric procedures can be verified [162]. A function is used to describe the failure behavior of the samples in a parametric analysis which is the life distribution of the samples. Although it is widely recognized that parametric approaches allow analysts more information based on the assumed distribution (such as the mean, median, variance, quantile, and the mean residual life), two reasons make parametric methods less preferred:

1. In many cases, the assumed distribution is only an approximation and, thus, any extrapolation for prediction purposes has to be managed carefully; in other words, very often, the extrapolation fails to provide good predictions.

2. The parameter(s) in the distribution must be estimated; however, they may not converge given the observed data due to the small sample size, imprecise measurements, or the existence of outliers.

We do not have these two problems in a nonparametric analysis, whose only assumption is that samples are from an identical population. Actually, many statistical methods can be regarded as nonparametric techniques, such as the Chebyshev's inequality, the Markov inequality, and the central limit theory, because they do not assume any particular distribution. Generally, nonparametric methods help reliability analysts in two areas: testing hypotheses and estimating the distribution function. A nonparametric method provides a model-free estimate of the reliability, $R(t)$, or the CDF of a sample. Many modifications, refinements, and extensions have been made since most of the fundamental nonparametric methods were proposed in the 1940s. Singpurwalla and Wong [387] provide a brief review on nonparametric non-Bayesian methods. The most famous nonparametric Bayesian method is achieved by using the Dirichlet process by Ferguson [140].

In this chapter, we focus on the nonparametric non-Bayesian method. A series of papers which analyze Nelson's insulating fluid data are introduced because they shed light on the general structure of nonparametric accelerating life testing by using the proportional hazard (PH) model. The life table estimator (LTE) and the Kaplan-Meier product limit estimator (KME) are reviewed and extended for censored data. The re-distribution algorithms proposed by Efron [112] and Dinse [106], which were originally applied on right-censored data, are modified for left-censored observations, and the revised KME is proposed and tested on some simulated data sets. The relationship between the KME and the Dirichlet process is presented in Section 7.3.6. Detailed descriptions on the Dirichlet process appear in Chapter 11. Goodness-of-fit (GOF) tests are applied to test whether the assumed distribution is suitable for the samples under a pre-specified confidence level if the life distribution of the samples is unknown. A brief review on basic GOF techniques is included in Section 7.4.

Throughout this chapter, we implicitly assume that all the samples are from a common population so that they all have the same failure mechanism.

7.1. The Proportional Hazard Rate Model

The more reliable a product is, the longer the testing time that is needed to measure the product's reliability. Therefore, higher stresses are applied on samples so that failures can be recorded in a shorter time. However, there are problems associated with this accelerated testing:

- The relationship between the parameter(s), α ($\vec{\alpha}$) of normal use condition, x_i, and those of the stressed environment, x_j, $g(x_i, x_j, \alpha)$, is usually unknown, although a linear model is usually assumed. Even if the linear model fits the observations, it may only hold up to a certain range.
- If we are only interested in the range where the linear model works, it is usually difficult to estimate the parameters, which may be different for different kinds of samples.
- If a linear relationship does not exist, other models have to be tested to find some suitable candidates.

The CDF under stress x_j at time t can be expressed as

$$\begin{aligned} \Pr\{T < t; x_i, x_j\} &= F_T(h_{ij}(t)) \\ h_{ij}(t) &= \begin{cases} g(x_i, x_j, \alpha) h_o(t) & \Rightarrow \text{multiplicative model} \\ g(x_i, x_j, \alpha) + h_o(t) & \Rightarrow \text{additive model} \end{cases} \end{aligned}$$

where $g(x_i, x_j, \alpha)$ is a factor between two stress levels x_i and x_j given that α, and $h_o(t)$, the population hazard rate at t under normal stress level, are non-negative and non-decreasing.

Two multiplicative models are discussed in Schmoyer [364].

1. The proportional hazard (PH) model
$$h_{ij}(t) = g(x_i, x_j, \alpha) h_o(t), \; i, j \in \{1, \cdots, n\}. \tag{7.1}$$
2. The accelerated failure time (AFT) model
Let $h_o(t) = t$.
$$\begin{aligned} h_{ij}(t) &= g(x_i, x_j, \alpha) t, \; i, j \in \{1, \cdots, n\} \\ &= \begin{cases} (x_j/x_i)^\alpha & \Rightarrow \text{the Arrhenius model} \\ \exp[\alpha(1/x_j - 1/x_i)] & \Rightarrow \text{the Eyring model} \end{cases} \end{aligned} \tag{7.2}$$

Schmoyer [364, pp 180-181] also compares the Weibull model, which is the unique model belonging to both the PH and the AFT families. In Table 7.1,

Table 7.1: The model comparison by Schmoyer.

Model	$F_o(t)$	$g(x)$	$h(t)$
PH	$1-e^{-t}$	sigmoid	
AFT		sigmoid	t
Cox PH	$1-e^{-t}$	$e^{\beta x}$	
Weibull	$1-e^{-t^{1/\sigma}}$	$e^{-(\alpha+\beta x)}$	t

$F_o(t)$ is the distribution function under the normal (or operating) stress level, and sigmoid refers to the function which is non-decreasing, defined on $[0, \infty)$, and either convex, concave, or convex to the left of a point M and concave to the right of M [364]. A more general formula is to use α_{ij} instead of α in Eqs. (7.1) and (7.2), which is considered by Shaked *et al.* [375]. A PH model whose g is a positive-coefficient polynomial in stress and $h(t) = t^K$ for some integer K is called the multi-stage model. The g in the PH model and the multi-stage model mentioned above are all sigmoid.

Suppose $i \neq j$ and k different stress levels are used; that is, i, j=1,$\cdots$,k. The time transformation factor, η_{ij}, can be estimated by, given failure times $t_{i1}, \ldots, t_{in_i}$ and $t_{j1}, \ldots, t_{jn_j}$,

$$\hat{\eta}_{ij} = \frac{g(x_j, x_o, \hat{\alpha})}{g(x_i, x_o, \hat{\alpha})} = \overline{t_i}/\overline{t_j},$$

where $\overline{t_i} = \frac{1}{n_i}\sum_{k=1}^{n_i} t_{ik}$, $\hat{\alpha} = \sum_{i\neq j} w_{ij}\hat{\alpha}_{ij}$, $\sum_{i\neq j} w_{ij}=1$, and x_o is the operating stress.

The PH approach can be considered semi-parametric because a function has to be assumed for $g(x_i, x_j, \alpha)$. We can estimate the expected life under normal operating $\hat{\mu}_o$ from Shaked and Singpurwalla [374]:

$$\hat{\mu}_o = \frac{1}{N}\sum_{i=1}^{k}\sum_{l=1}^{n_i} \hat{t}_{il} = \sum_{i=1}^{k} \frac{n_i}{N}\left(\frac{x_i}{x_o}\right)^{\hat{\alpha}}\overline{t_i} \tag{7.3}$$

where

$$N = \sum_{i=1}^{k} n_i,$$

$$\hat{t}_{il} = (x_i/x_o)^{\hat{\alpha}} t_{il}, \quad l = 1, \cdots, n_i, \quad i = 1, \cdots, k.$$

The estimated distribution function under operating condition, $\hat{F}_o(t)$, can be obtained by $\hat{t}_{il}$ in Eq. (7.3). We can use $\hat{\mu}_o$ in Eq. (7.3) to obtain the median, the mean, and the $100(1-\alpha)$ quantile (see Shaked *et al.* [375] for illustration).

The inverse power law is also considered in Basu and Ebrahimi [36] and Sethuraman and Singpurwalla [372] where censored data and competing risks are taken into consideration by using the KME and by transforming the observations by taking logarithms.

7.2. The Life Table Estimator (LTE)

One of the classical methods to estimate $R(t)$ is the LTE approach, which is developed under the conditional reliability p_i. The reliability estimate is

$$\hat{s}_i = \prod_{j=1}^{i} \hat{p}_j, \quad i = 1, 2, \cdots, k$$

where

$\hat{p}_i$	$1 - \hat{q}_i$
$\hat{q}_i$	$\frac{d_i}{n'_i}$
d_i	the number of deaths during I_i
I_i	the ith testing interval, which begins at the $(i-1)^{st}$ failure and ends at the ith failure
n'_i	the number of effective number of individuals at risk in I_i; $n'_i = n_i - \frac{r_i}{2}$
n_i	the number of samples at risk at time t_{i-1}, the beginning of I_i. (Let n_1 be n. The value of n_i is not necessarily the number of individuals still alive at time t_i of the original total n, but rather the number alive and which have not been censored prior to t_i. $n_i = n_{i-1} - d_i - r_i$, $i = 2, \cdots, k$)
r_i	the number of removals in I_i
t_i	the ending time of I_i, $i = 1, 2, \cdots, k$
p_i	Pr{ an individual survives beyond I_i | this individual survives beyond I_{i-1}} $= \frac{s_i}{s_{i-1}}$; $p_k = 0$
q_i	$1 - p_i$, $q_k = 1$
s_i	the reliability defined as Pr{ an individual survives beyond I_i}.

Variance can be estimated by Greenwood's formula,

$$\widehat{\text{Var}}(\hat{s}_i) = \hat{s}_i^{\,2} \sum_{j=1}^{i} \frac{\hat{q}_j}{n'_j \hat{p}_j}. \tag{7.4}$$

Modifications are needed for Eq. (7.4) if the samples are censored. The intervals in a life table need not be the same, and the number of intervals used will depend on the amount of available data and the aims of the analysis.

7.3. The Kaplan-Meier Product Limits Estimator

Incomplete data are common in practice; they can happen when, for example,

- the automatic detecting device breaks down and is not able to record exact failure times of the samples so that we can only inspect the sample manually when the test is interrupted; that is, we only know that the samples fail before certain times (left-censored),

- the test has to be terminated due to restricted budget; however, we know the samples fail after test termination time (right-censored) ,
- extra samples with known testing history are available and are put on test (left-truncation), or
- some samples have to be removed because of failure of the testing device (withdrawal).

These incomplete data are considered here for both the Kaplan-Meier product limits estimator (KME) and the re-distribution techniques. The extension of the re-distribution on preparing the failure plots similar to those by Leemis [244] is also discussed.

The KME [211] is the most popular nonparametric method for analyzing reliability. However, it is not able to handle censored data. Efron [112] proposes a re-distribute-to-the-right algorithm for right-censored data, which can also be re-distributed from right to left by a method by Dinse [106]. We extend the idea for left-censored data and other incomplete observations. Several interesting observations are identified as a guide for further investigation.

If there are no censored observations in a sample of size n, the reliability, or the empirical survival function (ESF) is defined as

$$\hat{R}(t) = \frac{\text{Number of observation} \geq t}{n} = \prod_{j:t_j<t} \frac{n_j - d_j}{n_j}, \; t \geq 0 \tag{7.5}$$

and its variance is given by

$$\widehat{\text{Var}}(\hat{R}(t)) = \hat{R}(t)^2 \sum_{j:t_j<t} \frac{d_j}{(n_j - d_j)(n_j)}$$

where $\hat{R}(t)$ is right continuous. Eq. (7.5) is essentially the same as that for the reliability estimators $\hat{s_j} = \prod_{i=1}^{j} \hat{p}_i$ [237, p 72] and is under the notion:

$$R(t) = \frac{\text{Number of observation } \geq t}{\text{Total number of samples}}, \; t \geq 0. \tag{7.6}$$

Eq. (7.6) can be applied to have the hazard rate in a specific time period:

$$h(t) = \frac{\text{number failed in the testing period}}{\text{(total number in test)(duration)}} = \frac{f}{N\Delta t}. \tag{7.7}$$

In Chapter 9, we use Eq. (7.7) to find the changing points t_1, the time from DFR to CFR, and t_2, the time from CFR to IFR under the U-shape hazard rate assumption.

Kaplan and Meier [211] give the product limit estimator (PLE) as an MLE. By following the notation in Section 7.2 and define s_i as the number of censoring times in I_i and L_j^i as the jth censoring time in I_i, $j = 1, \cdots, s_i$, then the surviving function $R(t)$ is a nonincreasing left-continuous function [237, p 75], and the observed likelihood function is

$$L(\mathbf{R}) = \prod_{i=1}^{k}\Big\{[\prod_{j=1}^{s_i} R(L_j^i)]\big[R(t_i) - R(t_i+0)\big]^{d_i}\Big\} \prod_{j=1}^{s_{k+1}} R(L_j^{k+1}) \tag{7.8}$$

for k intervals where $\big(R(t_i) - R(t_i+0)\big)$ is the probability that an individual dies at t_i. The next step is to maximize Eq. (7.8) with respect to all $R(t)$s that have the following properties:

- $\hat{R}(t_1) = \hat{R}(L_j^1) = 1,\ j = 1, \cdots, s_1$
- $\hat{R}(L_j^{i+1}) = \hat{R}(t_i+0) = \hat{R}(t_{i+1}) = 1,\ i = 1, \cdots, k;\ j = 1, \cdots, s_{i+1}$

Let $P_i = R(t_i+0)$ and $P_0 = 1$, we have the maximum likelihood function

$$\begin{aligned} L(\vec{P}) &= \prod_{i=1}^{k}(P_{i-1} - P_i)^{d_i} P_i^{s_{i+1}} \\ &= \prod_{i=1}^{k}(p_1 \cdots p_{i-1} q_i)^{d_i} (p_1 \cdots p_i)^{s_{i+1}} \\ &= \prod_{i=1}^{k} q_i^{d_i} p_i^{n_i - d_i} (\text{a term without } p_i). \end{aligned} \tag{7.9}$$

Since $p_i = \frac{P_i}{P_{i-1}}$, $q_i = 1 - p_i$, and $s_i = n_i - d_i$, Eq. (7.9) is maximized at $\hat{p}_i = \frac{n_i - d_i}{n_i}$. Hence,

$$\hat{R}(t_i+0) = \hat{R}(t_{i-1}+0)\frac{n_i - d_i}{n_i} \Rightarrow \hat{R}(t) = \prod_{i:t_i<t} \frac{n_i - d_i}{n_i}.$$

Define $c_{r,i:n}$ and $y_{i:n}$ as a realization of right-censored distribution and failure distribution, respectively. The KME given $x_{1:n}, \ldots, x_{n:n}$, which may be right-censored, is

$$\hat{R}(t) = \prod_{x_{i:n} \le t} (1 - \frac{1}{n-i+1})^{\delta_{i:n}} \tag{7.10}$$

where

$$\begin{aligned} \delta_{i:n} &= \begin{cases} 1, & \text{censored}: c_{r,i:n} < y_{i:n} \\ 0, & \text{uncensored}: c_{r,i:n} \ge y_{i:n} \end{cases} \\ x_{i:n} &= \min(c_{r,i:n}, y_{i:n}). \end{aligned}$$

The Greenwood's formula, which is similar to that in Eq. (7.4), can be applied for the variance of $\hat{R}(t)$ [284, p 51]:

$$\hat{\text{Var}}(\hat{R}(t)) = \hat{R}^2(t) \sum_{x_{i:n} \le t} \frac{1 - \delta_{i:n}}{(n-i)(n-i+1)}. \tag{7.11}$$

The (1-α)100% confidence interval (CI) can be derived by

$$\hat{R}(t) \pm z_{\alpha/2}\sqrt{\hat{\text{Var}}\hat{R}(t)} \tag{7.12}$$

where $\hat{\text{Var}}(\hat{R}(t))$ is in Eq. (7.11) and z_θ is the $100\theta^{th}$ percentile of the standard normal distribution.

The KME of $R_n(t)$ is defined as

$$R_n(t) = \begin{cases} 1, & 0 \le t < t_1 \\ R_n(t_i), & t_i \le t < t_{i+1},\ i = 1, \cdots, n-1 \\ R_n(t_n), & t = t_n \\ 0, & t > t_n \mid \delta_n = 1 \end{cases} \tag{7.13}$$

where

$$R_n(t_i) = \prod_{j=1}^{i} \left(1 - \frac{\delta_j}{n-j+1}\right), \quad i = 1, \cdots, n$$

$$\delta_i = \begin{cases} 0, & \text{if } t_i \text{ is a time to censoring} \\ 1, & \text{if } t_i \text{ is a time to failure} \end{cases}$$

Since the case $\{t > t_n \mid \delta_n = 0\}$ was left undefined by Kaplan and Meier, Efron and Gill modified the KME [161]:

KMEE Efron's version of KME
$R_n(t) = 0$, for $t > t_n$

KMEG Gill's version of KME
$R_n(t) = R_n(t_n)$, for $t > t_n$.

Geurts [161] finds that the robustness and lower value of the MSE of the KMEE give the KMEE an advantage over the KMEG only for moderate to low values of $R(t)$; at a high $R(t)$ the KMEG is preferred over the KMEE for its smaller bias and as a better estimator of (smaller) variance than KMEE.

Similar research that deals with the last sample is given in [302], where the reduced sample and the modified reduced sample methods are introduced. The reduced sample estimator is, for $x_{i:n} \le t < x_{i+1:n}$,

$$\hat{F}_r(t) = \frac{n_i - d_i}{m_i}$$

where m_i is the number of elements satisfying $\{i : c_i \ge t\}$ and c_i is the censoring time. Although this reduced sample estimator does not have the aforementioned undesirable property (at the largest observation) and it is exactly unbiased [302, p 39], it has other disadvantages. For example, it is not necessarily a monotone function. The modified reduced sample estimator is, for $x_{i:n} \le t < x_{i+1:n}$,

$$\hat{F}_m(t) = (n_i - d_i)\left[n_i + \sum_{j<i} \frac{d_j \hat{G}(x_{i:n})}{\hat{G}(x_{j:n})}\right]^{-1}$$

where

$$\hat{G}(c) = \prod_{x_i < c} \frac{l_i - m_i}{l_i}$$

$$l_i = n_i - d_i,$$

$$m_i \equiv \text{the number of these that are actually censored at } x_{i:n}.$$

Geurts [161] proves that the modified reduced sample estimator is identical to the KME. Similar to the above research, a nonparametric software reliability growth model is proposed by Sofer and Miller [392]. They use

$$\hat{M}(t) = n + \delta \frac{t - t_n}{T - t_n}, t_n \leq t \leq T,$$

where $\hat{M}(t)$ is the mean value function and δ can be chosen to be either 0, 0.5, or 1; the larger δ will give more conservative estimates. We should note that the use of $M(t)$ here is similar to the $\Lambda(t)$ used in the paper by Leemis [244].

7.3.1. Re-distribution Methods

Efron introduces another method of computing the PLE. Let $w_{i:n}$s be the weight of $x_{i:n}$s. Efron's re-distribution-to-the-right algorithm (for right-censored data) consists of the following steps:

E-method

Step 1 Arrange $x_1, \ldots, x_n$ in ascending order and call the ordered data $x_{1:n}, \ldots, x_{n:n}$.

Step 2 Start from $x_{1:n}$ and move toward $x_{n:n}$.

Step 3 Uniformly re-distribute the weight of any encountered right censored data to the xs at its right hand side, no matter if it is a failure or a censored observation.

Step 4 Repeat Step 1 $\sim$ Step 3 until reaching $x_{n:n}$.

Dinse [106] proposes another re-distributing method which follows the procedures below:

D-method

Step 1 Arrange $x_1, \ldots, x_n$ in ascending order and call the ordered data $x_{1:n}, \ldots, x_{n:n}$.

Step 2 Start from $x_{n:n}$ and move toward $x_{1:n}$.

Step 3 Re-distribute the weight of any encountered right-censored data to the failure times at its right hand side and until the distributed weights are proportional to the weight already put on these failures.

Step 4 Repeat Step 1 $\sim$ Step 3 until reaching $x_{1:n}$.

Redistribution allows for the contribution from the censored data to the reliability $R(t)$. If the largest observation is censored, we treat it as uncensored data so that we can proceed with the above two procedures.

The $w_{1:n}, \ldots, w_{n:n}$ are adjusted to be $w'_{1:n}, \ldots, w'_{n:n}$, and the $\hat{R}(t)$ can be estimated by

$$\hat{R}(t) = 1 - \sum_{x_{i:n} \leq t; \delta_{i:n} = 0} w'_{i:n}. \qquad (7.14)$$

It can be proved that the E and D methods result in the same $\hat{R}(t)$ if both $x_{1:n}$ and $x_{n:n}$ are failure times.

□ **Example 7.1**

Seven observations are available and the fourth, the sixth, and the seventh ones are right-censored as shown in Figure 7.1. We use a shaded circle for $t = 0$, circles for censored times, and an "X" for a failure. In Figure 7.1, where the

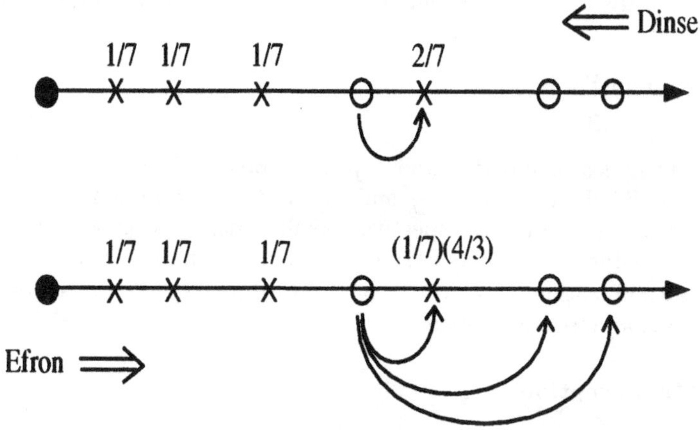

Figure 7.1: Analyzing right-censored data by the E- and the D-method.

weight of each failure is shown, under the E-method, $w'_{5:7} = w_{5:7} + \frac{1}{3}w_{4:7}$, which is smaller than $w'_{5:7} = w_{5:7} + w_{4:7}$ in Dinse format; therefore, the D-method gives us more conservative estimates in this example. Define

$$R(t-) = \lim_{\epsilon \to 0} R(t-\epsilon),$$

we have

$$\begin{array}{ll} \mathrm{E-method:} & \hat{R}(x_{6:7}) = \hat{R}(x_{7:7}-) = \hat{R}(x_{5:7}) = 1 - (\frac{1}{7} + \frac{1}{7} + \frac{1}{7} + \frac{4}{21}) = \frac{8}{21} \\ \mathrm{D-method:} & \hat{R}(x_{6:7}) = \hat{R}(x_{7:7}-) = \hat{R}(x_{5:7}) = 1 - (\frac{1}{7} + \frac{1}{7} + \frac{1}{7} + \frac{2}{7}) = \frac{6}{21} \end{array}$$

to estimate the reliability after the last failure and before the last observation, respectively. As for the $R(t)$ where $t \geq x_{n:n}$ and $x_{n:n}$ is a right-censored data, we can use

$$\hat{R}(t) = \hat{R}(x_{k:n}) - (n-k)\frac{\alpha}{n}, \quad 0 \leq \alpha \leq 1 \tag{7.15}$$

as a reliability estimator where $x_{k:n}$ is the last failure observed before test termination. Using $\alpha = \frac{1}{2}$ in Eq. (7.15), which is a very conservative choice, we have, for $t \geq x_{7:7}$,

$$\begin{array}{ll} \mathrm{E-method:} & \hat{R}(t) = \frac{5}{21} \\ \mathrm{D-method:} & \hat{R}(t) = \frac{3}{21}. \end{array}$$ □

In practice, it is possible that several tests, e.g., k tests, are performed; in other words, k results will be available. It is suggested by Srinivasan and Zhou [396] that we should pool these k tests (instead of using any convex combination estimators to have smaller asymptotic variance) and sort the observations for making inferences. Suppose we have independent censoring schemes for k data sets; the ith censoring can be described by a random variable $G_i(t)$. Thus,

$$\begin{aligned} Z_{ij} &= \min(Y_{ij}, C_{ij}) \\ \delta_{ij} &= I_{[Y_{ij} \leq C_{ij}]} \end{aligned}$$

where I is an indicator function and Z_{ij} is the observed life time of the ith sample in the jth data set and C_{ij} and Y_{ij} are the corresponding censored time and actual survival time, respectively [396]. Then it is reported that the pooled hazard estimator $\hat{\Lambda}_p(t)$ is smaller than the asymptotic variance of the best convex combination estimator $\hat{\Lambda}_c(t)$; the inequality is strict unless the censoring distributions are identical.

7.3.2. A Misconception

A nonparametric estimation of the cumulative hazard rate function for a nonhomogeneous Poisson process (NHPP) is developed by Leemis [244], who applies the idea of preparing the plot of the expected number of events by time t, $\Lambda(t)$. Supposing there are k realizations,

$$\hat{\Lambda}(t) = \frac{in}{(n+1)k} + \frac{n(t - t_{i:n})}{(n+1)k(t_{i+1:n} - t_{i:n})}, \quad t_{i:n} < t \leq t_{i+1:n}, \; i = 0, 2, \cdots, n.$$

Leemis derives an exact variance for $\hat{\Lambda}(t)$, the number of failures before and at t:

$$\text{Var}(\hat{\Lambda}(t)) = \frac{\hat{\Lambda}(t)}{k} \tag{7.16}$$

where k is the number of tests which are pooled. It is also shown in the same paper that, by the central limit theorem (CLT),

$$\frac{\hat{\Lambda}(t) - \Lambda(t)}{\sqrt{\frac{\hat{\Lambda}(t)}{k}}} \sim N(0,1). \tag{7.17}$$

From Eq. (7.17), the asymptotically exact 100(1-α)% CI for $\hat{\Lambda}(t)$ is

$$\hat{\Lambda}(t) \pm z_{\alpha/2}\sqrt{\frac{\hat{\Lambda}(t)}{k}} \tag{7.18}$$

where $z_{\alpha/2}$ is the $(1 - \alpha/2)100\%$ of the standard normal distribution. From Eq. (7.18), it can be seen that the more realizations (different data sets from the common NHPP) we have, the smaller the variance becomes. Two extensions are discussed [244].

- A period during which events cannot happen.
  A good example is the lunch break, where no new failure will be discovered because all workers are off duty. In this case, $\Lambda(t)$ will have a horizontal segment (as one can figure from intuition) with $\hat{\Lambda}(\frac{a+b}{2})$, which is the average of the cumulative hazard rates of the beginning and the termination of the lunch break, e.g., (a, b).

- Ties
  The author proposes to connect the midpoints of the two adjacent intervals, which does not work for the first and the last segments.

It must be noted that it is incorrect to use

$$\hat{R}(t) = 1 - \frac{\hat{\Lambda}(t)}{n} \Rightarrow \text{Var}\big(\hat{R}(t)\big) = \frac{1}{n^2}\text{Var}\big(\hat{\Lambda}(t)\big)$$

because the sample size changes in each test interval in our experiment. That is, the sample size is not always equal to n. However, it is handy to apply the re-distribution technique for preparing failure plots when censored data are present.

It is also possible that we assign more (less) weight to the observations from certain tests if the analysts believe these tests are more (less) representative than others; that is, let $w_i \geq 1$ ($w_i \leq 1$). This is an engineering approach and can be achieved by incorporating experts' opinions. Of course, we should discard the test results if the experimental procedures are wrong, which is equivalent to assigning zero weight to the samples of the discarded test(s).

7.3.3. Left-censored Data

We have a left-censored sample if $c_{l,i:n} > y_{i:n}$. As indicated by Miller [284, p 192], the weight on each "jump" for a KME is

$$\begin{aligned} \hat{\Delta}i : n = \;& \hat{R}(x_{i:n}-) - \hat{R}(x_{i:n}) \\ = \;& \tfrac{1-\delta_{i:n}}{n} \prod_{j=1}^{i-1} (\frac{n-j+1}{n-j})^{\delta_{i:n}} \\ = \;& \tfrac{1-\delta_{i:n}}{n}(1 + \tfrac{1}{n-j_1})(1 + \tfrac{1}{n-j_2}) \cdots (1 + \tfrac{1}{n-j_i}) \end{aligned} \tag{7.19}$$

where j_k is the index of the kth right-censored data. By Eq. (7.19), Miller proves that Efron's re-distribution algorithm gives the usual KME. Dinse uses the adjusting weight concept to verify that his procedures give the exact result as Efron's; therefore, the re-distribution method by Dinse also results in KME's.

Suppose we have a set of left-censored data, $x_1, \ldots, x_n$. Recall that the only difference between the left and the right-censored data is that we re-distribute the weights of the censored data to their left instead of to the right. From Eq. (7.19), it can be seen that we can apply both approaches mentioned earlier on left-censored data and still have the KME because the only difference will be the j_ks and, in the sequel, this makes Eq. (7.19) remain unchanged as we only re-arrange the j_ks.

We now propose the revised E-method and the revised D-method for left-censored data and call them E$'$-method and D$'$-method, respectively.

E'-method

Step 1 Arrange the data in ascending order to have $x_{1:n}, \ldots, x_{n:n}$.

Step 2 Start from $x_{1:n}$ and move toward $x_{n:n}$.

Step 3 Re-distribute the weight of any encountered left-censored data to the failure(s) at its left hand side so that the distributed weights are proportional to the weights already in these observations. If x_j is the first failure and $j > 1$, then $w_j' = w_j + \sum_{k=1}^{j-1} w_k$, so commence procedures from $j+1$.

Step 4 Repeat Step 1 $\sim$ Step 3 until reaching $x_{n:n}$.

D'-method

Step 1 Arrange the data in ascending order to have $x_{1:n}, \ldots, x_{n:n}$.

Step 2 Start from $x_{n:n}$ and move toward $x_{1:n}$.

Step 3 Uniformly re-distribute the weight of any encountered left-censored data to its left hand side (no matter whether a failure or a censored observation). If x_j is the first failure and $j > 1$, then $w_j' = w_j + \sum_{k=1}^{j-1} w_k$, so do not allocate any weight to the observation(s) with an index smaller than j.

Step 4 Repeat Step 1 $\sim$ Step 3 until reaching $x_{1:n}$.

We can see that the E$'$– and the D$'$-methods are defined according to the processing direction and use the re-distribution scheme in the D- and the E-methods, respectively. For left-censored data, we do not have the undetermined case mentioned above at the far right because, for example, if $x_{n:n}$ is censored, $w_{n:n}$ will be re-distributed to the data at its left hand side.

Similarly, the E$'$– and the D$'$-methods result in the same $\hat{R}(t)$ if both $x_{1:n}$ and $x_{n:n}$ are failure times. Figure 7.1 explains the reason why the match is broken when both $x_{1:n}$ and $x_{n:n}$ are not failures. As mentioned earlier, if $x_{j:n}$ is the first failure and $j > 1$, we should not allocate weight on the $x_{m:n}$, $m < j$. This is depicted in Figure 7.2.

Furthermore, we should add $w_{m:n}$ $(m = 1, \cdots, j-1)$ to $w_{j:n}$, which is the weight of the first observed failure, because we know the left-censored data $1 \sim (j-1)$ fail before $x_{j:n}$.

7.3.4. Handling Both Left- and Right-censored Data

We can also apply the D- and E-methods on both left and right-censored data. Eq. (7.14) can be slightly modified to estimate $R(t)$

$$\hat{R}(t) = 1 - \sum_{x_{i:n} \leq t; \delta_{i:n} = \gamma_{i:n} = 0} w_{i:n}'.$$

If neither $x_{1:n}$ nor $x_{n:n}$ is censored, and we consider both left- and right-censoring, we always have $\hat{R}_E(t) < \hat{R}_D(t)$, where $\hat{R}_E(t)$ and $\hat{R}_D(t)$ are the

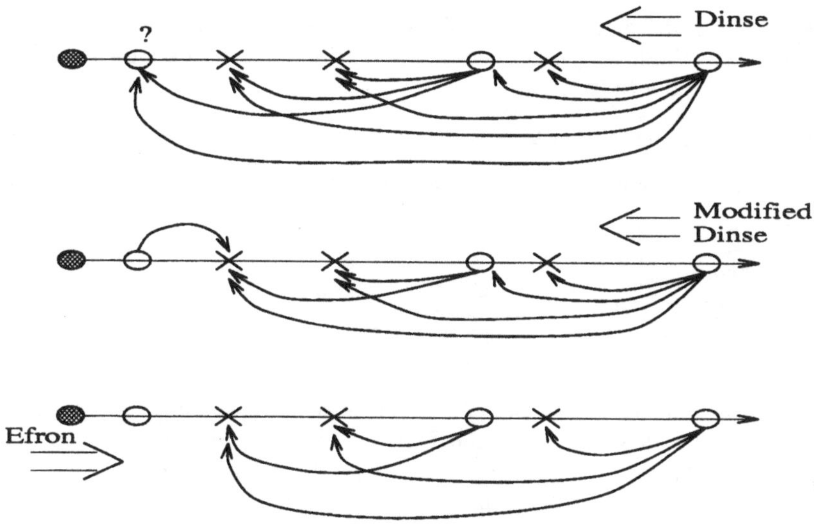

Figure 7.2: Using the E$'$- and the D$'$-methods for left-censored data.

estimated reliability from the E- (and E$'$) and the D- (and D$'$) method, respectively.

The weight on a failure observation i ($i \in \{1, 2, \cdots, n\}$) after re-distribution for the E-method ($w'_{i:n,E}$) and the D-method($w'_{i:n,D}$) are

$$\begin{aligned} w'_{i:n,E} &= w_{i:n} + \sum_{j<i} \delta_{j:n} \frac{w^*_{j:n,E}}{n-j} + \sum_{j>i} \gamma_{j:n} w_{j:n} \frac{(1-\delta_{i:n})(1-\gamma_{i:n})w^*_{i:n,E}}{\sum_{k<j}(1-\delta_{k:n})(1-\gamma_{k:n})w^*_{k:n,E}} \\ w'_{i:n,D} &= w_{i:n} + \sum_{j>i} \gamma_{j:n} \frac{w^*_{j:n,D}}{n-j} + \sum_{j<i} \gamma_{j:n} w_{j:n} \frac{(1-\delta_{i:n})(1-\gamma_{i:n})w^*_{i:n,D}}{\sum_{k>j}(1-\delta_{k:n})(1-\gamma_{k:n})w^*_{k:n,D}}, \end{aligned}$$

respectively, where $w^*_{i:n,E}$ and $w^*_{i:n,D}$ are the adjusted weight by the E-method and the D-method, respectively. Let $i = 1$ (since $x_{1:n}$ is a failure), then we have

$$\begin{aligned} w'_{1:n,E} &= w_{1:n} + \sum_{j>1} \gamma_{j:n} w_{j:n} \frac{(1-\delta_{1:n})(1-\gamma_{1:n})w^*_{1:n,E}}{\sum_{k<j}(1-\delta_{k:n})(1-\gamma_{k:n})w^*_{k:n,E}} \\ w'_{1:n,D} &= w_{1:n} + \sum_{j>1} \gamma_{j:n} \frac{w^*_{j:n,D}}{n-j}. \end{aligned}$$

To have the desired result, we need to show

$$\sum_{j>1} \gamma_{j:n} w_{j:n} \frac{(1-\delta_{1:n})(1-\gamma_{1:n})w^*_{1:n,E}}{\sum_{k<j}(1-\delta_{k:n})(1-\gamma_{k:n})w^*_{k:n,E}} > \sum_{j>1} \gamma_{j:n} \frac{w^*_{j:n,D}}{n-j}, \tag{7.20}$$

which can be achieved for the following reasons:

- For the D-method, $w_{j:n}$ is distributed to $(n-j)$ observations; whereas for the E-method, $w_{j:n}$ is only distributed to failures whose number is less than or equal to $(n-j)$.
- The closer the failures to $x_{1:n}$, the more weights they receive.

Hence, Eq. (7.20) is verified. The same method applies to other terms, and we will have

$$\begin{aligned} &w'_{i:n,E} > w'_{i:n,D}, \quad i = 1, 2, \cdots, k \\ &w'_{i:n,E} < w'_{i:n,D}, \quad i = k+1, k+2, \cdots, n \\ &\sum_{\forall i, \delta_i = \gamma_i = 0} w'_{i:n,E} = \sum_{\forall i, \delta_i = \gamma_i = 0} w'_{i:n,D} = 1 \end{aligned} \tag{7.21}$$

where $1 < k < n$. In words, Eq. (7.21) indicates that, before the kth ordered observation, the weights put on the failures according to the E-method are larger than those by the D-method; the opposite situation occurs for every failure after the kth observation so that the sums will both equal 1.

The above result does not always hold if either $x_{1:n}$ or $x_{n:n}$ (or both) is censored, although an example in a later section illustrates the same result (the cumulative number of failures from the E-method is larger than those from the D-method at each failure).

Tsai *et al.* [423] consider the PLE under right-censoring and left-truncation. Consider

T random left-truncation time

C random right-censoring time

X random life time under the reliability R, which is, as shown before, left continuous; T and C are independent of X

Y_i $\min(X_i, C_i)$ and $Y_i \geq T_i$,

then, as usual, the PLE of R is

$$\hat{R}(x) = \sum_{x_i < x} \frac{n_i - d_i}{n_i}$$

where d_i and n_i are defined previously; to be specific, $n_i = \sum_{\forall j} I_{[t_j \leq x_i \leq y_j]}$, the number of samples at risk. The $\hat{R}$ reduces to the KME for right-censored data, if $T_1 = \cdots = T_n = 0$. It is illustrated in a plot that the KME of R obtained by ignoring the truncation overestimates reliabilities considerably and is very close to the upper 90% confidence level. This result is not surprising because the sample at risk becomes smaller after considering left-truncation; that is, if the effect of truncation is ignored, the hazard rate decreases, i.e., the survival rate increases.

7.3.5. The Revised KME Method

In some cases, new samples which have survived for a certain time may be added to tests at certain stages. These types of samples are called left-truncation.

Meanwhile, it is also possible to withdraw samples from tests at any time. Unfortunately, the aforementioned re-distribution methods fail to reflect the change in the sample's size at risk at any time because the re-distribution only occurs at failure times. We have the same shortcoming when the re-distribution methods are applied to prepare the failure plots as described by Leemis [244], which can be calculated by modifying Eq. (7.14)

$$\hat{M}(t) = \sum_{x_{i:n} \leq t; \delta_{i:n} = \gamma_{i:n} = 0} W'_{i:n} \tag{7.22}$$

where $W'_{i:n}$ is the failure count at the ith observation. If the weight for each observation is $\frac{1}{n}$, we have $W'_{i:n} = nw'_{i:n}$.

The traditional KME in Eq. (7.10) should be used if samples are added or removed. However, from Eq. (7.10), we can see that the KME is not able to manage the left-censored observations. Because we know the left-censored samples fail before a certain time, we can make use of the re-distribution concept to allocate the weights of left-censored samples to the failures at its left hand side proportional to the weights already on them. Thus, the estimated reliability becomes

$$\hat{R}(t) = \prod_{x_{i:n} \leq t} \left(1 - \frac{d'}{n'_{i:n} - i + 1}\right)^{(1-\delta_{i:n})(1-\gamma_{i:n})} \tag{7.23}$$

In Eq. (7.23), d' is adjusted by distributing the weight of left-censored data to the failures at their left hand side proportional to the weights already put on them, and the $n'_{i:n}$ is the effective sample size in I_i. Let k be the number of segments separated by any incomplete observation between two failures, then $n'_{i:n}$ is obtained by

$$n'_{i:n} = \left(\sum_{j=1}^{k} t_j n_{i:n,j}\right) / \left(\sum_{j=1}^{k} t_j\right)$$

as shown in Figure 7.3, which has the following relationship

$$\begin{aligned} n_{i:n,1} - 1 = & \; n_{i:n,2} = n_{i:n,3} \\ n_{i:n,3} - 1 = & \; n_{i:n,4} = n_{i:n,5} + 1. \end{aligned}$$

From our experiments, this adjustment does provide better estimations, although they are not very significant.

Theoretically, the reliability should be a strictly monotonic decreasing function. It implies the impropriety of the "steps" in the reliability plots. To avoid these steps, we can connect the mid-points of the adjacent steps to provide estimates for the reliability for t which is between two failures; that is, e.g., $x_{i-1:n} < t < x_{i:n}$, $i = 1, 2, \cdots, k-1$. For I_k, the last test interval, we use

$$\hat{R}_m(t) = \hat{R}(x_{k:n}), \; x_{i-1:n} < t < x_{i:n} \tag{7.24}$$

where the subscript m is used to denote adjusting the reliability by connecting the mid-points.

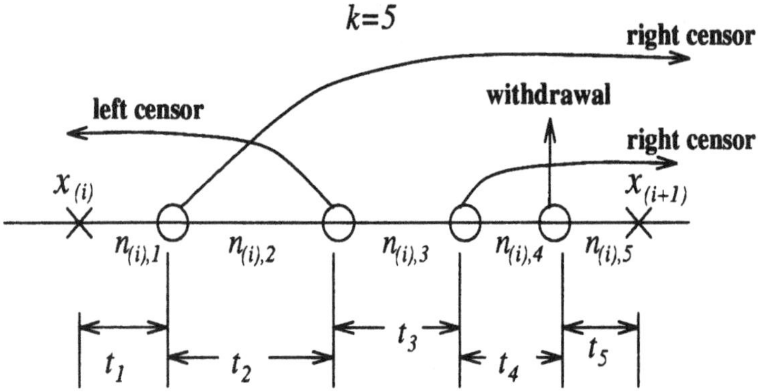

Figure 7.3: Calculating the effective sample size.

□ **Example 7.2**

Consider three types of censoring schemes: light, medium, and heavy censor; the values for λ_c and λ in each censoring scheme are given in Table 7.2. Simulation is applied to generate failure and censored times. Let

$$\begin{aligned} C_l &\sim \text{EXP}(\lambda_c) \text{ for left censoring,} \\ C_r &\sim \text{EXP}(\lambda_c) \text{ for right censoring, and} \\ Y &\sim \text{EXP}(\lambda). \end{aligned}$$

It is found that the pooled estimator will have more gain in the case of heavy censoring. Recall that for two exponentially distributed random variables, X

Table 7.2: The parameters for each censoring scheme.

Left/Right	Type	λ_c	λ
right	light	0.01	0.09
	medium	0.05	0.10
	heavy	0.10	0.10
left	light	0.90	0.10
	medium	0.20	0.10
	heavy	0.10	0.10

and Y, with parameter λ_x and λ_y, respectively,

$$\begin{aligned} \Pr\{X > Y\} &= \lambda_y/(\lambda_x + \lambda_y) \\ \Pr\{X < Y\} &= \lambda_x/(\lambda_x + \lambda_y). \end{aligned}$$

Thus, in average, the percentage of right and left-censored data can be derived by $\lambda_c/(\lambda + \lambda_c)$ and by $\lambda/(\lambda + \lambda_c)$, respectively. Table 7.2 indicates that, on average, for right-censor, the light, medium, and heavy censoring are equivalent to having 10%, 33%, and 50% of the data censored, respectively; similarly, for left-censor, the same average percentage for the three censoring schemes can be obtained. The error of an estimate $\hat{R}(t_i)$ for $R(t_i)$, e_i, is defined as e_i=| $\hat{R}$

$(t_i) - R(t_i) \mid$ and n_f as the number of observed failures. We define the following criteria to compare the errors:

- Absolute Error (AE) $= \sum_{\forall i} e_i$,
- Average Absolute Error (AAE)= AE$/n_f$,
- Relative Error (RE) $= \sum_{\forall i} \frac{e_i}{R(t_i)}$,
- Average Relative Error (ARE) = RE$/n_f$, and
- Maximum Error (ME) $= \max_{\forall i}\{e_i\}$.

The AAE and ARE depict the general fitting effects. The ME which can be treated as the worst case and can also be used for testing exponentiality under the corresponding Kolmogorov-Smirnov statistic is $(\sqrt{n}+0.26+\frac{0.5}{\sqrt{n}})(ME-\frac{0.2}{n})$. From the critical values under different αs [18, p 536], we have different critical values for ME as shown in Table 7.3.

Table 7.3: Critical values in each sample size.

Sample size (n)	$\alpha = 0.10$	$\alpha = 0.05$	$\alpha = 0.025$	$\alpha = 0.01$
30	0.1774	0.1944	0.2098	0.2294
90	0.1226	0.1327	0.1419	0.1535
100	0.1165	0.1261	0.1348	0.1459
1000	0.0375	0.0406	0.0434	0.0470

Eight data sets are generated. Table 7.4 illustrates the percentage of exact failures, the censored data, and the sample size. Data set #7 is generated by

Table 7.4: Parameters in each data set.

Data set	right-censored (%)	left-censored (%)	exact failure (%)	sample size	scenario
1	100	0	0	30	light
2	0	100	0	30	medium
3	30	30	40	30	heavy
4	10	10	80	30	heavy
5	10	10	80	100	heavy
6	10	10	80	1000	heavy

assuming $\lambda = 0.1$, $\lambda_c = 0.05$, $n = 100$. We consider the left-truncation and possible withdrawal: the percentages for right-censor, left-censor, left-truncation and withdrawal are all 20. Data set #8 is derived by pooling data sets #2 $\sim$ #4 (therefore, the k in Eqs. (7.16), (7.17), and (7.18) is 3). The results are shown in Tables 7.5, where KME refers to the estimator derived by Eq. (7.23) and E, D denotes the E- (& E$'$) and the D- (& D$'$) method, respectively. The adjusted

Table 7.5: Results of the simulated data sets.

Data set	Method	AE	AAE	RE	ARE	ME
#1	KME	0.609	0.022	2.707	0.097	0.049
	KME_m	0.585	0.020	2.817	0.097	0.053
	E	0.617	0.022	2.795	0.100	0.049
	E_m	0.574	0.020	2.606	0.090	0.053
	D	0.617	0.022	2.795	0.100	0.049
	D_m	0.574	0.019	2.606	0.090	0.053
#2	KME	0.655	0.028	3.590	0.156	0.058
	KME_m	0.762	0.032	3.538	0.147	0.159
	E	0.669	0.029	3.602	0.157	0.063
	E_m	0.623	0.026	3.328	0.139	0.086
	D	0.669	0.029	3.602	0.157	0.063
	D_m	0.623	0.026	3.328	0.139	0.086
#3	KME	1.600	0.089	4.845	0.269	0.226
	KME_m	1.380	0.073	4.554	0.240	0.193
	E	1.366	0.076	4.393	0.244	0.210
	E_m	1.289	0.068	4.233	0.223	0.167
	D	1.305	0.073	4.065	0.226	0.199
	D_m	1.315	0.069	4.014	0.211	0.148
#4	KME	0.595	0.027	2.769	0.126	0.132
	KME_m	0.934	0.041	5.179	0.225	0.148
	E	1.067	0.049	3.936	0.179	0.142
	E_m	1.585	0.069	7.445	0.324	0.158
	D	1.337	0.061	4.691	0.213	0.147
	D_m	1.859	0.081	8.459	0.368	0.162
#5	KME	2.375	0.027	6.824	0.078	0.097
	KME_m	2.640	0.030	8.339	0.094	0.102
	E	2.978	0.034	8.276	0.094	0.107
	E_m	3.352	0.038	10.444	0.117	0.111
	D	3.197	0.036	8.865	0.101	0.109
	D_m	3.580	0.040	11.201	0.126	0.113
#6	KME	6.586	0.007	29.991	0.033	0.024
	KME_m	6.372	0.007	27.714	0.030	0.023
	E	5.632	0.006	21.937	0.024	0.016
	E_m	5.667	0.006	20.114	0.022	0.016
	D	5.710	0.006	20.666	0.023	0.017
	D_m	5.777	0.006	18.952	0.021	0.017
#7	KME	4.412	0.077	8.091	0.142	0.155
	KME_m	4.079	0.072	7.681	0.135	0.136
	E	3.432	0.060	7.670	0.135	0.124
	E_m	3.182	0.056	8.398	0.147	0.107
	D	2.609	0.046	8.223	0.144	0.104
	D_m	2.426	0.043	9.159	0.161	0.091
#8	KME	1.736	0.028	8.231	0.131	0.068
	KME_m	1.479	0.023	6.618	0.103	0.074
	E	1.627	0.026	6.166	0.098	0.082
	E_m	1.753	0.027	5.462	0.085	0.088
	D	3.086	0.049	8.130	0.129	0.123
	D_m	3.359	0.052	8.022	0.125	0.129

$\hat{R}(t)$ by Eq. (7.24) for each method is marked by a subscript m. Figures 7.4 $\sim$ 7.11 show the reliability plot of each data set using three estimation techniques and the "true" curve is plotted from $R(t) = e^{-\lambda t}$. The 95% CIs are included in Figures 7.7 $\sim$ 7.9, and they are calculated by Eqs. (7.11) and (7.23). For data set #7, the re-distribution methods simply neglect the existence of the withdrawal and left truncated data. The adjusted reliability plots constructed by connecting the mid-points are given in Figures 7.10 and 7.11.

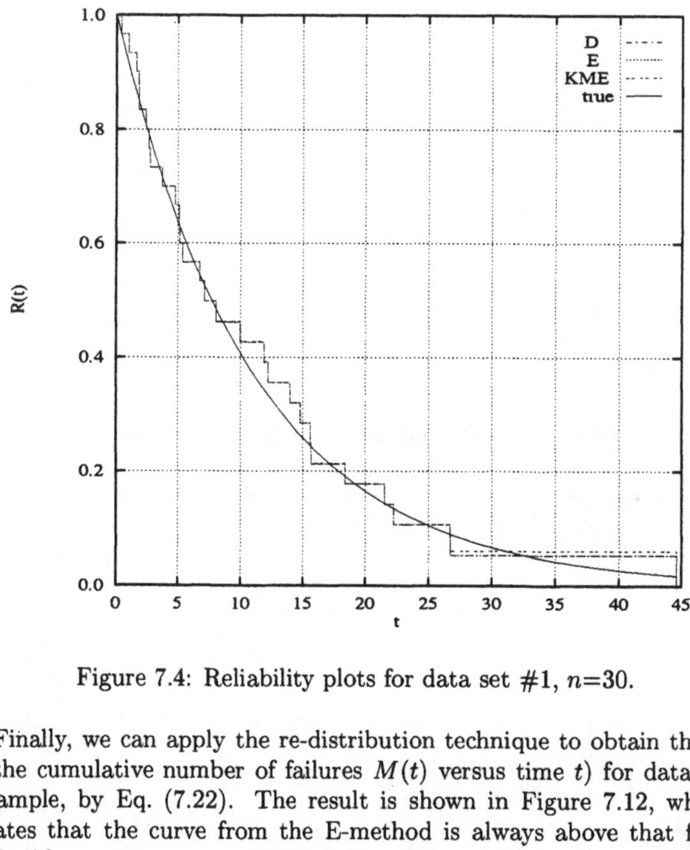

Figure 7.4: Reliability plots for data set #1, n=30.

Finally, we can apply the re-distribution technique to obtain the failure plot (the cumulative number of failures $M(t)$ versus time t) for data set #8, for example, by Eq. (7.22). The result is shown in Figure 7.12, which also illustrates that the curve from the E-method is always above that from the D-method because $x_{90:90}$ is a failure and $x_{1:90}$ is censored in data set #8, which has 63 failures (30% of the samples are censored). The 95% CIs are drawn in Figure 7.12 by Eq. (7.18), and the "true" curve is plotted by $M(t) = ne^{-\lambda t} = 90e^{-0.1t}$. From Figure 7.12, we can see a good fitting to the true cumulative failure curve by using the re-distribution method to account for the censored data.

Some modifications are suggested on the re-distribution methods to handle the incomplete data. In Figures 7.4 and 7.5, since the $x_{1:30}$ and $x_{30:30}$ in both data sets are failure times, the two re-distribution methods have the same result and the KME has more optimistic estimations at some intervals. Data set #3 is heavily censored and, due to loss of information, has the worst esti-

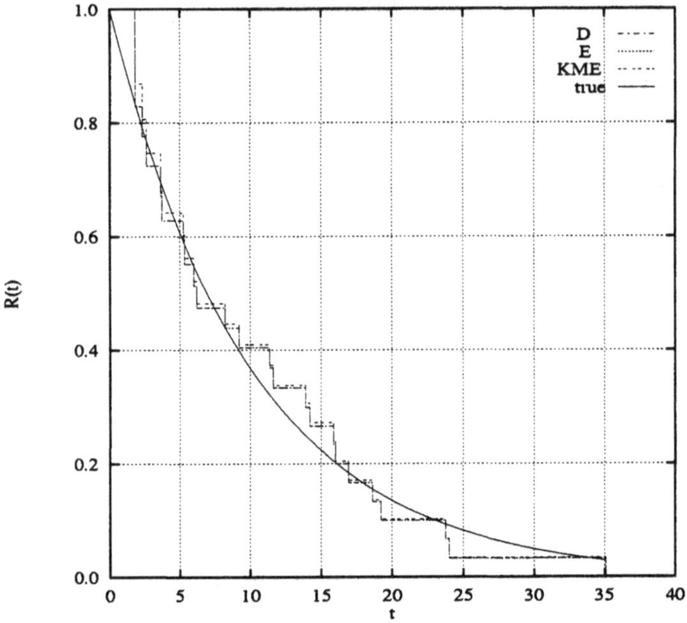

Figure 7.5: Reliability plots for data set #2, n=30.

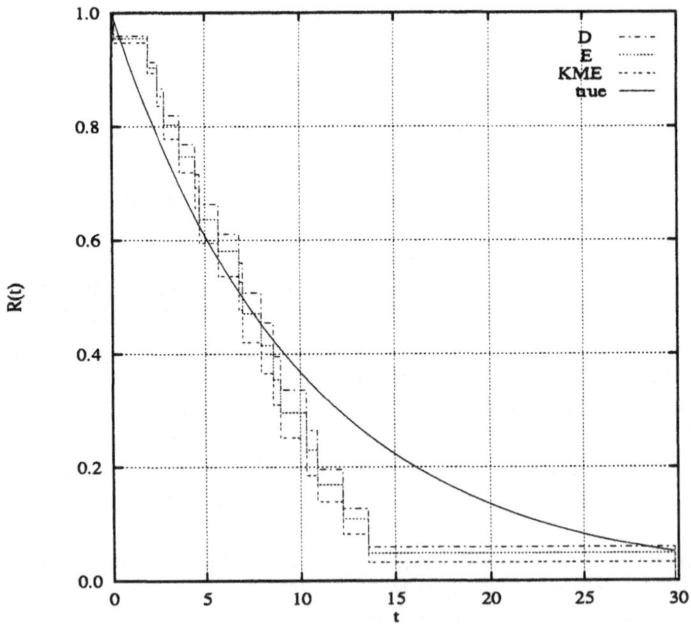

Figure 7.6: Reliability plots for data set #3, n=30.

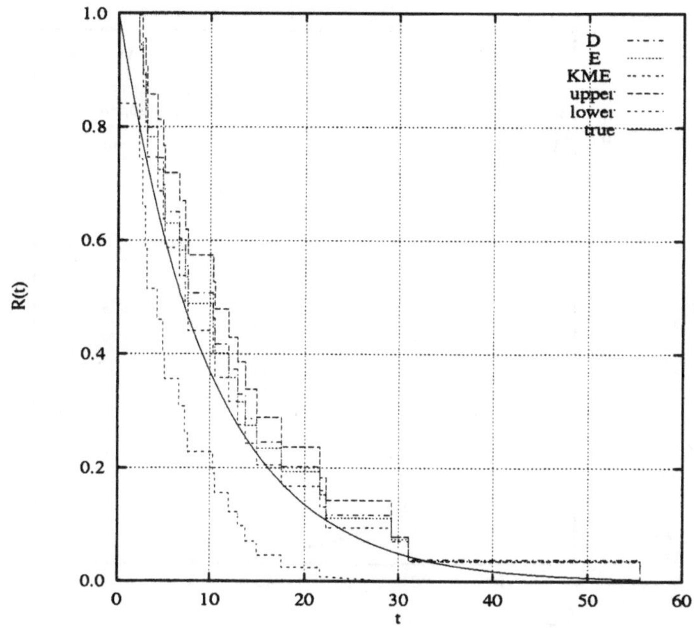

Figure 7.7: Reliability plots for data set #4 (with 95% CI), n=30.

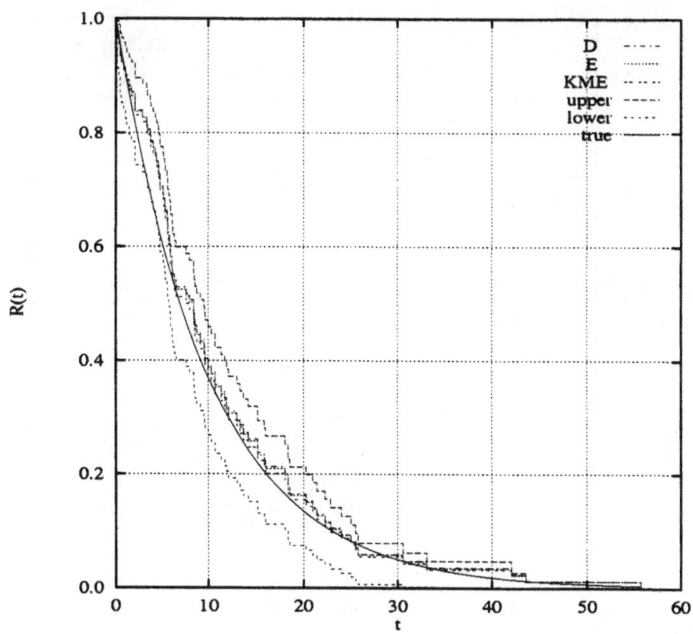

Figure 7.8: Reliability plots for data set #5 (with 95% CI), n=100.

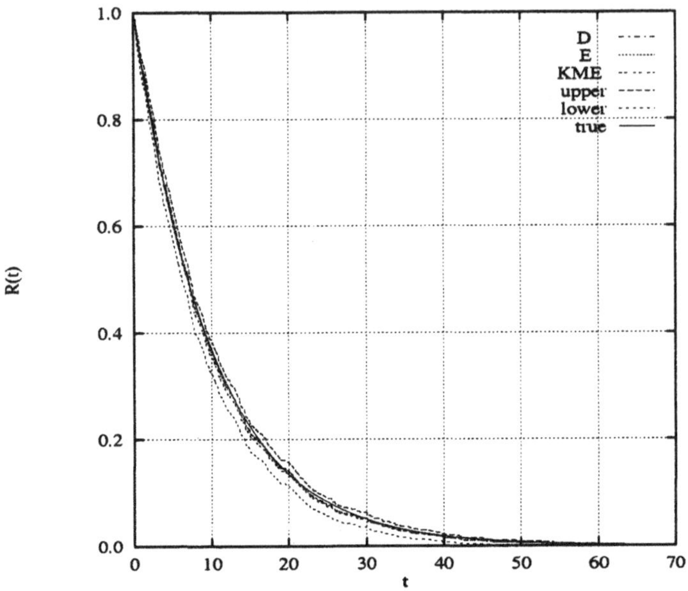

Figure 7.9: Reliability plots for data set #6 (with 95% CI), n=1000.

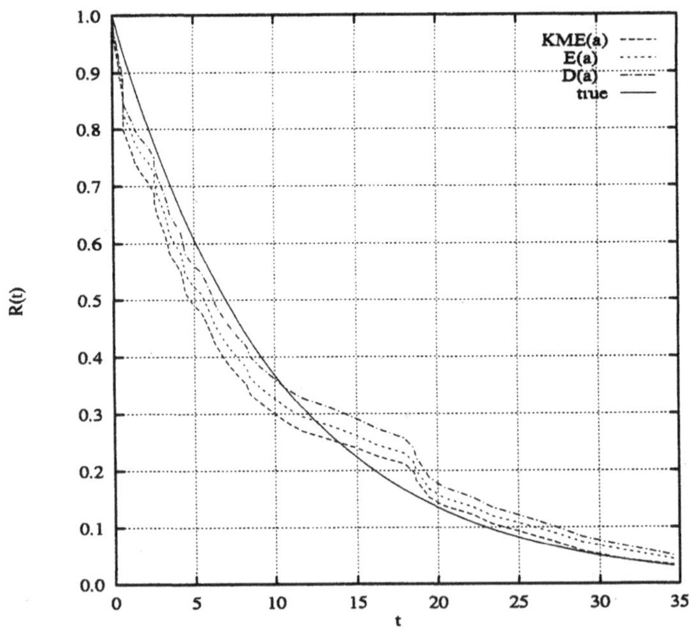

Figure 7.10: Adjusted reliability plots for data set #7, n=100.

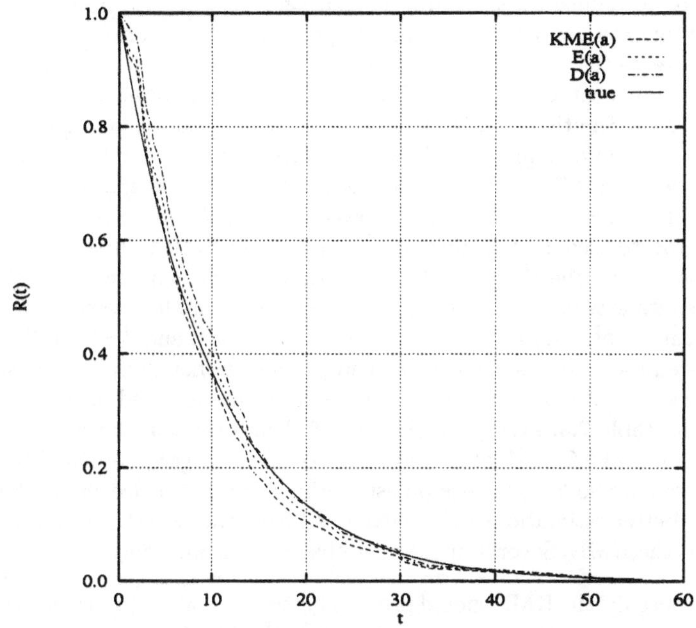

Figure 7.11: Adjusted reliability plots for data set #8, n=90.

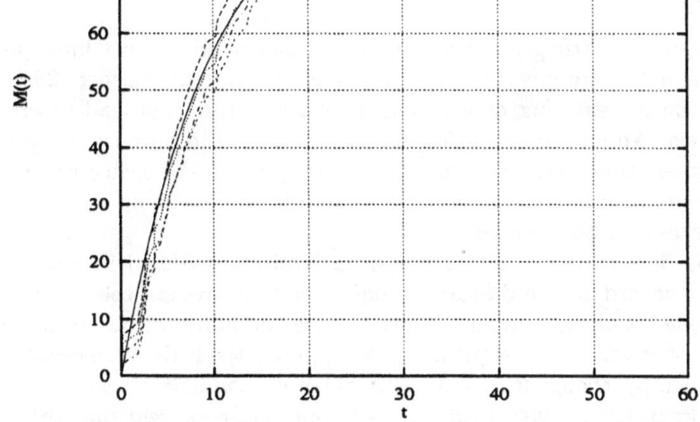

Figure 7.12: Cumulative number of failures for data set #8, n=90, k=3.

mation. If we reduce the censoring percentage, we will have better results as indicated by data set #4; the estimation gets worse as t increases. In data set #5, we increase the sample size from 30 in data set #4 to 100 and, as expected, better estimates can be obtained. The estimated curves are even closer to the true failure distribution if n is increased to 1,000 as shown in Figure 7.9. As expected, the more samples we have, the narrower the confidence bands become (from Figures 7.8, 7.8, and 7.9). Data set #7 depicts the situation when left-truncation and withdrawal are allowed. Figure 7.10 indicates that the D method (the revised KME) gives the most optimistic (the most pessimistic) estimation among the three. In fact, the same conclusion can be reached for all of the data sets used. For heavy censoring, the D method seems to produce better estimations than the E method. From the AAE and ARE in Table 7.5, the adjusted estimators show better fit in general, although in some cases, the worst cases (the ME) are worse than those of the un-adjusted ones.

From Table 7.3, except for the tests for data set #3 whose p-values are between 0.01 and 0.025, all other tests have p-values larger than .10. Therefore, the proposed methods do have good estimations. Table 7.5 also shows that as n increases, better estimations can be expected from the re-distribution methods than from the KME. Several interesting observations include:

- Supposedly, the KME should provide better estimation (in terms of smaller error measures) than the E- and the D-methods, which do not consider the changes in at-risk sample size, for the cases with extra data or withdrawal of data. However, from Table 7.5, this hypothesis can not be verified.

- For samples with both left- and right-censored data, the two re-distribution methods may result in different estimations, although they both give the traditional KME.

- As shown in Figures 7.4 $\sim$ 7.12, the D-method gives the most optimistic, and the KME the most pessimistic, estimation among the three approaches.

Where nonparametric generalized maximum likelihood product limit point estimates and CIs are given for a cure model, Laska and Meisner [235] report that too much censoring or insufficient follow-up time can lead to erroneous conclusions. An inadequate follow-up time will tend to produce overestimates of cure rates; thus, intensive care should be given to the choice of the length of follow-up. If censoring is heavy, reasonably long follow-up times and large sample sizes may be required.

The determination of the follow-up is also considered in Andersen [9], where two hazard rate models are evaluated: the additive and the multiplicative. Schoenfelder *et al.* [367] compare the life table and Markov chain techniques for follow-up studies. A time sequential plan applying the KME with censored data is developed by Zheng [461], where Brownian motion is mentioned.

Morgan [290, p 903] restates the finding of Schoenfeld that the number of deaths d^* required to detect a specific alternative hazard rate of Δ with power $1-\beta$ and level of significance α, under the PH model with equal random allocation and the hazard ratio between the two treatments A and B, is 1, $d^* = (\frac{2(Z_{1-\beta}+Z_{1-\alpha})}{ln\Delta})^2$ where d^* depends on the hazard rate and duration of patient accrual and length of follow-up. Morgan shows that the KME does not allow estimation of total study lengths at times where the Kaplan-Meier curve is undefined and thus prevents extrapolation beyond existing knowledge [290, p 909].

To sum up, we suggest using the revised KME method, which incorporates the concept of re-distribution for left-censored data, for preparing the reliability plots with incomplete observations. Greenwood's formula should be used for deriving CIs.

Once the reliability is derived, the curve can be smoothed by several smoothing techniques including the kernel estimator, the penalized maximum likelihood function, and the concept of completely monotone [281]. □

7.3.6. The Nonparametric Bayesian Approach

The advantages for deriving Bayesian decision rules in nonparametric statistical problems are that [141, p. 615]:

- the support of the prior with respect to some suitable topology on the space of probability measures should be large,
- the posterior distribution given a sample from the true probability measures should be manageable analytically, and
- Bayes rules are certainly desirable since generally they are admissible and have large sample properties.

One of the drawbacks of decision theory in general, and of the Bayesian approach in particular, is the difficulty of putting the cost of the computation into the model.

From the KME, we can re-arrange the terms in $\hat{R}(t)$ of Eq. (7.10) [284]:

$$\begin{aligned}\hat{R}(t) &= \prod_{x_{i:n}\leq t}\left(\frac{n-i}{n-i+1}\right)^{\delta_{i:n}}\\ &= \prod_{x_{i:n}\leq t}\left(\frac{n-i+1}{n-i}\right)^{-\delta_{i:n}}\frac{1}{n}\left(\frac{n}{n-1}\cdot\frac{n-1}{n-2}\cdots\frac{N_x(t)+1}{N_x(t)}\right)\frac{N_x(t)}{1} \quad (7.25)\\ &= \prod_{x_{i:n}\leq t}\left(\frac{n-i+1}{n-i}\right)^{1-\delta_{i:n}}\frac{N_x(t)}{n}\end{aligned}$$

where $N_x(t) = \#\ (x_{i:n} > t)$. Susarla and Van Ryzin [410] show that the Bayes estimator of $R(t)$ under square loss function and with a Dirichlet process prior (see Chapter 11 for Dirichlet process) with parameter α on the family of all possible distributions is

$$\hat{R}_\alpha(t) = \frac{\alpha(t,\infty)+N(t)}{\alpha(0,\infty)+n}\prod_{x_{i:n}\leq t}\left(\frac{\alpha(x_{i:n},\infty)+(n-i+1)}{\alpha(x_{i:n},\infty)+(n-i)}\right)^{1-\delta_{i:n}} \quad (7.26)$$

where the parameter α is a finite non-negative measure on $(0, \infty)$.

Consider a reliability of a random variable T with parameter Λ_0 [209, p 201]. Assume finite k disjoint partitions of $[0, \infty)$, $[t_0 = 0, t_1), [t_1, t_2), \cdots, [t_{k-1}, t_k = \infty)$; then the probability of falling in the interval $[t_{i-1}, t_i)$ is

$$q_i = \Pr\{T \in [t_{i-1}, t_i) \mid T \geq t_{i-1}, \Lambda_0\},$$

and the cumulative hazard rate function can be expressed as

$$\Lambda_0(t_i) = \sum_{j=1}^{i} -\log(1 - q_i),$$

which is also the parameter of the reliability of interest. Under the application of the Dirichlet process to describe the q_is, we have $q_i \sim \text{BETA}(\gamma_{i-1} - \gamma_i, \gamma_i)$, $i = 1, \cdots, k$.

Most of the nonparametric Bayesian analyses apply the Dirichlet distribution or process as the core model except Robinson and Dietrich [350]. Colombo *et al.* [87] use the Dirichlet process in their failure rate analysis where the $\hat{F}(t)$ is smoothed to make it continuous so that it is suitable for deriving the quantile. The Dirichlet family of distributions is useful for Bayesian analysis largely because it is the natural conjugate family to the multinomial [55, 270]. Another reason for using the Dirichlet process is that it leads to the usual KME as a limiting case.

The Dirichlet distribution and process defined in Ferguson [140] are summarized in Section 10.1. The most important factor in a Dirichlet process is α, which can be interpreted as the prior sample size [373]: small $\alpha(X)$ implies small "prior sample size" and corresponds to limited prior information. Furthermore, if $\alpha(X)$ is made to tend to zero, then the Bayes estimator mathematically converges to the classical estimator, namely, the sample mean. However, it is misleading to think of $\alpha(X)$ as the sample size and the smallness of $\alpha(X)$ as no prior information. In fact, very small values of $\alpha(X)$ mean that the prior has a lot of information concerning the unknown true distribution and is of a form that would be generally unacceptable to a statistician.

Many properties and the corresponding Bayes estimates of population parameters of the Dirichlet prior are derived in Ferguson [140]. We will apply the Dirichlet process to determine burn-in times in Chapter 10.

7.4. Goodness-of-fit (GOF) Tests

The χ^2 test can be used to test if a set of samples comes from a fully specified distribution, $H_o : X \sim F(x)$. Suppose n samples $x_1, \ldots, x_n$ are available and they are divided into k disjoint intervals, $I_1, \ldots, I_k$. Let $p_i = \Pr\{X \in I_i\}$ where $X \sim F(x)$ and let n_i be the number of samples in I_i. Thus, we have $\sum_{i=1}^{k} p_i = 1$, $\sum_{i=1}^{k} n_i = n$, and the expected number of samples in I_i is np_i. We now have a multinominal problem, and if m_i is the number of samples in I_i,

$$\chi^2 = \sum_{i=1}^{k} \frac{(m_i - np_i)^2}{np_i}. \tag{7.27}$$

The null hypothesis H_o will be rejected if the χ^2 in Eq. (7.27) is larger than $\chi^2_{1-\alpha}(k-1)$ at the α significance level. The χ^2 table is available in most statistical text books, such as [18, 192]. For large k, we can refer to Figure 6.3 or use the approximation

$$\chi^2_{\alpha}(k) \approx k(1 - \frac{2}{9k} + z_{\alpha}\sqrt{\frac{2}{9k}})^3. \tag{7.28}$$

Let V and D be the Kuiper and the Kolmogorov-Smirnov statistic, respectively. They are given by

$$\begin{aligned} V &= D^+ + D^- \\ D &= \max(D^+, D^-). \end{aligned} \tag{7.29}$$

where

$$\begin{aligned} D^+ &= \max_i(\frac{i}{n} - z_i), \\ D^- &= \max_i(z_i - \frac{i-1}{n}). \end{aligned}$$

The exact distributions for V and D can be found in much of the literature, for example, in Durbin [111], which is a great monograph on the basic theories of goodness-of-fit tests. Let $F_o(x)$ be the specified continuous CDF and $F_n(x)$, the sample CDF. Besides the statistics mentioned above, the other type of frequently used statistic is in the form

$$CR_n^2 = \int_{-\infty}^{\infty} \left(F_n(x) - F_o(x)\right)^2 dx,$$

which was first proposed by Cramér in 1928 and is also called the quadratic statistic. In 1931, von Mises generalized CR_n to

$$vM_n^2 = \int_{-\infty}^{\infty} g(x)\left(F_n(x) - F_o(x)\right)^2 dx$$

with a suitably chosen weight function, $g(x)$. Smirnov, in 1936, further extended vM_n to

$$SM_n^2 = n\int_{-\infty}^{\infty} \psi(F_o(x))(x)\left[F_n(x) - F_o(x)\right]^2 dF_o(x)$$

so as to have a distribution-free statistic [111]. By using the probability integral transformation, i.e., set $t = F_o(x)$, we have

$$\mathcal{W}_n^2 = n\int_0^1 \psi(t)\left(F_n(x) - t\right)^2 dt. \tag{7.30}$$

When $\psi(t) = 1$, we have the Cramér-von Mises statistic (W^2), and when $\psi(t) = 1/t(1-t)$, we have the Anderson-Darling statistic (A^2). The Watson statistic, U^2, is defined by

$$U_n^2 = n\int_0^1 \left(F_n(t) - t - \int_0^1 (F_n(t) - t)dt\right)^2 dt. \tag{7.31}$$

The subscript n in all the notation used above indicates the sample size, and the superscript 2 reflects the fact that we take the square of the difference between the sample and the specified CDFs.

Substituting $F_n(t)$ by i/n for $t_i \leq t < t_{i+1}$, Eq. (7.30) can be expressed as [111]

$$\mathcal{W}_n^2 = 2\sum_{i=1}^{n}(\phi_1(t_i) - \frac{i-0.5}{n}\phi_2(t_i)) + n\int_0^1 (1-t)^2\psi(t)dt \tag{7.32}$$

where

$$\phi_1(t) = \int_0^t u\psi(u)du, \quad \phi_2(t) = \int_0^t \psi(u)du.$$

From Eq. (7.32), we have

$$W_n^2 = \sum_{i=1}^{n}(t_i - \frac{i-0.5}{n})^2 + \frac{1}{12n}, \tag{7.33}$$

the usual representation of the Cramér-von Mises statistic.

One important issue in a goodness-of-fit test is to determine the critical values given sample size n and confidence level α. The methods for testing U-shaped hazard rates and exponentiality are to be mentioned in the next two sections.

7.4.1. Testing U-shaped Hazard Rates

Let n be the number of samples. For a TTT plot (see Section 6.4.3), an area-based statistic is proposed to test the exponentiality

$$W_n^2 = \frac{1}{12n} + \sum_{i=1}^{n}(TT_{i:n} - \frac{i-0.5}{n})^2 \tag{7.34}$$

where

$$TT_{i:n} = \frac{\sum_{j=1}^{i} t_{j:n} + (n-i)t_{i:n}}{\sum_{j=1}^{n} t_{j:n}} = TT_{i-1:n} + (n-i+1)(t_{i:n} - t_{i-1:n})$$
$$t_{m+1:n} = \cdots = t_{n:n} = t_T.$$

It is easy to see that W_n^2 is the Cramér-von Mises statistic in Eq. (7.33). The statistic G_n, which is originally proposed by Bergman [39], is defined as

$$G_n = V_n + n - M_n$$

where

$$\begin{aligned} V_n &= \min\{i \geq 1 : TT_{i:n} \geq \frac{i}{n}\} \\ M_n &= \max\{i \leq n-1 : TT_{i:n} \leq \frac{i}{n}\}. \end{aligned} \tag{7.35}$$

This statistic is used to test H_0: the distribution is exponentially distributed, and H_a: otherwise. The null hypothesis is rejected if G_n is large. Define the Durbin transformation as

$$\begin{aligned} d_i &= (n-i)(w_i - w_{i-1}), \quad i = 1, 2, \cdots, n-1 \\ w_i &= \frac{t_{n-i:n}}{t_{n:n}}, \end{aligned} \tag{7.36}$$

and a statistic, V, based on the Durbin transformation as

$$V = \sum_{i=1}^{n-2} u_i, \text{ or } \sum_{i=1}^{n-2} v_i \tag{7.37}$$

where

$$\begin{aligned} u_i &= (n-i)(d_{i:n-1} - d_{i-1:n-1}), \ d_0 = 0 \\ v_i &= (n-i)(w_{i:n-1} - w_{i-1:n-1}), \ w_0 = 0. \end{aligned}$$

Since the area under the TTT-plot, A, should be close to 0.5, if the process is exponentially distributed, then, $A = (V-0.5)(n-1)$; that is, $V \sim N(\frac{n-2}{2}, \frac{n-2}{12})$.

7.4.2. Testing Exponentiality

In general, reliability and its related parameters are not likely to remain the same throughout a series or sequence of experiments due to such factors as [267, p. 205]

- changing test conditions,
- component modifications, and
- varying system environment.

Besides, it is a common research topic that, during life testing, the shape parameter shifts from an exponential to a gamma [253]. Lingappaiah [253] only considers a shift in the shape parameter from 1 to 2. Therefore, the nonparametric tests for exponentiality developed in Wang and Chang [436] are helpful to find change points from CFR to DFR or IFR. Changing point problems are important in many fields, including time series analysis and reliability prediction; Zacks [456] provides good references for the changing point problems. The sequential testing method suggested in Lindley [252] is applied in El-Sayyad [123] to the exponential model and in Bury [57] to the Weibull failure model. Testing for constant versus monotone increasing failure rate based on the ranks of the normalized spacing between the ordered observations was considered in Proschan and Pyke [336]. The asymptotic normality of their statistics was proven for fixed alternatives; later, it was proven that this asymptotic normality also holds for a sequence of alternatives, F_{θ_n}, that approach the H_o distribution $1 - e^{-\lambda t}$, $t \geq 0$, as $n \to \infty$ [46].

There are three main approaches to the problem of obtaining a test statistic with a distribution independent of the unknown scale parameter θ of an exponential distribution: the ratio method, the Kolmogorov-Smirnov type method, and the rank test type method. Hollander and Proschan [193] give a thorough description of these methods. In Chapter 9, the Anderson-Darling statistic for testing exponentiality as suggested by D'Agostino and Stephens [99, p 110] is used.

7.4.3. Cumulative Hazard Plots

Figure 7.13 contains the TTT plots of some distributions. Note that the lognor-

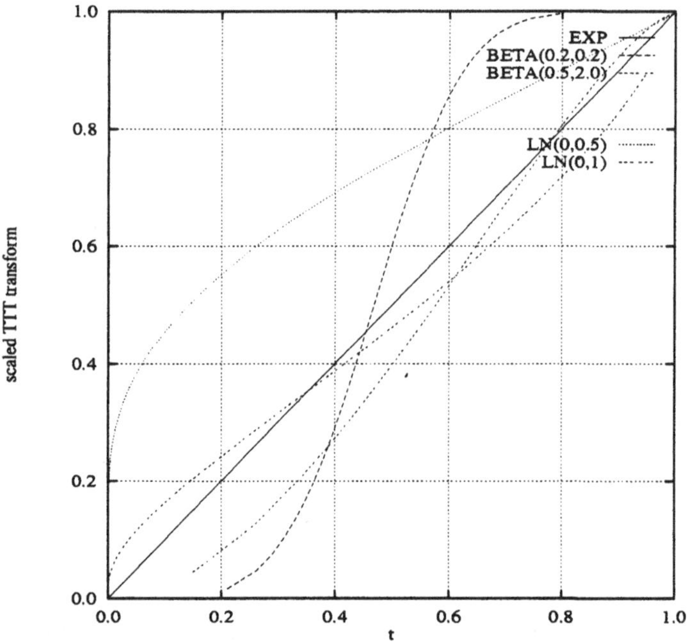

Figure 7.13: TTT plots of some distributions.

mal distribution with parameter 0 and 1 has an up-side-down U-shaped hazard rate. TTT plots of some beta distributions indicate that, very possibly, we will have trouble in detecting the "S"-shaped TTT curve; i.e., these curves do not have a clear trend of passing the diagonal line as shown in Figures 7.14 ~ 7.17.

We propose using the scaled cumulative hazard rate plots as shown in Figures 7.18 and 7.19, which are scaled by dividing by $\log(1 - \frac{n+0.5}{n+1})$.

We have concave (convex) curves for distributions which have DFR (IFR). Notice that the exponential distributions have a straight line in this plot. Take BETA(0.9, 0.1) for example; its TTT plot is almost concave and the scaled cumulative hazard (SCH) plot is convex. And for BETA(0.1, 0.9) and BETA(0.2, 2), both have convex TTT plots, and they have "S-shaped" SCH's. The "S-shaped" SCH plot is a promising property. From the SCH plots above, we know that the largest curvature for the U-shaped hazard rates happens closer to 1 than those for IFR distributions.

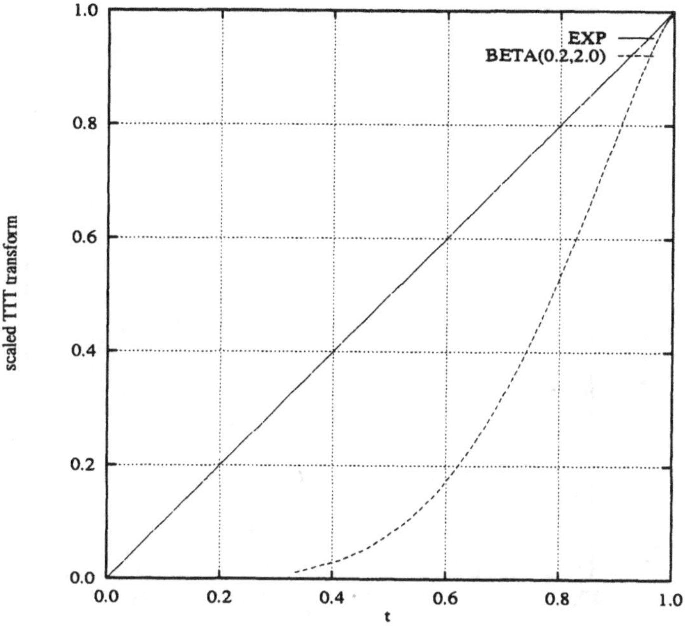

Figure 7.14: TTT plot of BETA(0.2, 2).

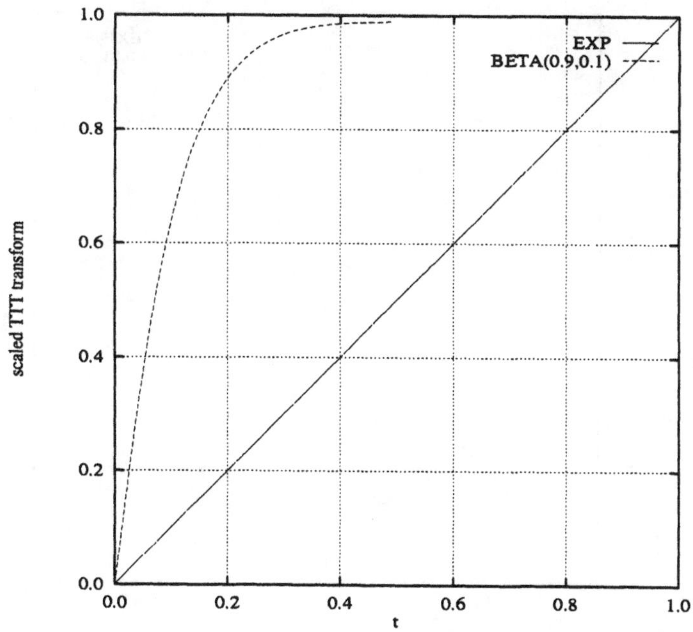

Figure 7.15: TTT plot of BETA(0.9, 0.1).

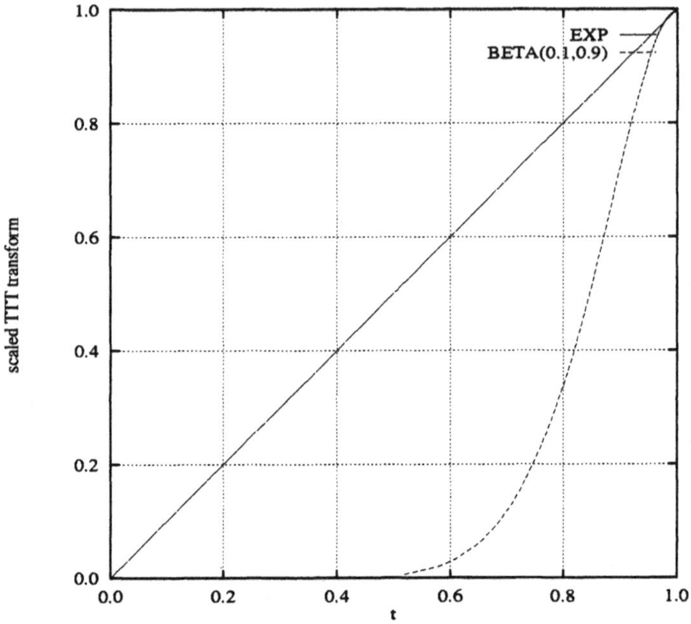

Figure 7.16: TTT plot of BETA(0.1, 0.9).

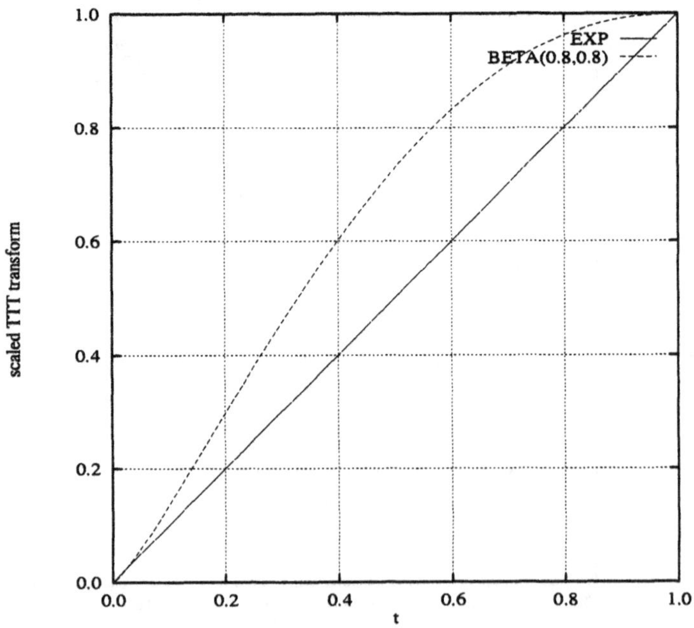

Figure 7.17: TTT plot of BETA(0.8, 0.8).

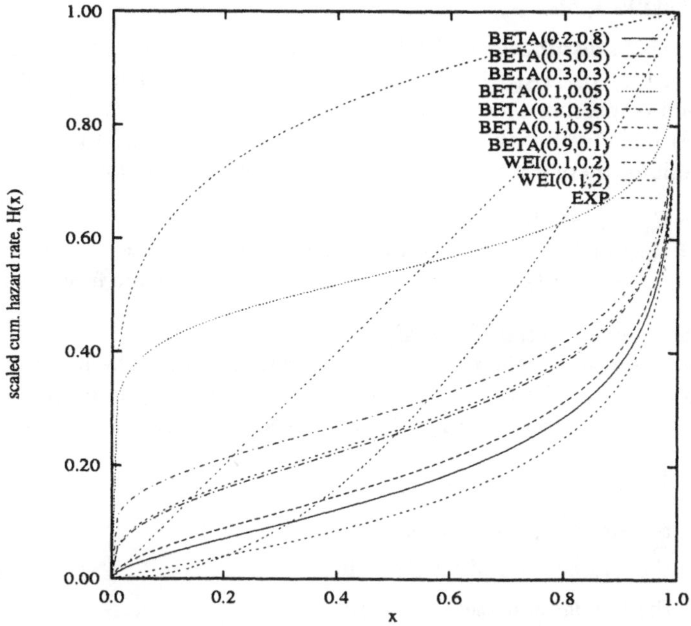

Figure 7.18: Scaled cumulative hazard plots - 1.

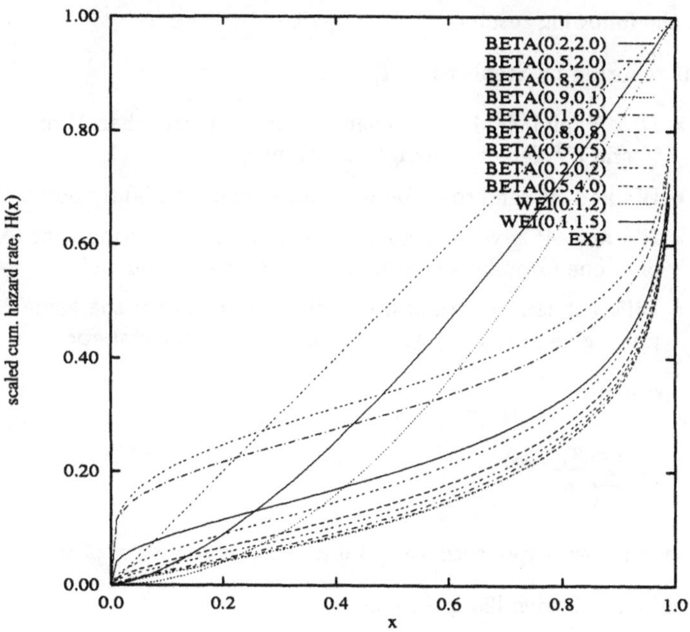

Figure 7.19: Scaled cumulative hazard plots - 2.

7.4.4. Testing GOF

Since some beta distributions with U-shaped hazard rates (e.g., BETA(0.1, 0.9), BETA(0.9, 0.1), BETA(0.2, 2.0), BETA(0.8, 0.8), and others) do not have a clear trend of passing the diagonal, very possibly, the test similar to that in Eq. (7.36) by substituting the $TT_{i:n}$ of Eq. (7.35) with $\Phi(\frac{i}{n})$ will fail in these cases. Besides, there are two unclear points about this statistic:

1. Are V_n and M_n both integers?
   The answer is YES if they are defined on the is; however, the example used by Kunitz [222] shows that these two variables are fractional.

2. What are the initial values of V_n and M_n?
   We do not have this information from Kunitz [222]. The initial numbers are important for the cases when there is no point passing the diagonal as mentioned above so that G_n remains well-defined.

Recall that

D	the Kolmogorov statistic
W^2	the Cramér-von Mises statistic
V	the Kuiper statistic
U^2	the Watson statistic
A^2	the Anderson-Darling statistic.

We have the following results:

1. It is reported by Stephens [721] that

   - D, W^2, and A^2 detect a change in mean better than V and U^2; and V and U^2 detect a change in variance,
   - W^2 and A^2 tend to be better than D and U^2 slightly better than V,
   - W^2 and A^2 give very good performance considering that for all n, only one formula is needed for each statistic, and
   - Different true distributions with approximately the same (β_1, β_2) point give nearly the same powers for any one statistic.

2. The statistic

$$T^2 = \sum_{i=1}^{n} \frac{x_i^\alpha}{n} / \overline{x}^\alpha$$

   is used to test exponentiality [243] where $\overline{x}$ is the mean of $x_1, \ldots, x_n$.

Groeneveld and Meeden [20, p 114] use

$$\beta_2(\alpha, F) = \frac{F^{-1}(.75+\alpha) + F^{-1}(.75-\alpha) - 2F^{-1}(.75)}{F^{-1}(.75+\alpha) - F^{-1}(.75-\alpha)}, \ \alpha \in (0, 0.25)$$

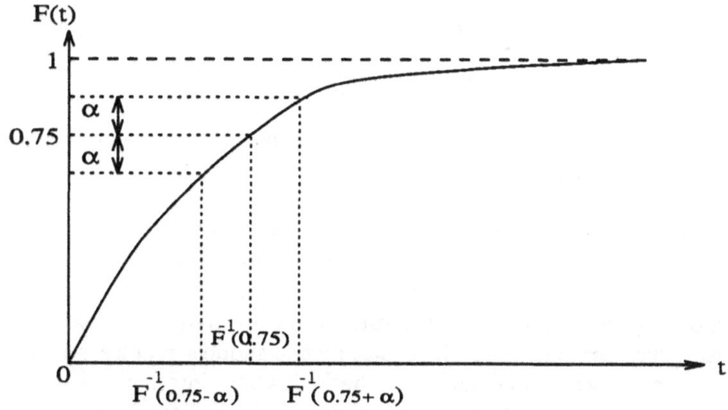

Figure 7.20: The quantile-based measure.

to measure the kurtosis of a symmetric distribution F, and they conclude that $\beta_2(\alpha, F) \in$ (-1, 1). The idea is depicted in Figure 7.20.

Other quantile-based measures are

$$t_p(F) = \frac{F^{-1}(.5+p) - F^{-1}(.5-p)}{F^{-1}(.75) - F^{-1}(.25)}$$

$$t_{\tau,\delta} = \frac{U_\tau(F) - L_\tau(F)}{U_\delta(F) - L_\delta(F)}$$

$$U_\epsilon = E[X \mid X > F^{-1}(1-\epsilon)]$$

$$L_\epsilon = E[X \mid X < F^{-1}(1-\epsilon)]$$

where $X \sim F$. However, from several CDF plots, it is seen that these quantile-based measures are not powerful enough to distinguish the U-shaped hazard rate distribution from others. For example, refer to the CDF curves of some beta distributions and of the exponential distribution. Several ordering schemes on symmetric distributions are also available in Balanda [20]; they are Van Zwet's ordering, Lawrence's ordering, Loh's ordering, and Oja's ordering.

7.5. Smoothing Techniques

Knowing how to set up the reliability enhancement programs given certain data is very important. To get this, we have to know the estimated hazard rate from the data set. There are several possible ways of doing this:

- Use the methods developed by Tanner [417, 416] and Tanner and Wong [418].

- Apply the spline smoothing method and derive $\hat{R}(t)$ & $\hat{H}(t)$.

- Obtain $\hat{f}(t)$ from the kernel estimator.

- Estimate the $\hat{R}(t)$ or $\hat{F}(t)$ by the spline regression techniques as described in [136, 137, 167].

Several asymptotic and exact CI estimations have been proposed in [136, 137, 167], and the bootstrap and the jackknife approaches are introduced by Efron [116].

Smoothing techniques for curve estimation are frequently used for nonparametric curve estimations. Since parametric models require assumptions which are often unwarranted and not checked when entering a new field in the empirical sciences, nonparametric analysis can be an attractive alternative. The development of computer and computing techniques allows one to avoid unrealistic assumptions and let the data speak for themselves [323].

Linear regression is one of the most widely used statistical smoothing techniques. The usual linear regression is a method to fit n observations (t_i, y_i), $i = 1, \cdots, n$ into the model

$$Y = a + bt + \text{error.}, \tag{7.38}$$

The summary or reduction of the observed data and prediction are the main purposes for using regression. The linear form in Eq. (7.38) can then be extended to a model of polynomial regression

$$Y = f(t) + \text{error} \tag{7.39}$$

where $f(t)$ is a polynomial function.
The classical approach is to use a low order polynomial for f. This approach is widely used in practice and is easily implemented using a multiple regression approach. Polynomial regression is a popular technique but suffers from two major drawbacks [167]:

- Individual observations can exert an influence, in unexpected ways, on remote parts of the curve.

- The model elaboration implicit in increasing the polynomial degree happens in discrete steps and can not be controlled continuously.

The roughness penalty approach is a method for relaxing the model assumptions along lines in classical linear regression and is little different from polynomial regression. Given a curve f defined on an interval [a,b], there are many different ways of measuring how rough the curve is. One attractive motivation arises from a formalization of a mechanical device that was often used for drawing smooth curves. If a thin piece of flexible wood, called a spline, is bent to the shape of a graph of f, then the leading term in the strain energy is proportional to the integrated squared second derivative.

Given a smoothing parameter $\lambda > 0$ and any twice-differentiable function f on $[a, b]$, we want to minimize the penalized sum of squares

$$R(f) = \sum_{i=1}^{n} (Y_i - f(t_i))^2 + \lambda \int_a^b (f''(t))^2 dt. \tag{7.40}$$

The penalized least squares estimator $\hat{f}$ is defined to be the minimizer of the function $R(f)$ over the class of all twice-differentiable functions f.

The addition of the roughness penalty term

$$\lambda \int_a^b (f''(t))^2 dt$$

in Eq. (7.40) ensures that for a given value λ, minimizing $R(f)$ will give the best compromise between smoothness and goodness-of-fit [167]. If λ is large, then the main component in $R(f)$ of Eq. (7.40) will be the roughness penalty term, and hence the minimizer $\hat{f}$ will display very little curvature. On the other hand, if λ is relatively small, then the main contribution to $R(f)$ will be the residual sum of squares.

7.6. Conclusions

We present nonparametric methods in this chapter. In nonparametric reliability analysis, we can avoid both the extrapolation of distribution and the estimation of parameters. Thus, nonparametric techniques help reliability analysts in testing hypotheses and estimating parameters and distribution functions.

Among the nonparametric techniques, PH, LTE, and KME are reviewed and described. The KME is the most popular nonparametric technique for analyzing the reliability.

When we want to know whether the assumed distribution is suitable for the samples under a prespecified confidence level, goodness-of-fit tests can be used. One important feature is to determine the critical values given sample size n and confidence level α. Also in this chapter, the method for testing U-shaped hazard rates and exponentiality and smoothing techniques are mentioned.

8. Parametric Approaches To Decide Optimal System Burn-in Time

The incompatibility factor introduced in Chapter 6 is elaborated in detail in this chapter. Two different models, the time-independent and the time-dependent models, are presented by applying "compatibility factors," which are not widely considered elsewhere.

Although both the lognormal and Weibull distributions are often applied to describe the early failure property of electronic components, to make use of the generic data available in the Bellcore manual [37], we assume in this chapter that early failures of all components follow a Weibull distribution with the pdf and the CDF given by

$$\begin{aligned} f(t) &= \alpha t^{-\beta} \exp[-(\tfrac{\alpha}{1-\beta})t^{1-\beta}] \\ F(t) &= 1 - \exp[-(\tfrac{\alpha}{1-\beta})t^{1-\beta}] \end{aligned} \tag{8.1}$$

respectively. These functions can be obtained using parameter transformation from Eqs. (3.14) and (3.15). The hazard rate function is

$$h(t) = \begin{cases} \alpha t^{-\beta}, & \text{when } 0 < t \leq t_L \\ \lambda_L = \alpha t_L^{-\beta}, & \text{otherwise} \end{cases} \tag{8.2}$$

where β is usually between 0.6 and 0.9 for electronic components. When $\beta = 0$, the Weibull distribution becomes the exponential distribution.

The three level systems, the stress factors, the cost factors, and the Arrhenius equation mentioned in the previous chapters are used hereafter. Without loss of generality, we also assume that the components are in series in subsystems; the series configuration dominates other structures in real world products. A black box, which is exponentially distributed with parameters $\lambda_{i,u_j+1,L}$, is used to represent the behavior of all remaining non-IC components.

A one-year-warranty is used by most electronic equipment manufacturers. Very few electronic equipment manufacturers use a warranty duration of longer than two years, 17,520 hrs. To make some allowances, observation point, t_o, is set at 20,000 hrs (2.28 years) for the example in Section 8.1. A larger t_o (30,000 hrs) is used in Section 8.2 to account for the fact that most highly reliable computer systems tend to have longer warranty periods.

As mentioned in Chapter 6, the costs considered are burn-in set-up cost (B^s, B_j^u, and B_i^c), time-dependent variable cost (c^s, c_j^u, c_i^c), shop repair cost ($c_{ij,s}$), field repair cost ($c_{ij,f}$), and loss in future sales $(1+l)e_{ij,t_o}c_{ij,f}$. If we fix t_o and consider both failure rate and costs for a specific subsystem, the relationship between them can be expressed by Figure 8.1. From time 0 to t^*,

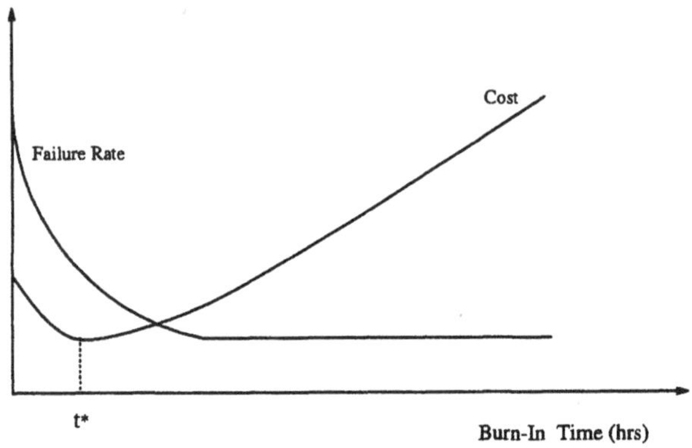

Figure 8.1: Cost and failure rate relation.

the time with minimum cost, the curve goes downward until it reaches t^*. This result is caused by the cost of future loss. Unless field repair cost $c_{ij,f}$ is much larger than shop repair cost $c_{ij,s}$, t^* will be near zero. The same argument should hold for component and system costs.

8.1. A Time-independent Model

In this section, we assume that the removal of incompatibility is independent of the burn-in times. For a three-level system, eight burn-in policies can be chosen as shown in Table 6.1. For each policy, we specify a range of incompatibility. During a certain level of burn-in (eg. subsystem level), we can remove a significant amount of incompatibility at and below this level (namely subsystem and component levels). An optimization program is formulated to maximize the highest level of a given burn-in policy (eg. maximize the system reliability for the component- subsystem burn-in policy) subject to system cost and reliability requirements.

Considerations concerning setting burn-in temperature and stress factors are investigated again in Section 8.1.1. The compatibility factors are then described in detail before the formulation of cost and reliability functions. Parameters used in the model are assigned suitable values in an example that is solved by the Hooke and Jeeves pattern search method [420].

8.1.1. Stress Curves

Ten temperature stress curves and eleven electric stress curves are provided in Figures 8.2~8.4 by Bellcore [37]. Temperature curve #7 (Figure 8.3) is used to find burn-in time according to the Arrhenius equation. If another ambient temperature, T_p, is preferred, make a shift by reading in curve #7 = (Operating Temperature) + $(40-T_p)$. Bellcore obtains these data by placing a temperature probe in the air 0.5 inch above the unit(s) while they are operating under normal conditions. Except for temperature stress curve #7 (see Figure 8.3), which is

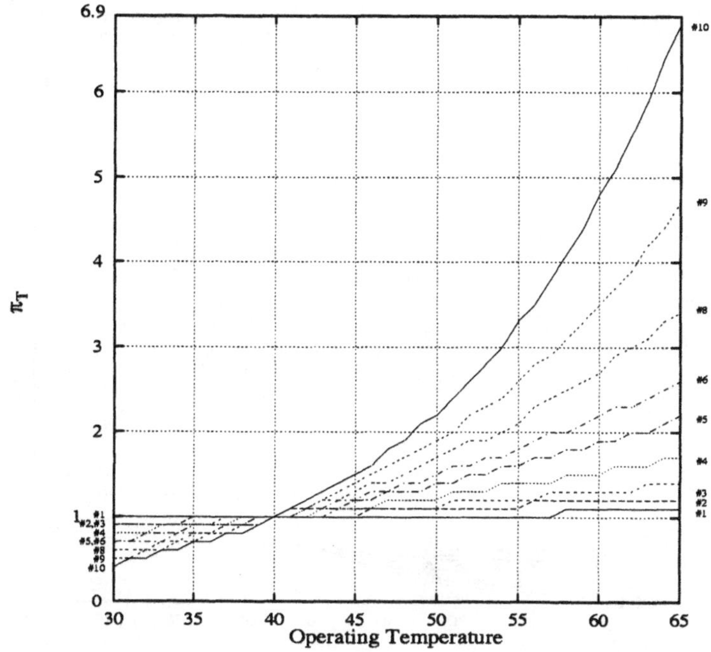

Figure 8.2: Temperature stress curves.

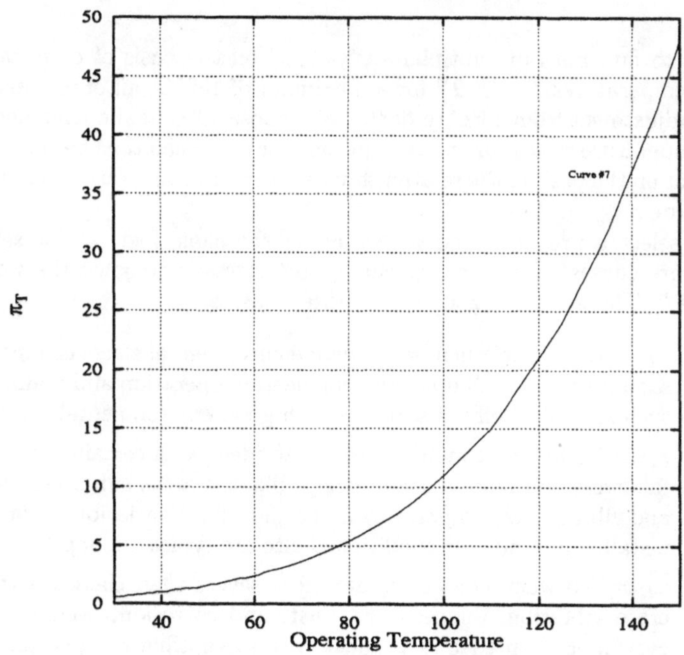

Figure 8.3: Temperature stress curves (cont.).

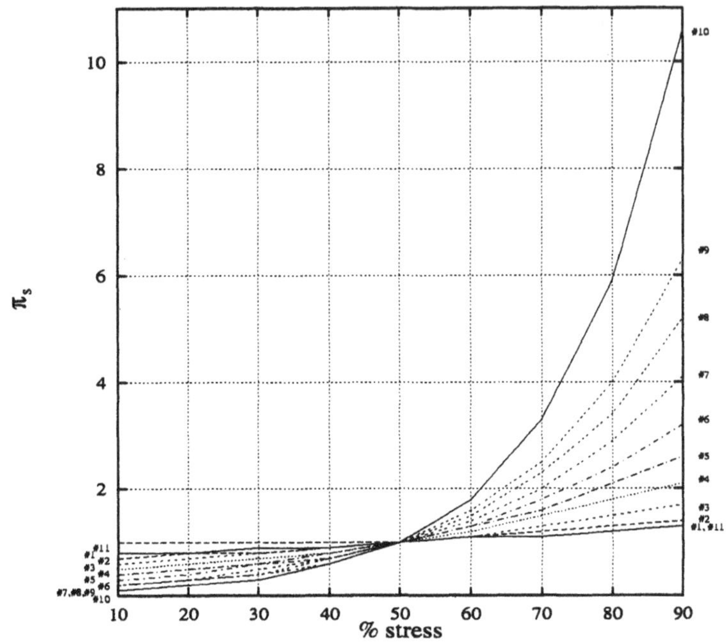

Figure 8.4: Electric stress curves.

also used to find burn-in multipliers $\varphi^s, \varphi^u_j, \varphi^c_i$ on the basis of corresponding burn-in temperatures T^s, T^u_j, T^c_i for a maximum of 150^oC, all other curves are for a π_T adjustment from 30^oC to 65^oC. We choose 40^oC as the reference point for all temperature stress curves. An appropriate curve can thus be chosen for a component on the basis of the relationship between its failure rate and operating temperature.

To select a proper π_S curve, we apply the same logic as for selecting temperature curves. For every π_S curve, 50% stress is assigned the weighted value of 1.0. The π_E is categorized into three classes:

Class 1 $\pi_E = 1.0$ is used when nearly zero environmental stress is imposed on samples along with optimum engineering operation and maintenance such as central offices and labs with good environmental control.

Class 2 $\pi_E = 1.5$ for the conditions less than ideal with certain environmental stress and limited maintenance like the areas subjected to shock and vibration or temperature and atmospheric variations. Computers installed in a non-air conditioned cite are typical examples.

Class 3 $\pi_E = 5.0$ when conditions are more severe than class 2 with moderate vibration, temperature, dust, and corrosion. Maintenance is even more limited and products are susceptible to operator abuse. Typical applications are portable CD players, portable PCs, testing equipment, and mobile telephones.

As the value of π_E increases, higher quality is required. Finally, π_Q reflects the degree of manufacturing control and tests conducted by the producers before products are shipped to customers. The military standards MIL-HDBK-217C, MIL-STD-883, and MIL-M-38510 have detailed tables for π_Q. To simplify the analysis (as compared to the references mentioned above), we classify three levels for π_Q: $\pi_Q = 1.5$ for commercial-grade devices, $\pi_Q = 1.0$ for average parts, and $\pi_Q = 0.5$ for loosest, lowest-quality-required parts. It is reported in O'Connor [304] that, since the defective percentage values are supported by many published data on part screening and reliability, the relationship between the different quality levels and the π_Q may be represented by Figure 8.5. Hence, once the defective percentage is known, π_Q can be determined accordingly [37].

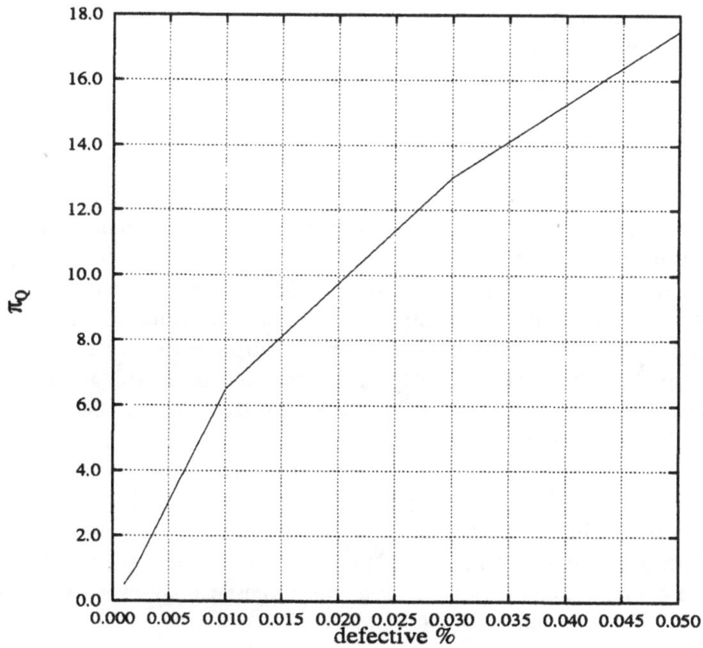

Figure 8.5: Use of the defective percentage to set π_Qs.

The components' generic failure rates are adjusted by Eq. (5.19). The adjusted (by the aforementioned stress factors) α, λ_L are used hereafter.

8.1.2. System and Subsystem Burn-in Temperature

Arsenault and Roberts [10] provide a guideline to determine component burn-in temperatures for equipment in commercial air lines. The external surface or case temperature of component, T_c, can be found by

$$T_c = \begin{cases} 0.67 \times T_M & \text{tantalum and sintered anode capacitors} \\ 0.72 \times T_M & \text{other types of capacitors} \\ \min(120^oC,\ 0.68 \times T_M) & \text{resistors} \\ 0.6 \times T_{MJ} - (\theta_{JC} \times P) & \text{IC, semiconductors, and diodes} \\ 0.75 \times T_M & \text{relays and switches} \\ T_M - 35 & \text{transformers, coils, and chokes} \end{cases} \quad (8.3)$$

where

T_{MJ} the component manufacturer's maximum permissible semi-conductor junction temperature at zero power; in ^oC

T_M the component manufacturer's maximum permissible body temperature at zero power; in ^oC

θ_{JC} the component manufacturer's thermal resistance from junction to case; in $^oC/$Watt

P the normal power dissipation for the specific circuit application; in Watts.

Jones [204] lists tables for T_M, T_{MJ} and θ_{JC}. Plugging them into Eq. (8.3) and choosing the minimum value allows one to determine the system burn-in temperature as 70°C. Since Eq. (8.3) is designed for commercial air lines, an industry requiring high-reliability, 70^oC should be suitable for most of the applications. Therefore, the system burn-in temperature can be set at 70°C based on Arsenault *et al.* [10, 204]. In a example in a later section, we set the system burn-in time to be 75°C under the assumption that components are carefully selected to be more temperature-durable; this approach is very often applied in industry to increase the burn-in efficiency. From the previous discussion, in Eq. (5.6), we use E_a=0.4, $T_1 = 70 + 273 = 343$, $T_2 = T_p + 273$ for system burn-in. Subsystems without components that are vulnerable at high temperatures can be burned-in for a longer time to reduce failure rate. To sum up, unless special components are used, we suggest the system and subsystem burn-in temperatures be no higher than 80°C.

After the burn-in temperatures are set, we assume that component i will be burned-in for the same duration, x_i^c, for any subsystem using it. Furthermore, all components in subsystem j will be burned-in for x_j^u regardless of whether they have already been burned-in at the component level or not. In each iteration of a "pattern" search (a pattern is a set of the combination of x_i^c, x_j^u, and x^s, which is abbreviated as X), we set $t'_{ij,L}$ to $t_{ij,L}$ at the beginning and subsequently reset it by the following algorithm:

for $(k = 1)$ to $(k = 3)$
$t'_{ij,L} \leftarrow t'_{ij,L} - h_k$
if $(t'_{ij,L} < 0)$ $t'_{ij,L} = 0$

where

$$h_1 = \frac{\varphi_i^c}{\varphi_{T_p}} x_i^c, \quad h_2 = \frac{\varphi_j^u}{\varphi_{T_p}} x_j^u, \quad h_3 = \frac{\varphi^s}{\varphi_{T_p}} x^s.$$

φ_{T_p} is the multiplier at ambient temperature. T_p is found from temperature curve #7. $t'_{ij,L} = 0$ means that the CFR has been reached. It might still be economical and worthwhile to burn-in longer if the trade-offs among failure rate, cost, capacity and time are considered.

8.1.3. Compatibility Factors

In practice, the actual system failure rate after burn-in is greater than what is normally predicted. ESD [230], improper manufacturing processes, weak solder joints caused by vibration (shipping and handling), component incompatibility from bad selection of components and so forth [48] should also be considered. Whitbeck and Leemis [751] use a "pseudo-component" to represent assembly defects. In a similar manner, we use the compatibility factors κ^s, κ^u, and κ^c to simulate these phenomena. Large values of κ indicate less incompatibility. No study has been performed on incompatibility. For simplicity let us assume, as suggested by IBM, that compatibility is uniformly distributed between $(\kappa^s - \varepsilon^s, \kappa^s + \varepsilon^s)$ for system level, $(\kappa^u - \varepsilon^u, \kappa^u + \varepsilon^u)$ for subsystem level, and $(\kappa^c - \varepsilon^c, \kappa^c + \varepsilon^c)$ for component level. $\kappa^s, \kappa^u, \kappa^c, \varepsilon^s, \varepsilon^u$ and ε^c are normalized to be unitless.

The overall compatibility factor, Λ, is larger in the system with system-level burn-in than the one without system-level burn-in. The reason is that when system level burn-in is implemented, a large portion of failures due to component incompatibility are weeded out during burn-in. If only component burn-in is used, some potential failures are not completely weeded out, thus a smaller compatibility factor will occur. The same argument holds between system and subsystem and between subsystem and component. If several levels of burn-in are used, even more potential failures will occur during burn-in. Another assumption is that burn-in at any level will get rid of 50% of the incompatibility but still will not exceed the maximum original compatibility level $(\kappa + \varepsilon)$; that is, the incompatibility range will be shrunk in half. The new average is also shifted upward by $\frac{\varepsilon}{2}$; that is, if only component burn-in is used,

$$\kappa^c_{new} = \kappa^c + \frac{\varepsilon^c}{2}.$$

Higher level burn-in can compress the ε. To explain this, suppose both component and system burn-in is implemented. The new compatibility factors for each level become

$$\kappa^c_{new} = \kappa^c + \frac{\varepsilon^c}{2} + \frac{\varepsilon^c}{4}, \quad \kappa^u_{new} = \kappa^u + \frac{\varepsilon^u}{2}, \quad \kappa^s_{new} = \kappa^s + \frac{\varepsilon^s}{2}.$$

Now if we burn-in at every level, the compatibility factors are

$$\kappa^c_{new} = \kappa^c + \frac{\varepsilon^c}{2} + \frac{\varepsilon^c}{4} + \frac{\varepsilon^c}{8}, \quad \kappa^u_{new} = \kappa^u + \frac{\varepsilon^u}{2} + \frac{\varepsilon^u}{4}, \quad \kappa^s_{new} = \kappa^s + \frac{\varepsilon^s}{2}.$$

Λ is assumed to be equal to the multiplication of the compatibility factors of the three levels; that is,

$$\Lambda = \kappa^s_{new}\kappa^u_{new}\kappa^c_{new}.$$

If any burn-in is done, the adjusted κ should be used.

8.1.4. Cost Consideration

To derive total burn-in costs, we have to know $t_{ij,b}$, $t_{ij,L}$, $e_{ij,b}$, and e_{ij,t_o}. The total burn-in time equivalent to burn-in at T_p for component i in subsystem j is

$$t_{ij,b} = \frac{1}{\varphi_{T_p}}(\varphi^s x^s + \varphi_j^u x_j^u + \varphi_i^c x_i^c).$$

Time to reach the CFR, $t_{ij,L}$, can be easily derived from Eq. (8.2):

$$t_{ij,L} = (\frac{\alpha_{ij}}{\lambda_{ij,L}})^{\frac{1}{\beta_{ij}}}.$$

Under the Weibull failure law,

$$e_{ij,b} = \int_0^{t_{ij,b}} \alpha_{ij} w^{-\beta_{ij}} dw = \frac{\alpha_{ij}}{1-\beta_{ij}}(t_{ij,b})^{1-\beta_{ij}} \tag{8.4}$$

which is used to evaluate shop repairing cost. Similarly, to consider loss in future sales, e_{ij,t_o} is calculated as

$$e_{ij,t_o} = \begin{cases} \frac{\alpha_{ij}}{1-\beta_{ij}}\left((t_o + t_{ij,b})^{1-\beta_{ij}} - (t_{ij,b})^{1-\beta_{ij}}\right), & \text{if } 0 \le t_o \le t_{ij,L} - t_{ij,b} \\ \frac{\alpha_{ij}}{1-\beta_{ij}}\left((t_{ij,L})^{1-\beta_{ij}} - (t_{ij,b})^{1-\beta_{ij}}\right) + \lambda_{ij,L}(t_o - t_{ij,L} + t_{ij,b}), & \text{o.w.} \end{cases} \tag{8.5}$$

The s-expected total cost therefore becomes

$$C_T(t_o) = (B^s + c^s x^s) + \sum_{j=1}^{n}\{B_j^u + c_j^u x_j^u + \sum_{i=1}^{u_j+1}[e_{ij,b}c_{ij,s} + (1+l)e_{ij,t_o}c_{ij,f}]\}$$
$$+ \sum_{i=1}^{v}(B_i^c + c_i^c x_i^c),$$

where $e_{v+1\ j,t_o}$ is the s-expected fraction of failure at time t_o for all non-IC components in subsystem j. The reliability of subsystem j at time t_o is

$$R_j^u(t_o) = \prod_i R_{ij}^c = \exp(-\sum_{i=1}^{u_j+1} e_{ij,t_o}), \quad R_{ij}^c = \exp(-H_{ij}),$$

which must be less than or equal to

$$R_{j,max}^u(t_o) = R_j^u(t_o \mid t_{ij,b} = t_{\max}, \ \forall\, i,\ j) = \exp(-\sum_{i=1}^{v+1} \lambda_{ij,L}{}'t) \tag{8.6}$$

and $t_{\max}$ is defined as

$$t_{\max} = \min(t_o, \max_{\forall i,j}\ t_{ij,L}).$$

We take a minimum in the above equation to prevent the case of any component having large $t_{ij,L}$. Eq. (8.6) indicates that the maximum attainable subsystem reliability at time t_o occurs when all the components are burned-in for $t_{\max}$ hrs.

From previous discussion, we know that there is usually an incompatibility at each level. Under the given parameters and assumptions [76], the relation between the eight burn-in options and compatibility levels discussed in Section 6.1.3 is shown in Figure 8.6 for the system configuration in Figure 3.6. Besides the system shown in Figure 3.6, two other system configurations are

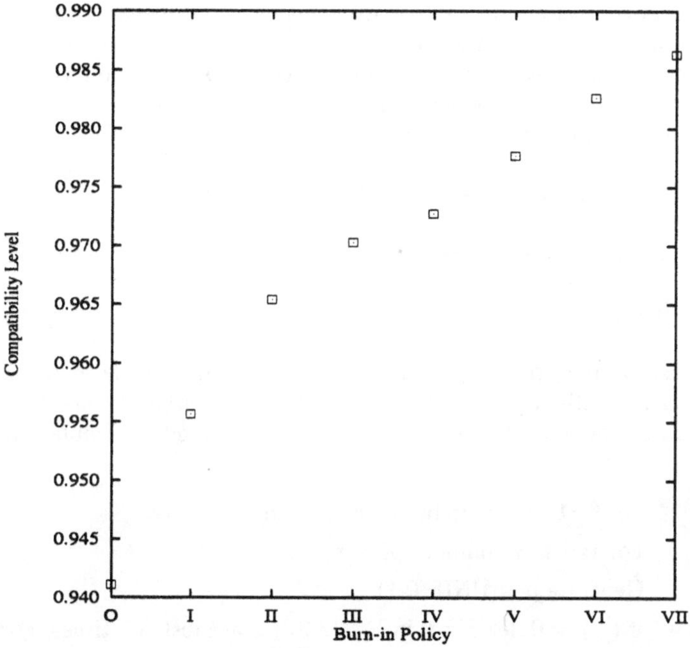

Figure 8.6: Compatibility growth at various burn-in levels.

also formulated here: parallel and composite. Then, the system reliability is

$$R^s(t_o \mid X) = \begin{cases} \Lambda \times \sum_{j=1}^{n} R_j^u(t_o) & \text{for serial structures} \\ \Lambda \times \{1 - \prod_{j=1}^{n}[1 - R_j^u(t_o)]\} & \text{for parallel systems} \\ \Lambda \times R^s_{complex} \text{ (in Eq. (3.24))} & \text{for a complex system} \\ \Lambda \times R^s_{bridge} \text{ (in Eq. (3.23))} & \text{for a bridge system.} \end{cases} \tag{8.7}$$

All α_{ij} and $\lambda'_{ij,L}$ are adjusted by Eq. (5.19).

The problem of interest becomes

$$\begin{array}{rll} \text{Minimize} & (-1.0) \times R^s(t_o \mid X) & \\ \text{s.t.} & R_j^u(t_o) - R^u_{j,min}(t_o) & \geq 0 \\ & R^u_{j,max}(t_o) - R_j^u(t_o) & \geq 0 \\ & C_{T,max} - C_T(t_o) & \geq 0 \\ 0 & \leq x^s,\ x_j^u,\ x_i^c \leq t_{\max}, & \forall\, i,j \end{array} \tag{8.8}$$

The nonlinear programming used to solve this problem, Pattern Search, is available in [149, 328, 420].

8.1.5. Evaluation

Component failure times are generated after burn-in times are found in order to evaluate how the system works. We apply inverse transformation techniques to generate random failure times. Let Eq. (8.1) be p, a random number uniformly distributed between zero and one. Note that $(1-p)$ is also uniform on $(0,1)$ and thus $-\ln(1-p)$ has the same distribution as $-\ln(p)$.
For Weibull distribution, set

$$\text{Random time } t = [-\frac{1-\beta}{\alpha}(\ln p)]^{\frac{1}{1-\beta}}. \tag{8.9}$$

For exponential distribution, set

$$\text{Random time } t = \frac{\ln p}{-\lambda}. \tag{8.10}$$

One can easily verify that Eqs. (8.9) and (8.10) are both monotonic functions. The antithetic method [51] is applied to reduce the variance since Eq. (8.9) is a monotonic function. The following algorithm is used to simulate a failure time:

Step 1: For j=1 to j=number of subsystem.

Step 2: For i=1 to i=number of component.

Step 3: Generate $p \sim$ UNIF(0,1).

Step 4: If $t'_{ij,L} = 0$, let $\beta = 0$ in Eq. (8.9) for at most N_{ij} times; go to Step 7.

Step 5: Generate $U_1 \sim$ UNIF(0,1).

Step 6: If $U_1 \leq p_2/(1-p_1)$, use Eq. (8.9) for at most N_{ij} times; otherwise, $\beta = 0$ in Eq. (8.9) for at most N_{ij} times where

$$p_1 = \int_0^{t_{ij,b}} \alpha_{ij} t^{-\beta_{ij}} \exp\left[-\left(\frac{\alpha_{ij}}{1-\beta_{ij}}\right) t^{1-\beta_{ij}}\right] dt$$

$$p_2 = \int_{t_{ij,b}}^{t_{ij,L}} \alpha_{ij} t^{-\beta_{ij}} \exp\left[-\left(\frac{\alpha_{ij}}{1-\beta_{ij}}\right) t^{1-\beta_{ij}}\right] dt.$$

Step 7: Continue.

The reason we divide by $(1-p_1)$ in Step 6 is that the cumulative distribution function for all components after burn-in at time infinity is $(1-p_1)$, which is less than 1. So we "normalize" it by dividing $(1-p_1)$. Simpson's rule is applied to derive p_1 and p_2. A failure (system failure or non-system failure) is found once a failure time is less than T_o; the algorithm is terminated at this point, and another new system simulation proceeds. Thus, the maximum number of simulations in Steps 4 and 6 is N_{ij}. We define a system failure as the point when a critical component fails.

□ Example 8.1

A computer program that is user-interactive and user-friendly is designed to simulate real situations described in the previous sections. Complete screen messages are provided by the program as guidelines and manuals. Efficiency is another important concern of this program; it takes less than one minute to get a local minimum for Eq. (8.8).

The system configuration is shown in Figure 3.6. There are six types of ICs used in five subsystems. Assume properties of IC components in different subsystem are the same, i.e.,

$$\alpha_{i1} = \alpha_{i2} = \alpha_{i3} = \alpha_{i4} = \alpha_{i5}, \qquad \beta_{i1} = \beta_{i2} = \beta_{i3} = \beta_{i4} = \beta_{i5},$$
$$\lambda_{i1} = \lambda_{i2} = \lambda_{i3} = \lambda_{i4} = \lambda_{i5}, \qquad B_1^c = B_2^c = B_3^c = B_4^c = B_5^c = B^c,$$
$$c_{i1,f} = c_{i2,f} = c_{i3,f} = c_{i4,f} = c_{i5,f}, \quad c_{i1,s} = c_{i2,s} = c_{i3,s} = c_{i4,s} = c_{i5,s},$$
$$c_1^c = c_2^c = c_3^c = c_4^c = c_5^c = c^c, \qquad c_1^u = c_2^u = c_3^u = c_4^u = c_5^u = c^u,$$

and the subsystem burn-in set-up costs are the same, i.e., $B_1^u = B_2^u = B_3^u = B_4^u = B_5^u = B^u$. The number of each type of IC in each subsystem and specifications of IC and non-IC black boxes are shown in Tables 8.1 and 8.2, respectively. All αs and λs are in FIT, a failure in 10^9 device hrs, and IC #3 is the critical component in subsystem #3, #4, and #5. The stress curve 12 in Table 8.2 uses 1.0 factor for all stress levels. Specifications for each subsystem are in Table 8.3. The burn-in temperatures and costs in Tables 8.2 and 8.3 are in ^{o}C and in \$/(unit-hr), respectively. The program will calculate the $R_{j,max}^u(t_o)$s, and we set $R_{j,min}^u(t_o) = 0.95$, $\forall\ j$. Other parameters are shown in Table 8.4. To start the pattern search, we set the initial values to be 168 hrs (1 week) for all the components' burn-in times, 60 hrs (2.5 days) for all subsystem burn-in times, and 24 hrs (1 day) for the system burn-in time. The compatibility levels for all possible burn-in combinations (policy I ~ policy VII) and output data are summarized in Table 8.5.

Table 8.1: Number of ICs in each subsystem.

Type	Subsystem				
	#1	#2	#3	#4	#5
IC#1	3	10	6	3	3
IC#2	12	5	3	13	11
IC#3	2	12	4	2	5
IC#4	5	8	10	5	5
IC#5	2	6	8	2	2
IC#6	3	4	7	3	4

Also consider the bridge system in Figure 3.5, whose reliability can be derived by Eq. (3.23), by using the same settings in Tables 8.1 ~ 8.4. The results appear in Table 8.6. The unit for the burn-in time is hours in Tables 8.5 and 8.6. System evaluation is made under policy VII. The components, the subsystems, and the system are burned-in for the corresponding durations in Table 8.5 at their burn-in temperatures. After 400,000 simulations, no fatal error (which occurs when critical components fail) and minor failure (which

Table 8.2: IC burn-in specifications.

Type	β	α	λ	$c_{ij,f}$	$c_{ij,s}$	Curve #		Burn-in	
						Temp.	Stress	Cost	Temp.
IC#1	0.86	80000	35.3	32.71	19.35	5	12	0.15	150
IC#2	0.80	75000	52.5	70.68	6.29	8	12	0.15	135
IC#3	0.90	50000	28.3	16.12	10.00	5	12	0.25	150
IC#4	0.89	60000	18.3	56.90	8.62	5	12	0.20	125
IC#5	0.82	23000	7.2	12.12	2.10	5	12	0.18	130
IC#6	0.83	17000	9.2	15.20	3.11	5	12	0.22	125
NonIC#1	0.00	4.6	4.6	8.26	1.03	7	10	-	-
NonIC#2	0.00	7.0	7.0	9.50	2.90	9	8	-	-
NonIC#3	0.00	10.5	10.5	10.25	8.00	8	7	-	-
NonIC#4	0.00	4.5	4.5	8.25	1.00	7	10	-	-
NonIC#5	0.00	4.6	4.6	8.05	0.90	7	10	-	-

Table 8.3: Subsystem specifications.

Type	Subsystem				
	#1	#2	#3	#4	#5
Burn-in cost	3.0	2.5	2.3	3.1	3.2
Burn-in temperature	85	75	90	85	85

Table 8.4: Parameters associated with ICs used in Example 8.1.

Cost	B^s=120.0, B^u=60.0, B^c=20.0, c^s=3.0, $C_{T,max}$ =800
Temperature	$T_p = 40°C$, $T^s = 75°C$
Stress Level	$\pi_E = 1.5$, $\pi_s = 1.0$, $\pi_{Q_1} = \pi_{Q_2} = \pi_{Q_3} = 1.0$
Compatibility	κ^s=0.99, ε^s=0.0099999 κ^u=0.98, ε^u=0.0199999 κ^c=0.97, ε^c=0.0299999
Others	l=3, t_o=20,000

Table 8.5: Results for the complex system in Example 8.1.

Policy #	I	II	III	IV	V	VI	VII
Compatibility (Unitless)	0.956	0.965	0.970	0.973	0.978	0.983	0.986
Com. #1(x_1^c)	468.2			60.5	183.2		150.67
Com. #2(x_2^c)	16.8			60.5	0.0		104.8
Com. #3(x_3^c)	166.5			60.5	0.0		131.9
Com. #4(x_4^c)	500.6			60.5	0.0		155.2
Com. #5(x_5^c)	440.0			60.5	0.0		378.0
Com. #6(x_6^c)	50.0			67.2	0.0		228.9
Sub. #1(x_1^u)		0.5		22.2		0.6	0.6
Sub. #2(x_2^u)		149.4		21.6		92.7	4.8
Sub. #3(x_3^u)		62.5		21.6		60.7	4.2
Sub. #4(x_4^u)		0.5		21.6		0.6	1.2
Sub. #5(x_5^u)		0.4		21.6		0.8	1.7
System(x^s)			92.7		48.0	0.4	0.7
Total Cost	374.2	755.7	402.0	798.6	551.5	798.4	798.5
Reliability (20,000 hrs)	0.956	0.965	0.970	0.973	0.978	0.983	0.986

happens when non-critical components fail) is found. From the results, system burn-in (policies III, V, VI and VII) is usually better than only component burn-in (policy I) or only subsystem burn-in (policy II) because system burn-in can get rid of a large portion of the potential defects that can not be eliminated by policies I, II and IV. Another interesting observation is that for policy V, some component burn-in times are zero for both examples (Tables 8.5 and 8.6). If reliability is the primary concern, all level burn-in (policy VII) is recommended; it can weed out most of the incompatibilities. But, the disadvantage is its high cost. If cost is really a restriction, policies III and V are good candidates for the complex system and, policies V and VI, for the bridge system. □

8.2. A Time-dependent Model

The removal of incompatibility in the previous section is independent of the burn-in time. We can also say that an expected value is used for estimating the removed incompatibility if the compatibility level is a uniform random variable with parameters κ and 1. In this section, a model is developed where the removed incompatibility is dependent on the burn-in times.

For a very high system reliability requirement, even if we perform the system, subsystem, and component burn-in, the overall system's reliability often does not achieve the requirement. In this case, redundancy is a good way to further increase system reliability if improving component quality is expensive. This section proposes a nonlinear model to (1) estimate the optimal burn-in times for all levels and (2) determine the optimal amount of redundancy for each subsystem.

For illustration purposes, we consider a bridge system whose structure is

Table 8.6: Results for the bridge system in Example 8.1.

Policy #	I	II	III	IV	V	VI	VII
Compatibility (Unitless)	0.956	0.965	0.970	0.973	0.978	0.983	0.986
Com. #1(x_1^c)	805.4			60.5	286.1		77.8
Com. #2(x_2^c)	457.1			60.5	0.0		73.0
Com. #3(x_3^c)	572.1			60.5	0.0		67.5
Com. #4(x_4^c)	580.0			60.5	0.0		84.1
Com. #5(x_5^c)	682.4			60.5	0.0		174.1
Com. #6(x_6^c)	1154.7			67.2	0.0		151.9
Sub. #1(x_1^u)		24.9		21.9		23.8	9.1
Sub. #2(x_2^u)		51.2		21.6		27.7	22.3
Sub. #3(x_3^u)		32.8		21.6		23.5	5.0
Sub. #4(x_4^u)		69.9		21.6		43.7	10.6
Sub. #5(x_5^u)		0.5		21.6		2.6	0.2
System(x^s)			225.8		24.7	45.0	3.1
Total Cost	740.0	799.4	799.1	799.8	582.8	703.2	799.3
Reliability (20,000 hrs)	0.955	0.964	0.969	0.971	0.978	0.981	0.984

in Figure 3.5. But it should be noted that this model can be applied to any other system configuration. For simplicity, all non-IC components are 100% compatible; that is, $\kappa_{v+1,j}^c = 1, \forall\ j$. And we will not burn-in these black boxes because they have a CFR [225] and are 100% compatible. As mentioned at the beginning of the chapter, components, i.e., ICs, in each subsystem are in series. Newly developed ICs with unknown parameters are compared to other similar ICs with known properties to determine suitable parameters. Components (subsystems) fail independently of each other in order to simplify analysis. Although it is possible that the deterioration of a component (subsystem) will increase the burden of other components (subsystems), that mechanism is too complicated and, thus, will not be included here.

8.2.1. Field and Shop Repair Costs

An allocated scheme can be used to set the field and shop repair costs, which is different from the one used in Section 8.1. Suppose the cost of each component is known. Cost analysts tend to disaggregate the maintenance cost, inspection cost, and repair cost to all components (or subsystems and system as well) and then calculate the repair cost per unit [159]. If the buying prices (or the procurement costs) for all components p_i^cs are known, the component field and shop repair costs can be expressed as

$$\begin{aligned} c_{ij,f} &= \xi_{i,f}^c p_i^c \\ c_{ij,s} &= \xi_{i,s}^c p_i^c \end{aligned} \tag{8.11}$$

where $\xi_{i,f}^c$ and $\xi_{i,s}^c$ are field and shop repair cost factors for component i in subsystem j. Similar definitions apply for $\xi_{j,f}^u$, $\xi_{j,s}^u$, $\xi_{j,f}^s$ and $\xi_{j,s}^s$. The subsystem cost p_j^u is

$$p_j^u = \sum_{i=1}^{v} N_{ij} p_i^c + p_j$$

where p_j is the price for all non-IC components, such as resistors, transistors, capacitors and diodes, used in the jth subsystem. By the same logic, the system price p^s is

$$p^s = \sum_{j=1}^{n} p_j^u.$$

No non-IC cost will be considered for p^s. Similar to Eq. (8.11), $c_{j,f}$, $c_{j,s}$, $c_{s,f}$, and $c_{s,s}$ can be derived by $\xi_{j,f}^u p_j^u$, $\xi_{j,s}^u p_j^u$, $\xi_f^s p^s$, and $\xi_s^s p^s$, respectively.

8.2.2. Time-dependent Compatibility Factors

Compatibility factors may be used to incorporate the reliability loss due to poor manufacturability. In Section 8.1, we introduce compatibility factors whose incompatibility range will be compressed by 50% given that any level of burn-in is performed. From experience and intuition, we know that the compatibility of a subsystem should be related to the number of components that compose it. Accordingly, the more subsystems in a system, the less compatible the system will be. It is also reasonable that if a subsystem (system) is burned-in longer, more incompatibility can be removed. Without loss of generality, we assume components in different subsystems have the same compatibility factor; that is, we use κ_i^c instead of κ_{ij}^c. On the basis of these observations, the following compatibility factors are proposed:

$$\begin{aligned} \kappa^{s\prime} &= m^s(u^s)\kappa^s + w^s(x^s)[1 - m^s(u^s)\kappa^s] \\ \kappa_j^{u\prime} &= m_j^u(u_j)\kappa_j^u + w_j^u(x_j^u)[1 - m_j^u(u_j)\kappa_j^u] \\ \kappa_{ij}^{c\prime} &= m_{ij}^c(N_{ij})\kappa_i^c + w_i^c(x_i^c)[1 - m_{ij}^c(N_{ij})\kappa_i^c] \end{aligned} \tag{8.12}$$

where $m^s(u^s)$, $m_j^u(u_j)$ and $m_{ij}^c(N_{ij})$ are strictly decreasing functions of u^s, u_j, and N_{ij}, respectively. They reflect the fact that more subsystems (components) result in lower compatibility, so we can use

$$\begin{aligned} m^s(u^s) &= 1 - \tfrac{1}{\rho^s}(u^s - n) \\ m_j^u(u_j) &= 1 - \tfrac{1}{\rho_j^u}(u_j - 1) \\ m_{ij}^c(N_{ij}) &= 1 - \tfrac{1}{\rho_i^c}(N_{ij} - 1) \end{aligned} \tag{8.13}$$

if $m_{ij}^c(N_{ij})$ remains the same in different subsystems. In addition, κ_i^c is the compatibility level for one unit of component i. If a subsystem contains more than one component i, the component compatibility level will decrease in an amount of $1/\rho_i^c$. This probably results from incompatible manufacturing processes, or, in other words, quality variations. The decrease in compatibility for an additional subsystem j is $1/\rho_j^u$. This decrease can also be treated as the result of quality variation among the jth subsystem. A similar explanation can be applied to $\kappa^{s'}$. The κs can be set by the defective weighting fraction in a lot. The defective weighting fraction can be derived by assigning a certain weight to some measurable specifications (current flow, energy consumption, and voltage drop) with respect to a reference level. The ρ^s, ρ^u_j, and ρ^c_i are positive real numbers. If the system reliability is greater than R_{req}, we should use $\rho > 1,000$; otherwise, the systems reliability is bounded due to the loss of compatibility from an additional subsystem (component). We can replace u^s by

$$u^s = \sum_{j=1}^{n} \zeta_j y^u_j$$

where ζ_j is a weighting factor that depicts the incompatibility influence of increasing one unit of subsystem j.

We assume $\zeta_j \geq 1$ since compatibility does not increase when the number of subsystems increases. The $w^s(x^s)$, $w^u_j(x^u_j)$, and $w^c_i(x^c_i)$ are strictly increasing functions of x^s, x^u_j and x^c_i, respectively, since the longer the burn-in time, the higher the compatibility will be. The following are used

$$\begin{aligned} w^s(x^s) &= 1 - \exp(-\tfrac{\varphi^s x^s}{\psi^s}) \\ w^u_j(x^u_j) &= 1 - \exp(-\tfrac{\varphi^u_j x^u_j}{\psi^u_j}) \\ w^c_i(x^c_i) &= 1 - \exp(-\tfrac{\varphi^c_i x^c_i}{\psi^c_i}) \end{aligned} \tag{8.14}$$

where ψ^s, ψ^u_j, and ψ^c_i are factors relating to the burn-in efficiency of the corresponding level, which can be decided by historical data.

Figure 8.7 indicates some possible "compatibility increasing curves" for different levels. Table 8.7 summarizes the parameters for each curve. We let

Table 8.7: Parameters of possible compatibility increasing curves.

	$m(\mu)(\kappa)$ (unitless)	φ (unitless)	Burn-in temperature (°C)	ψ (hrs)
System	0.993	4.5	75	3,000.0
Subsystem (1)	0.985	5.4	80	5,000.0
Subsystem (2)	0.985	5.4	80	7,500.0
Subsystem (3)	0.985	11.0	100	5,000.0
Component	0.992	48.0	150	100,000.0

a subsystem have the lowest initial compatibility. The efficiency of removing incompatibility is controlled by φ and ψ; the smaller the ψ, the more efficient the removal. The larger the φ, which means higher burn-in temperature, the more efficiently we can remove the incompatibility. Note that the vertical axis in Figure 8.7 represents the increase in compatibility. Figure 8.7 shows that system burn-in can remove incompatibility more efficiently than other levels of burn-in. Any function that satisfies the above requirements for functions m and w can be chosen; regression is applied to estimate the parameters. After taking

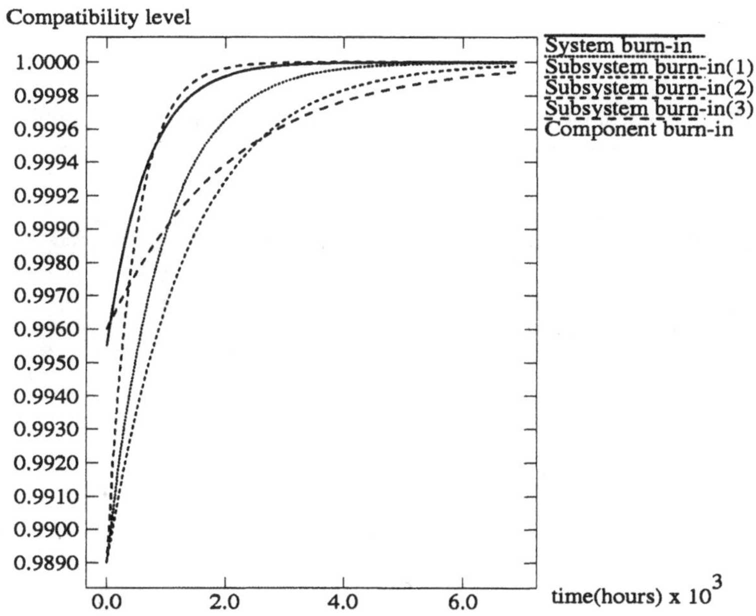

Figure 8.7: Possible compatibility increasing curves.

compatibility factors and redundancy into consideration, we have

$$R_j^{u\prime} = 1 - (1 - \kappa_j^{u\prime} \prod_{i=1}^{v} \kappa_{ij}^{c\prime} R_j^u)^{y_j^u}.$$

Since κ^s, κ_j^u, and κ_{ij}^c are unitless, the amount $(-\ln\ \kappa)$ represents the number which can not be removed due to incompatibility. Moreover, the expected number of incompatible items removed during burn-in is $(\ln\ \kappa' - \ln\ \kappa)$. Given parameters in Table 8.8, Figures 8.8 and 8.9 depict this phenomenon. The gap between curves # 1 and # 2 in Figure 8.8 after 48 hrs indicates the hazard rate increases because of the amount of incompatibility remaining. From Figure 8.8, we find that a smaller hazard rate increase will incur (Curve # 3) if there is higher compatibility. Curve # 4 indicates that if higher burn-in temperature can be used, all incompatibility will be removed if the burn-in time is long enough. In Figure 8.9, the λ is smaller than that in Figure 8.8, so almost all incompatibilities in the four curves will be removed under different rates (efficiencies) after 78 hrs of burn-in.

8.2.3. The Nonlinear Mixed-integer Optimization Model

Some components used in different subsystems will be burned-in for the same duration to save setup costs. Similarly, the same subsystem will be burned-in for the same duration. When adjustment costs are high, we burn-in the components

Table 8.8: Parameters for increases in hazard rate due to incompatibility.

Curve	κ (unitless)	φ (unitless)	Burn-in temperature (°C)	ψ	β	α (FIT)
#1	0.9999990	1.0	(no burn-in)	5,000	0.85	120,000
#2	0.9999990	3.7	70	5,000	0.85	120,000
#3	0.9999995	3.7	70	5,000	0.85	120,000
#4	0.9999995	11.0	100	5,000	0.85	120,000

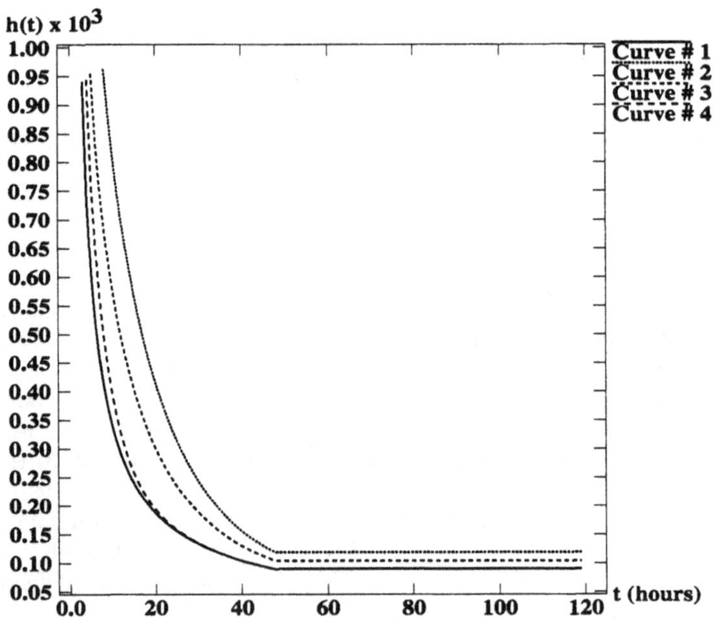

Figure 8.8: Increases in hazard rate due to incompatibility, $\lambda = 90$.

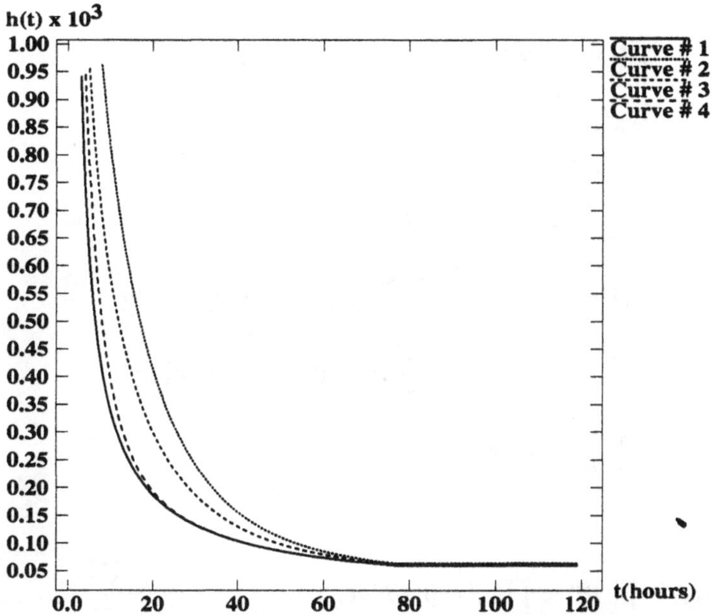

Figure 8.9: Increases in hazard rate due to incompatibility, $\lambda = 60$.

longer than the time when they reach the CFR. No component-redundancy within a subsystem will be considered because allowing component-redundancy can increase the design cost dramatically. Likewise, system redundancy is very expensive and hence not considered.

Subsystems are designed as a module to reduce maintenance cost; if a subsystem failure is detected, the damaged subsystem will be simply replaced by a new one. Unlike a subsystem, which contains many non-IC components like printed circuit boards (PCB), a system is formed by connecting these modules, and very few extra components are needed. Experience indicates that compatibility levels for subsystems are lower than those for systems, which depends only on the quality of the module connections.

Suppose the required system reliability is R_{req}; the problem becomes

$$\begin{array}{rlll} \text{Maximize} & \pi = \kappa^{s'} R^s(t_o) = R^{s'}(t_o) & & \\ \text{s.t.} & g_1 = R_{req} - \kappa^{s'} R^s(t_o) & \leq & 0 \\ & g_2 = C_T(t_o) - C_{\max} & \leq & 0 \\ & g_3 = -x^s & \leq & 0 \\ & g_{3+j} = -x_j^u & \leq & 0, \; j = 1,2,3,4,5 \\ & g_{8+i} = -x_i^c & \leq & 0, \; i = 1,2,3,4,5,6 \end{array} \tag{8.15}$$

where

$$
\begin{aligned}
C_T(t_o) &= (c^s x^s + B^s) + \sum_{j=1}^{n} [y_j^u c_j^u x_j^u + (y_j^u - 1) c_{j,r}^u + B_j^u] \\
&+ \sum_{j=1}^{n} \sum_{i=1}^{v} N_{ij} c_{ij,s} e_{ij,b} + (1+l) \sum_{j=1}^{n} \sum_{i=1}^{v+1} N_{ij} c_{ij,f} e_{ij,t_o} \\
&+ \sum_{i=1}^{v} (c_i^c x_i^c \sum_{j=1}^{n} N_{ij} + B_i^c) \\
&-(1+l)[c_{s,f} \ln \kappa^{s\prime} - \sum_{j=1}^{n} (c_{j,f} \ln \kappa_j^{u\prime} - \sum_{i=1}^{v} c_{ij,f} \ln \kappa_{ij}^{c\prime})] \\
&+ c_{s,s} (\ln \kappa^{s\prime} - \ln \kappa^s) \sum_{j=1}^{n} \{[c_{j,s} (\ln \kappa_j^{u\prime} - \ln \kappa_j^u) \\
&+ \sum_{i=1}^{v} c_{ij,s} (\ln \kappa_{ij}^{c\prime} - \ln \kappa_{ij}^c)]\}
\end{aligned}
$$

and

$$
R_j^u(t_o) = \prod_i R_{ij}^{c\prime} = \prod_i \kappa_{ij}^{c\prime} \exp(-N_{ij} e_{ij,t_o}) = \prod_i \kappa_{ij}^{c\prime} \exp(-\sum_{i=1}^{v+1} N_{ij} e_{ij,t_o}).
$$

The $e_{ij,b}$ and e_{ij,t_o} are defined in Eqs. (8.4) and (8.5), respectively.

□ Example 8.2

A bridge system (Figure 3.5), which has five subsystems ($n = 5$) and six different components ($v = 6$), is considered. The number of each type of component is tabulated in Table 8.1. To achieve a hazard rate less than 10 PPM (parts per million) at $t = 30{,}000$ hrs (≈ 3.4 years), instead of using the components in Table 8.2, the manufacturer modifies the designs and switches to other vendors for the components listed in Tables 8.9 and 8.10 because, from Table 8.2, it is obvious that the components' burn-in costs and prices in the previous section are too high. The shape parameters for all non-IC components are 0 since they have CFR. All αs and λs are in FIT. We also set

$$
\zeta_1 = 1.000233,\ \zeta_2 = 1.000278,\ \zeta_3 = 1.000270,
$$
$$
\zeta_4 = 1.000250,\ \zeta_5 = 1.000004.
$$

From Tables 8.9 and 8.10, we can obtain the time to reach the CFR region for components #1 ~ #6 to be 3,890.47, 9,395.25, 5,983.64, 6,893.66, 26,607.97, and 10,899.58 (in hr), respectively. Component and subsystem incompatibility characteristics are listed in Tables 8.11.

The set-up costs, loss of credibility, and stress factors are the same for those in Table 8.4. System level burn-in variable cost is \$3.00/(unit-time), $\xi_f^s = 1.25$, and $\xi_s^s = 0.05$; it is calculated that $p^s = 344.71$ before considering any redundancy. The system is designed to be used at the ambient temperature of 40°C. And the system burn-in temperature is set to be 75°C, which results in

Table 8.9: Component and subsystem burn-in specifications in Example 8.2.

Type	β	α	λ	p	ξ_s	ξ_f
IC #1	0.88	58,000	40.2	1.50	0.20	1.05
IC #2	0.81	19,000	11.5	1.85	0.15	1.12
IC #3	0.79	100,000	103.8	1.05	0.15	1.25
IC #4	0.81	90,000	70.0	1.35	0.16	1.28
IC #5	0.75	65,000	31.2	1.25	0.21	1.01
IC #6	0.85	30,000	11.1	1.60	0.20	1.40
NonIC #1	0.00	8.0	8.0	20.50	-	-
NonIC #2	0.00	9.4	8.0	16.45	-	-
NonIC #3	0.00	7.2	9.4	18.62	-	-
NonIC #4	0.00	10.1	7.2	20.13	-	-
NonIC #5	0.00	6.9	10.1	20.71	-	-
Subsys. #1	-	-	-	63.35	0.25	1.11
Subsys. #2	-	-	-	78.00	0.21	1.12
Subsys. #3	-	-	-	72.07	0.19	1.10
Subsys. #4	-	-	-	64.83	0.11	1.20
Subsys. #5	-	-	-	66.46	0.09	1.15

Table 8.10: Component and subsystem burn-in specifications in Example 8.2 (cont.).

Type	Curve #		Burn-in	
	Temperature	Stress	Cost	Temperature(°C)
IC #1	8	12	0.0080	125
IC #2	5	12	0.0090	135
IC #3	5	12	0.0097	130
IC #4	8	12	0.0102	135
IC #5	5	12	0.0117	125
IC #6	8	12	0.0081	140
NonIC #1	6	10	-	-
NonIC #2	7	11	-	-
NonIC #3	6	10	-	-
NonIC #4	6	10	-	-
NonIC #5	8	10	-	-
Subsys. #1	-	-	0.1600	85
Subsys. #2	-	-	0.1620	85
Subsys. #3	-	-	0.1710	90
Subsys. #4	-	-	0.1805	75
Subsys. #5	-	-	0.1018	80

the time transformation factors for components #1 ~ #6, subsystems #1 ~ #5, and the system to be 24, 32, 28, 32, 24, 37, 6.5, 6.5, 7.7, 4.5, 5.4, and 4.5, respectively. At last, $C_{\max}$ is set to be $1,500.

Without considering redundancy under these given parameters, subsystem reliabilities with no burn-in are

$$R_1^{u'} = 0.8824532,\ R_2^{u'} = 0.7822939,\ R_3^{u'} = 0.8220069,$$
$$R_4^{u'} = 0.8814274,\ R_5^{u'} = 0.8631047$$

where the system reliability is 0.9389660. If all components are burned-in to their CFRs, the reliability for each subsystem is

$$R_1^{u'} = 0.9680984,\ R_2^{u'} = 0.9275791,\ R_3^{u'} = 0.9463924,$$
$$R_4^{u'} = 0.9678018,\ R_5^{u'} = 0.9602288.$$

The corresponding system reliability is 0.9908025. It is clear that we have to allocate redundancies to achieve the required 0.999990 system reliability.

The revised XKL algorithm (see [450]) is used to solve Eq. (8.15). The initial values for the 14 Lagrange multipliers are:

$$\lambda_1 = 5,\ \lambda_2 = 10,\ \lambda_3 = 10,$$
$$\lambda_4 = 100, \ldots, \lambda_8 = 100, \qquad \lambda_9 = 500, \ldots, \lambda_{14} = 500.$$

All combinations with system reliability higher than 0.999990 and with cost less than or equal to $C_{\max}$ (= $ 1,500.00) are recorded to provide analysts more options because sometimes re-designs are necessary; if minor changes are made, it might not be necessary to run the program again, especially when there are many subsystems in a system (say, more than 15).

Results are listed in Table 8.12. The elements in vector $\vec{x}$ are in hours and the unit of $C(t_o)$ is dollars. The $R_j^{u'}$ included in Table 8.12 represents the jth subsystem's reliability under the burn-in policy $\vec{x}$. Five redundancy and burn-in combinations, which achieve the 0.999990 system reliability, are found. The optimal burn-in times for each component, subsystem and system are also in Table 8.12.

From Table 8.12, Solution #4 is the one with the highest system reliability; there are 4 units of subsystem #1, 4 of #2, 1 of #3, 2 of #4, and 1 of #5. Solution #5 can be considered since the difference in reliability from Solution #4 is only 1.8 PPM but can save more than $40 per system. The total burn-in time for the type-1 components in subsystem #1 for Solution #4 is

$$x_1^c\varphi_1^c + x_1^u\varphi_1^u + x^s\varphi^s = 3,877.81 < 3,890.47. \tag{8.16}$$

These components have not reached the CFR region. Sometimes, it is possible to "over" burn-in a component; for example, component #1 in Solution #1, from Eq. (8.16), which is burned-in for 4,728.76 hrs (> 3,890.47). However, we have to keep in mind that the longer the burn-in time, the more incompatibility will be removed, which can be verified by comparing Solution #1 and #2; Solution #2 suggests longer system burn-in time than Solution #1 and results in a higher resultant system reliability, although the reliability of each subsystem in Solution #2 is less than those in Solution #1. Thus, Eq. (8.16) cannot be used

Table 8.11: Data concerning the characteristics of incompatibility for components, subsystems, and the system in Example 8.2.

	κ (unitless)	ψ (hrs)	ρ (unitless)	Initial value (hrs)
IC #1	0.985	2,100	900	168
IC #2	0.980	2,750	850	168
IC #3	0.979	3,000	1,000	168
IC #4	0.981	2,700	4,400	168
IC #5	0.973	2,900	1,900	168
IC #6	0.989	3,100	1,200	168
Subsystem #1	0.975	110	11,000	60
Subsystem #2	0.995	90	7,500	60
Subsystem #3	0.990	90	5,000	60
Subsystem #4	0.985	100	6,000	60
Subsystem #5	0.970	100	4,000	60
System	0.999997	40	11,000	24

Table 8.12: Results for Example 8.2.

	Solutions				
	#1	#2	#3	#4	#5
y_1^u	4	4	4	4	3
y_2^u	4	4	4	4	3
y_3^u	1	1	1	1	2
y_4^u	2	2	2	2	2
y_5^u	1	1	1	1	1
x^s	22.69	25.22	23.46	43.15	37.89
x_1^u	59.62	59.65	59.58	60.46	59.14
x_2^u	59.64	59.61	59.57	60.06	59.15
x_3^u	59.61	59.58	59.55	59.82	58.97
x_4^u	59.99	60.00	59.99	59.98	59.99
x_5^u	1.92	0.51	1.74	6.23	0.36
x_1^c	176.63	171.47	175.38	137.11	170.39
x_2^c	176.65	173.12	175.41	136.90	164.76
x_3^c	178.50	171.47	176.42	121.48	165.53
x_4^c	175.78	173.22	174.76	141.74	166.26
x_5^c	177.05	170.11	174.90	123.85	167.53
x_6^c	180.46	170.73	178.99	122.46	165.13
$R_1^{u\prime}$	0.9459	0.9448	0.9456	0.9505	0.9438
$R_2^{u\prime}$	0.9066	0.9055	0.9063	0.9113	0.9046
$R_3^{u\prime}$	0.9253	0.9241	0.9249	0.9306	0.9232
$R_4^{u\prime}$	0.9453	0.9442	0.9450	0.9500	0.9432
$R_5^{u\prime}$	0.9131	0.9100	0.9125	0.9250	0.9088
$R^{s\prime}$	0.9999912	0.9999929	0.9999918	0.9999962	0.9999944
$C(t_o)$	1499.95	1500.00	1500.00	1490.74	1449.62

as the only criterion to decide the profitability of burn-in. If burn-in costs are reduced, the burn-in times will increase because it will become more beneficial to burn-in items (components, subsystems, and system) than to increase the subsystem number.

For Solution #4, since the burn-in times for subsystems #1 ~ #4 are almost the same, we can burn-in these four subsystems for the same duration, say, 60.1 hrs, which is the average of x_1^u, x_2^u, x_3^u, and x_4^u; that is, we can burn-in these four subsystems for about two and half days. The same action can be applied for components if adjustment cost is high. The five solutions indicate that subsystem #5 is the one with the least contribution to system reliability; in other words, it has the lowest reliability importance. □

□ Example 8.3

In the IC industries, one of the most important purposes for an HTOL test is to derive the operating life (Op-Life) curve. If the number of failures is large (e.g., more than 30), it is not convenient to find the critical value under the given degree of freedom and confidence level. The chi-square plot in Figure 6.3 can serve this purpose. For most cases in obtaining the Op-Life curve and in estimating E_a and β of the acceleration models, large sample sizes are important as shown in Example 6.5. This example shows how field engineers apply the parametric method to analyze the test data. The Weibull distribution, however, is not the only choice to describe the infant mortality behavior. The lognormal distribution is also used in industry.

Ten thousand 5V-memory-ICs are sampled from 20 lots for an HTOL test so that the samples can well represent the overall quality of the present production line. The HTOL test condition is described in Example 5.3. Reading points are 48-hr, 96-hr 168-hr, 500-hr, and 1,000-hr. After testing samples at the test points on a memory tester, cumulative failure percentages at the points are 0.19%, 0.27%, 0.35%, 0.60%, and 0.88%. These five data points are plotted with the Y-axis ($y=\ln\{\ln(1/[1-F(t)])\}$) and the X-axis ($x=\ln(t)$), and a regression line is derived based on these five points. Thus, if we use $F(t) = 1 - \exp[-(t/\alpha)^\beta]$, the slope of the regression line is the estimated β and the α can be obtained by $\alpha = \exp(-y_o/\beta)$ where y_o is the Y-intersection when $x = 0$.

The Weibull plot is shown in Figure 8.10, where the regression line is $y = 0.5x - 8.2$ with $R^2 = 0.999$. Therefore, $\beta = 0.5$ and $\alpha = \exp(8.2/0.5) = 1.326 \times 10^7$ (hrs). The mean is 3,026 years, which is the expected life without considering the acceleration effect (from temperature and bias). The Op-Life curve is shown in Figure 8.11. After enough samples are collected and if the testing capacity (both the stressing chambers and the E-test equipment) permits, the E_a and β of this IC can be estimated (see Example 6.5). If the overall acceleration factor is 200, the corresponding failure rate (in FIT) for each "step" is shown in the parenthesis. The LTFR can be derived by $0.88\%/1,000/200 \times 10^9 = 44$ FIT (the 0.88% is the accumulative failure percentage in the 1,000-hr HTOL test). From Eq. (6.7) and Figure 6.3, the 60% and the 90% UCL is 46.5 and 54.0 FIT, respectively; both of them can meet

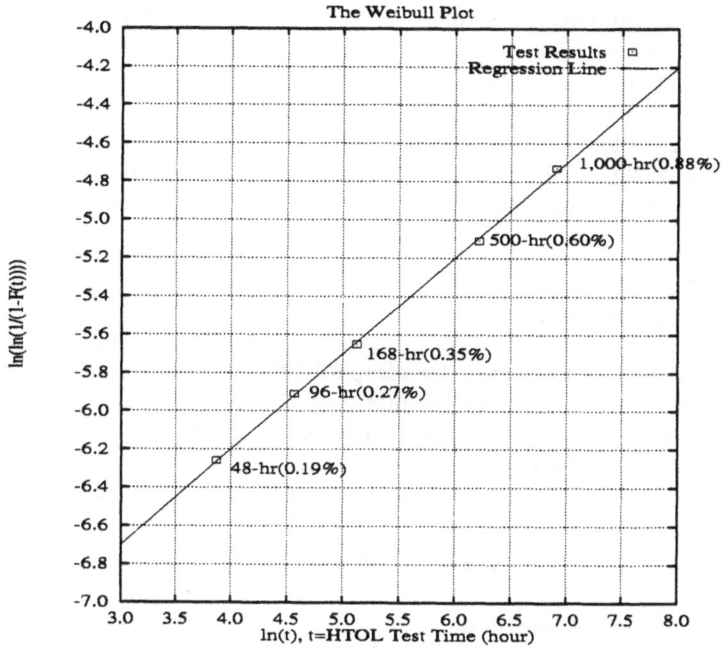

Figure 8.10: Weibull plots with the test results and regression line.

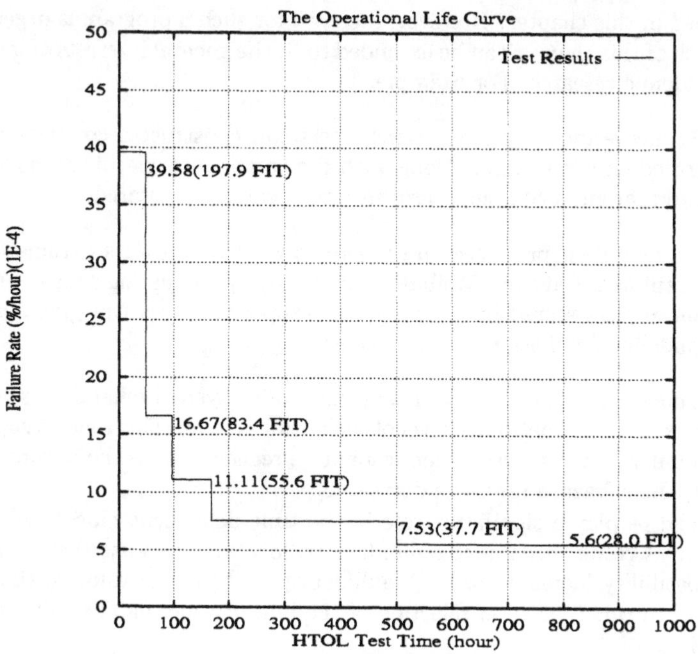

Figure 8.11: The Op-Life curve of the test data.

the LTFR requirements by most customers: less than 100 FIT under either 60% (α=0.4) or 90% (α=0.1) UCL. The $\chi^2_{1-\alpha}(df)$, where *df* denotes degree of freedom, critical values are shown in Figure 6.3.

The techniques in Section 7.5 can be used to "smooth" the five "steps" in Figure 8.11. Besides, if we believe the DFR should be valid for the HTOL failure behavior of this IC and the calculated hazard rates of 2 adjacent "steps", for example, are not decreasing, the PAV algorithm in Section 9.2.2 can be applied to adjust the hazard rates. □

8.3. Conclusions

The time-independent model for compatibility is only a surmise. More information should be collected to modify or re-model it. Also because of the lack of practical reports and data and suggestions made by field engineers, compatibility is assumed to be uniformly distributed $\sim \Upsilon(\kappa^s, \kappa^u, \kappa^c)$. If we can find a better model, then it becomes a stochastic programming problem:

$$\begin{aligned} \text{Minimize} \quad & (-1.0) \times \Phi \times \sum_{j=1}^{n} R_j^u(t_o) \\ \text{s.t.} \quad & (\text{Constraints}) \ \geq \ 0 \\ & \Phi \sim (\text{mean of a distribution}). \end{aligned}$$

There is no appropriate burn-in analysis program that considers all the factors described in this chapter. However, the need for such a program is urgent. The approach of this chapter can be extended to fit the general interests of industries and academic research. For example:

- By answering simple questions, users can construct a complicated integrated circuit system. Along with the component specifications and cost information, a complete burn-in report can be generated.

- If a new IC is produced and designers want to know its parameters (if it is still assumed to be Weibull distributed) by comparing similar products' parameters stored in the program; proposed values can give analysts a guideline for these new components.

Under the proposed time-dependent model, system reliability is very sensitive to κ^s, the compatibility level of the initial system. This sends a signal that the reliability of the target system cannot be reached unless the incompatibility or component hazard rates are decreased.

The ρs play a significant role in the time-dependent model; selection of different ρs result in distinct redundancy allocations because they control the incompatibility increase rates. One difficulty is the initial value settings for $\vec{x}$, the burn-in times. A good selection of the initial values can expedite program execution.

Since compatibility factors are included in the model, many derivatives must be calculated, making the execution time longer than that of the same system in Xu *et al.* [450]. It should take less time if the system configuration is simpler than the one shown in Figure 3.5.

When the burn-in cost increases, we tend to have a shorter execution time because most of the burn-in time combinations violate the cost restriction (g_2), so we do not have to solve the minimization problem of Eq. (8.8).

9. Nonparametric Approach and Its Applications to Burn-in

Two kinds of models, the time-independent and time-dependent models, were introduced in Chapter 8 to decide all-level burn-in times. The major disadvantage of the parametric approaches, like the ones in the Chapter 8, is the difficulty of assigning suitable distributions for the failure mechanisms (for the components, the subsystems, and the system) and estimating the parameters in these distributions. A fairly simple nonparametric method is described in this chapter. The Anderson-Darling statistic and the TTT transformation are used to find the two change points: t_{L_1} (from DFR to CFR) and t_{L_2} (from CFR to IFR) if the system is believed to have a bathtub hazard rate. Then, the pooled-adjacent-violator (PAV) algorithm [25] is used to unimodalize the hazard rate. The system burn-in time can be decided by calculating the mean residual life (MRL). An example is given to illustrate the procedures.

In this chapter, a "component" can be viewed as a module which is composed of many electronic items. Thus, it will be proper to assume the module has the U-shaped hazard rate. A series system is analyzed for illustration purposes. For demonstration purposes, simulation is applied to generate failure times for a U-shaped hazard rate.

Some preliminaries are summarized in Section 9.1. Section 9.2 depicts the methods. Section 9.3 provides the possible applications. An example and its results are given in Section 9.3.2 along with extensive discussions given in Section 9.4. The flow chart in Figure 9.1 demonstrates the analysis procedure of the proposed approach.

9.1. Introduction

A U-shaped hazard rate function is assumed for electronic components. This has been suggested by many statisticians as well as field engineers. Characteristics of the U-shaped hazard rate function are given in Chapter 6. A Weibull distribution is usually assumed to describe the infant mortality function for electronic components [37]; however, little research has been done on the electronic wearout process. To simplify the analysis, Weibull is used to portray the wearout behavior. From Eqs. (3.14), (3.15), and (3.16), the pdf, CDF, and

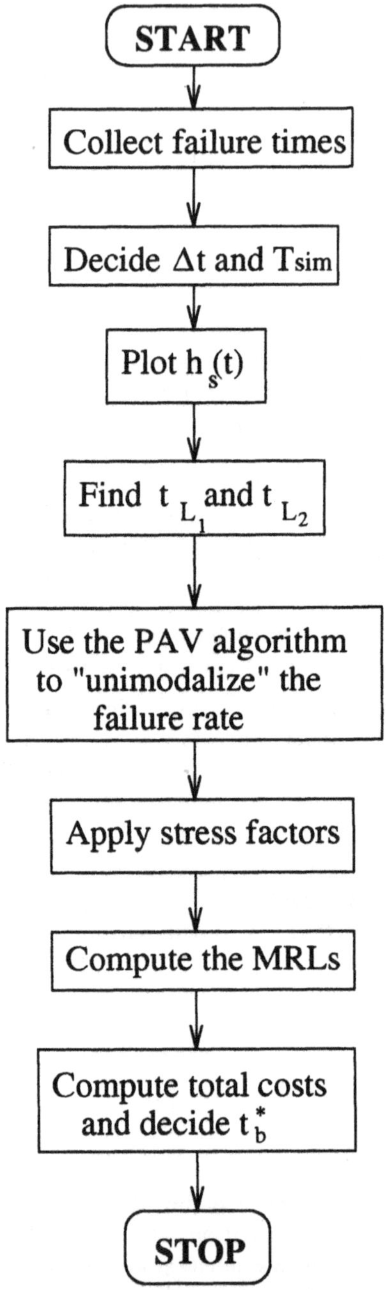

Figure 9.1: Analysis flow chart.

hazard rate of the Weibull distribution used in this chapter are

$$\begin{aligned} f(t) &= \lambda\beta(\lambda\, t)^{\beta-1}e^{-(\lambda\, t)^{\beta}}, \\ F(t) &= 1 - e^{-(\lambda\, t)^{\beta}}, \\ h(t) &= \lambda\beta(\lambda\, t)^{\beta-1}. \end{aligned}$$

Figures 9.2, 9.3, and 9.4 show the $f(t)$s, $F(t)$s, and $h(t)$s of a 5-component series system, respectively. The parameters of the five components are given in Table 9.1, where T^D and T^C are the lengths of the DFR and CFR regions, respectively, and both λ_i^D and λ_i^I are in the units of FIT, a failure in 10^9 device hrs, and T_i^D and T_i^C are in 1,000 hrs (Data from [225]).

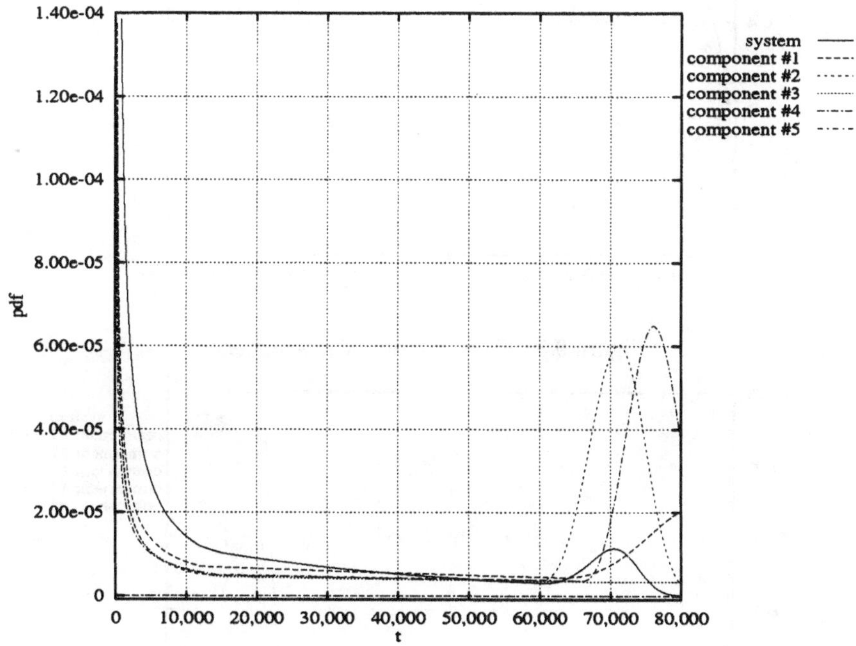

Figure 9.2: The pdf of a series system.

Table 9.1: Parameters of the components in a series system.

Component (i)	λ_i^D	β_i^D	T_i^D	T_i^C	λ_i^I	β_i^I
1	3,000	0.33	12	51	41,000	3.3
2	700	0.25	12	48	80,000	3.5
3	1,000	0.30	15	42	7,200	3.6
4	1,500	0.40	15	51	87,000	3.3
5	20	1.00	75	0	20	1.0

It is reported that the series system structure dominates most of an electronic apparatus. Suppose that a series system has n components, and compo-

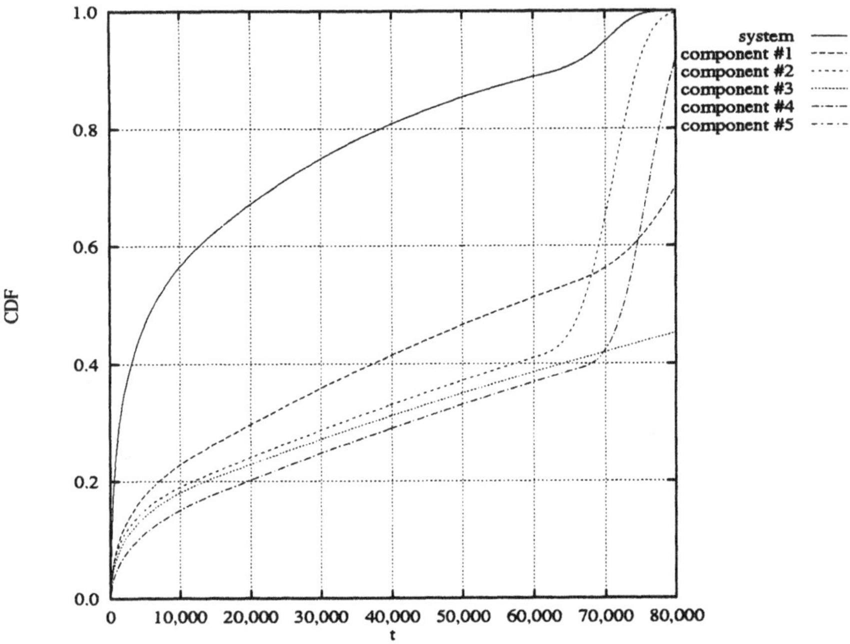

Figure 9.3: The CDF of a series system.

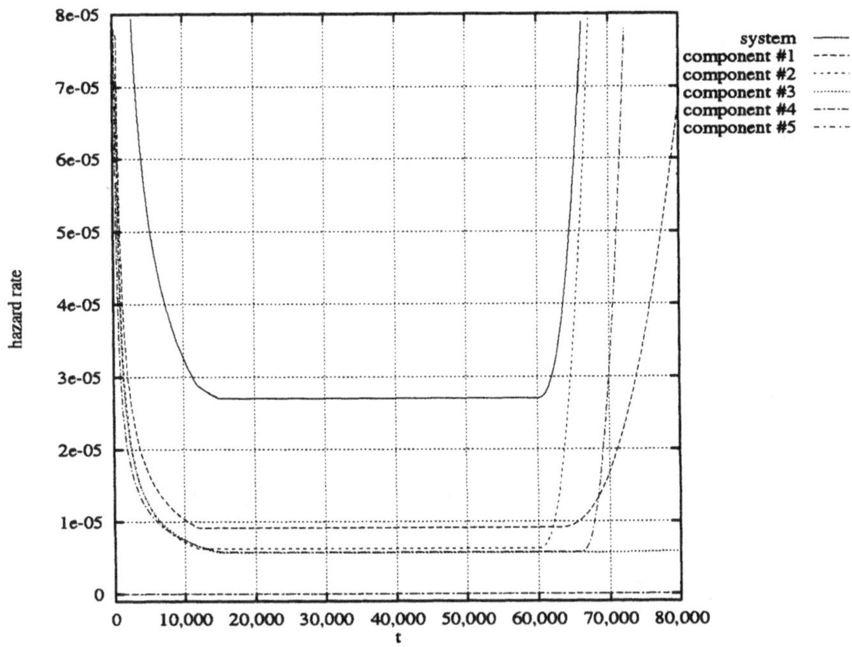

Figure 9.4: The hazard rates of a series system.

nent functions independently, then

$$R_s(t) = \exp\Big(-\sum_{i=1}^{n} h_i(t)\Big). \tag{9.1}$$

If a system contains components that follow the Weibull distribution with common shape parameter β, then the series system hazard rate becomes

$$\beta\, t^{\beta-1}(\sum_{i=1}^{n} \lambda_i)^{\beta}.$$

The life distribution of a series system will not be Weibull if the β_is are not the same.

9.2. Methods

A U-shaped system hazard rate is generated by Monte-Carlo simulation to illustrate the proposed method (see Appendix D). The Anderson-Darling statistic and the TTT plot can be used to test exponentiality in order to find t_{L_1} and t_{L_2}. Finally, the total costs and the MRLs under different burn-in times are calculated.

9.2.1. Testing Exponentiality

D'Agostino and Stephens [99] compared the following empirical distribution function (EDF) statistics for different distributions: the Cramér-Von Mises statistic (W^2), the Kolmogorov-Smirnov statistic (D), the Anderson-Darling statistic (A^2), the Watson statistic (U^2), and the Kuiper statistic (V) to test exponentiality. When we tested these statistics, the Anderson-Darling statistic was shown as the most powerful one in our model. This agrees with the findings by D'Agostino and Stephens [99]. The superiority of A^2 has also been documented by various power studies based on Monte Carlo sampling [99]. The Anderson-Darling statistic is thus used to check exponentiality. In order to test $H_o : X \sim F(x; \alpha)$ against X from other populations, where X is the life of the series system and α is an unspecified parameter, the Anderson-Darling statistic for exponential distribution (A_e^2) is

$$\begin{aligned} A_e^2 &= (1+\tfrac{0.6}{n})\widehat{A}^2 \\ \widehat{A}^2 &= -n - \tfrac{1}{n}\sum_{i=1}^{n}(2i-1)\{\ln F(x_{(i)};\hat{\nu}) + \ln[1-F(x_{(n+1-i)};\hat{\nu})]\} \\ \hat{\nu} &= \sum_{i=1}^{n}\frac{x_i}{n}. \end{aligned} \tag{9.2}$$

For the Weibull distribution, the statistic (A_w^2) is

$$\begin{aligned} A_w^2 &= (1+\tfrac{0.2}{\sqrt{n}})\widehat{A}^2 \\ \widehat{A}^2 &= -n - \tfrac{1}{n}\sum_{i=1}^{n}(2i-1)\{\ln F(x_{(i)};\hat{\lambda},\hat{\beta}) + \ln\,[1-F(x_{(n+1-i)};\hat{\lambda},\hat{\beta})]\} \end{aligned}$$

where $\hat{\lambda}$ and $\hat{\beta}$ satisfy

$$\begin{aligned} g(\hat{\beta}) &= \sum_j y_j^{\hat{\beta}} \ln\, y_j / \sum_j y_j^{\hat{\beta}} - 1/\hat{\beta} - \sum_j \ln\, y_j/n = 0 \\ \hat{\lambda} &= \left(n/\sum_j y_j^{\hat{\beta}}\right)^{1/\hat{\beta}}. \end{aligned} \tag{9.3}$$

This can be solved by the Newton-Raphson method through iteration by using

$$\hat{\beta}_m = \hat{\beta}_{m-1} - \frac{g(\hat{\beta}_{m-1})}{g'(\hat{\beta}_{m-1})}$$

for the mth iteration.

Although it is reported in D'Agostino and Stephens [99] that A^2 is more powerful than other statistics in general, we use other three statistics here for comparisons.

1. The Kolmogorov-Smirnov statistic (D):

$$\begin{aligned} D_e &= (\sqrt{n} + 0.26 + \tfrac{0.5}{\sqrt{n}})(\hat{D} - \tfrac{0.2}{n}) \\ D_w &= \sqrt{n}\hat{D} \\ \hat{D} &= \max(D^+, D^-) \end{aligned}$$

where

$$\begin{aligned} D^+ &= \max_i(\frac{i}{n} - F(x_{i:n})) \\ D^- &= \max_i(F(x_{i:n}) - \frac{i-1}{n}). \end{aligned}$$

2. The Kuiper statistic (V):

$$\begin{aligned} V_e &= (\sqrt{n} + 0.24 + \tfrac{0.35}{\sqrt{n}})(\hat{V} - \tfrac{0.2}{n}) \\ V_w &= \sqrt{n}\hat{V} \\ \hat{V} &= D^+ + D^-. \end{aligned}$$

3. The Cramér-Von Mises statistic (W^2):

$$\begin{aligned} W_e^2 &= (1 + \tfrac{0.16}{n})\widehat{W}^2 \\ W_w^2 &= (1 + \tfrac{0.2}{\sqrt{n}})\widehat{W}^2 \\ \widehat{W}^2 &= \tfrac{1}{12n} + \sum_{i=1}^{n}\left(F(x_{i:n} : \hat{\lambda}, \hat{\beta}) - \frac{i-0.5}{n}\right)^2. \end{aligned}$$

The parameters are estimated by Eqs. (9.2) and (9.3). Table 9.2 shows the critical values for the significance levels $\alpha = 0.10, 0.05, 0.025$, and 0.01. That is, we reject H_o if the test statistic is greater than the corresponding critical value under a pre-specified α.

Table 9.2: Critical values for the tests under different significance levels.

	$\alpha = 0.10$	$\alpha = 0.05$	$\alpha = 0.025$	$\alpha = 0.01$
A_e^2	1.062	1.321	1.591	1.959
A_w^2	0.637	0.757	0.877	1.038
D_e	0.995	1.094	1.184	1.298
D_w	0.803	0.874	0.939	1.007
V_e	1.527	1.655	1.774	1.910
V_w	1.372	1.477	1.557	1.671
W_e^2	0.177	0.224	0.273	0.337
W_w^2	0.102	0.124	0.146	0.175

Sometimes, the available samples in a segment may not be enough to give us a good testing result. In such cases, the TTT plot can be used to find the CFR region. As shown in Figure 6.15, there is an almost straight line between the convex (for small t) and the convex (for large t) sections. We can apply regression analysis to test the linearity of an empirical TTT plot. If a large coefficient of determination (R^2) can be obtained, the CFR region can then be determined. The example in the next section makes this idea more transparent.

9.2.2. The PAV Algorithm

The PAV algorithm is commonly used in isotonic regression. It can also be applied to smooth curves and to study the maximum likelihood estimation of completely ordered parameters [25, 241]. Suppose there are k segments. For $\theta_i > 0$, consider **P1**:

$$\begin{aligned}\mathbf{P1}: \quad \text{Maximize} \quad & \prod_{i=1}^{k} h_i \\ \text{s.t.} \quad & \sum_{i=1}^{k} \theta_i h_i = k \\ & h_1 \leq h_2 \leq \ldots \leq h_{k-1} \leq h_k.\end{aligned}$$

If $(h_1^o, h_2^o, \ldots, h_{k-1}^o, h_k^o)$ satisfies **P1**, then they become the $(h_1^*, h_2^*, \ldots, h_{k-1}^*, h_k^*)$. If not, perform the PAV algorithm:

Step 1 Without loss of generality, assume $h_2^o > h_3^o$; then construct $h_2^1 = h_3^1 = 2/(\theta_2 + \theta_3)$; $h_i^1 = h_i^0$, $i \neq 2, 3$.

Step 2 If h_i^1, $i \in [1, k]$, satisfy **P1**, then these become the h_i^*.

Step 3 If not, perform another PAV step; thus, say that $h_2^1 = h_3^1 > h_4^1$; then construct $h_2^2 = h_3^2 = h_4^2 = 3/(\theta_1 + \theta_2 + \theta_3)$; $h_i^2 = h_i^0$, $i \neq 2$, 3, 4.

Step 4 If h_i^2, $i \in [1, k]$, satisfy **P1**, then they become the h_i^*.

Step 5 Otherwise, go to Step 1.

To start the algorithm, from Eq. (D.2) and **P1**, θ_i is set to $N_i \Delta t / f_i$. This algorithm can be applied for the more general cases.

□ **Example 9.1**

$$\begin{aligned}\textbf{P2}: \quad & \text{Maximize} \quad \prod_{i=1}^{9} h_i \\ & \text{s.t.} \quad 3h_1 + 7h_2 + 5h_3 + 6h_4 + 8h_5 + 9h_6 + 6h_7 + 4h_8 + 7h_9 = 9 \\ & \qquad h_1 \geq h_2 \geq h_3 \geq h_4 = h_5 = h_6 \geq 0 \\ & \qquad h_6 \leq h_7 \leq h_8 \leq h_9\end{aligned}$$

The solution for **P2** is summarized in Table 9.3. □

Table 9.3: The PAV algorithm: solution for Example 9.1.

Iteration		Period (i)								
		1	2	3	4	5	6	7	8	9
0	θ_i	3	7	5	6	8	9	6	4	7
	h_i^0	$\frac{1}{3}$	$\frac{1}{7}$	$\frac{1}{5}$	$\frac{1}{6}$	$\frac{1}{8}$	$\frac{1}{9}$	$\frac{1}{6}$	$\frac{1}{4}$	$\frac{1}{7}$
1	θ_i	3	(12)	(12)	6	8	9	6	4	7
	h_i^1	$\frac{1}{3}$	$\frac{2}{12}$	$\frac{2}{12}$	$\frac{1}{6}$	$\frac{1}{8}$	$\frac{1}{9}$	$\frac{1}{6}$	$\frac{1}{4}$	$\frac{1}{7}$
2	θ_i	3	(12)	(12)	(23)	(23)	(23)	6	4	7
	h_i^2	$\frac{1}{3}$	$\frac{1}{6}$	$\frac{1}{6}$	$\frac{3}{23}$	$\frac{3}{23}$	$\frac{3}{23}$	$\frac{1}{6}$	$\frac{1}{4}$	$\frac{1}{7}$
3	θ_i	3	(12)	(12)	(23)	(23)	(23)	6	(11)	(11)
	h_i^3	$\frac{1}{3}$	$\frac{1}{6}$	$\frac{1}{6}$	$\frac{3}{23}$	$\frac{3}{23}$	$\frac{3}{23}$	$\frac{1}{6}$	$\frac{2}{11}$	$\frac{2}{11}$

9.3. Applications

After generating a U-shaped hazard rate of the series system, we perform an exponentiality test on the failure times of each segment so that t_{L_1} and t_{L_2} can be found. The PAV algorithm is applied to unimodalize the hazard rate because the simulated hazard rates usually are not well U-shaped. The modified hazard rate between t_{L_1} and t_{L_2} is $\hat{h^C}$.

9.3.1. Total Cost and Warranty Plan

Through field justifications, Kuo [225] formulated the s-expected cost, $C_B(t;\ t_{b,T_b})$, at time t given Eq. (6.13) where t_{b,T_b} is system burn-in time if burned-in at T_b^oC. The s-expected number of failures (e.g., e_b and e_t) can be obtained from (time)$\times(h_k^{'})$.

The warranty duration is normally set to be one year by most electronic manufacturers under the assumption that the product will not deteriorate before it is shipped out to the customers. If the product wears out after t_b within a year, say, $t_{L_2} - t_b < 8000$, it is better to choose a shorter warranty duration. If there is any trade-in policy for the product, a used system will have less salvage value if the time that expired since it is sold is greater than $t_{L_2} - t_b$. The expected cost incurred in the warranty duration hence becomes

$$C_B'(t;\ t_{b,T_b}) = C_B(t;\ t_{b,T_b}) + \eta e_w c_w \tag{9.4}$$

where

c_w repair cost if a failure is reported by a customer

e_w s-expected number of failures occurred during warranty period

η percentage of a failure to be claimed during the warranty period.

9.3.2. Optimal Burn-in Time

The MRL of a system surviving time t_b, the corresponding burn-in time if burn-in occurs at system ambient temperature, is defined as

$$\mu(t_b) \equiv \text{E}[X - t_b \mid X > t_b] = \frac{\int_{t_b}^{\infty} \exp[-H(x)]dx}{\exp[-H(t_b)]}. \tag{9.5}$$

A larger MRL is preferred. However, more has to be invested to achieve a larger MRL. The optimal burn-in time can be determined by subtracting the MRL weighted by a time-proportional gain, g, from the total cost; that is, the optimal burn-in time can be found from

$$\min_{t_b} \Big(C_B(t;\ t_{b,T_b}) - g\mu(t_b)\Big). \tag{9.6}$$

□ Example 9.2

Suppose that the lives of five components ($n = 5$) in a series system follow Weibull distributions as shown in Table 9.1. The U-shaped hazard rate curves of each component and the series system are plotted in Figure 9.4. Note that component #5, which is used to take care of all non-IC components [225], is the one with CFR throughout its lifetime. The failure times are listed in Table 9.4, where times start from the beginning of the first segment (that is, time 0) and $\Delta t = 3,000$ hrs and $T_{\text{SIM}} = 75,000$ hrs; that is, $k = 25$.

The N_k, f_k, $h_s(t)$, and $h_s'(t)$ are summarized in Table 9.5, where t_S and t_E represent the starting and the end time (in 1,000 hrs) of the corresponding segment, respectively. In Table 9.5, the '0' and '1' in columns 7 ~ 10 (columns 12 ~ 15) are used for not rejecting and rejecting the hypothesis that the exponential (Weibull) distribution holds for the segment, respectively; they are tested under a 95% confidence level. We mark '2' if no failure is detected. Figure 9.5 depicts the system hazard rates from simulation and after PAV adjustment.

Chapter 6 investigates the TTT transformation and plots to detect CFR, DFR, and IFR. If we use the data in Table 9.4, the corresponding TTT plot is in Figure 9.6. The diagonal line depicts the CFR case. It is clear that a straight line can be drawn in a portion of the plot. From simple linear regression, the fitted line is found to be

$$y = 2.470109x - 0.83678 \tag{9.7}$$

with R^2=0.98707. Under this high R^2, we can find that the CFR begins at segment 6 and ends at segment 21 (which is included in the CFR region), which is more accurate than the Anderson-Darling's testing results, according to the parameters shown in Table 9.1.

Table 9.4: Failure times of each segment used in Example 9.2.

Segment	Failure times
1	0.4 0.8 0.9 2.6 2.9 3.7 5.6 7.1 11.1 12.6 17.4 23.2 26.0 32.3 64.9 88.9 106.1 109.8 175.9 186.2 199.8 208.3 285.3 287.5 293.1 312.5 466.7 644.8 738.4 786.2 881.0 1014.1 1053.6 1250.4 1293.0 1416.3 1616.4 1696.1 1783.0 1820.8 1978.4 1989.4 1993.3 2038.6 2119.9 2347.0 2723.9 2907.6
2	4053.1 4149.8 4207.4 4332.8 4372.0 4376.6 4840.1 4856.9 5069.5 5848.7 5976.1
3	6284.3 7583.6 7814.1 8110.3 8333.9 8399.2 8416.4 8435.8 8672.6 8949.6
4	9137.2 9165.6 9170.0 9202.6 9393.0 9426.9 10097.0 10189.5 10977.6 11395.3
5	12256.3 12376.4 13211.0 13353.5 13855.9 13962.5 14205.7 14479.7 14778.9
6	17970.7
7	
8	21407.3
9	24034.7
10	28505.4 29031.3
11	30650.4 31344.9 31410.6
12	33105.3
13	36015.8
14	39120.4 39215.9 39651.2 40174.5 40454.2
15	42449.2
16	45270.5 46450.4 47922.5
17	50637.3
18	53101.2
19	55711.0 56017.06
20	57007.9 57166.36 58179.78
21	61736.0
22	63817.9 63857.7 64669.1 64762.7 65561.2 65607.7 65756.0 65918.3 65929.4
23	66053.2 66218.4 66479.3 66532.5 66556.1 66612.1 66612.2 66876.5 67136.5 67221.6 67233.1 67497.8 67880.0 67993.1 68011.1 68094.0 68126.1 68401.2 68422.8 68433.0 68555.4 68704.7 68727.6 68947.4
24	69008.4 69047.9 69096.7 69125.5 69158.6 69182.3 69231.9 69285.1 69304.8 69558.2 69600.6 69645.4 69764.3 69825.6 69828.5 69901.4 69967.8 70040.9 70176.2 70237.0 70280.1 70480.1 70537.5 70627.0 70652.3 70726.4 70748.9 70769.3 70786.2 70838.6 70909.1 71004.0 71048.9 71151.5 71158.2 71165.7 71396.4 71429.5 71485.6 71625.0 71786.5 71799.3 71846.0 71877.5 71949.0
25	72009.9 72025.1 72027.6 72056.6 72115.2 72116.7 72130.1 72218.5 72224.1 72353.3 72416.1 72550.8 72575.4 72610.0 72610.1 72650.1 72657.5 72664.7 72678.1 72719.3 72752.0 72804.5 72834.4 72841.1 72869.8 72878.0 72975.3 72979.1 72995.0 73012.9 73064.8 73069.3 73125.0 73126.5 73228.0 73270.3 73271.8 73327.8 73361.2 73383.8 73405.5 73435.7 73553.0 73580.7 73587.5 73617.7 73680.6 73726.6 73740.4 73840.7 73844.5 73877.0 73888.3 73971.2 74021.0 74050.3 74072.9 74080.9 74118.4 74166.6 74190.5 74196.5 74220.5 74243.5 74272.5 74341.1 74367.0 74372.4 74519.5 74574.4 74593.9 74640.3 74711.1 74746.3 74768.3 74947.5

Table 9.5: Simulation results for Example 9.2.

Segment (k)	t_S (10^3h)	t_E (10^3h)	N_k	f_k	$h_s(t)$ (10^{-6})	A_e^2	D_e	V_e	W_e^2	$h'_s(t)$ (10^{-6})	A_w^2	D_w	V_w	W_w^2
1	0	3	100	48	160	1	1	1	1	160	1	1	1	1
2	3	6	100	11	38	1	1	1	1	37	0	1	1	0
3	6	9	100	10	33	1	1	1	1	33	1	1	1	1
(t_{L_1}) 4	9	12	100	10	33	0	0	0	0	22	0	0	0	0
5	12	15	80	9	38	0	0	0	0	22	0	1	0	0
6	15	18	30	1	11	0	0	0	0	22	0	0	0	0
7	18	21	20	0	0	2	2	2	2	22	2	2	2	2
8	21	24	20	1	17	0	0	0	0	22	0	0	0	0
9	24	27	10	1	33	0	0	0	0	22	0	0	0	0
10	27	30	20	2	33	0	0	0	0	22	0	0	0	0
11	30	33	20	3	50	0	0	0	0	22	0	0	0	0
12	33	36	20	1	17	0	0	0	0	22	0	0	0	0
13	36	39	10	1	33	0	0	0	0	22	0	0	0	0
14	39	42	20	5	83	0	0	0	0	22	0	0	0	0
15	42	45	20	1	17	0	0	0	0	22	0	0	0	0
16	45	48	20	3	50	0	0	0	0	22	0	0	0	0
17	48	51	20	1	17	0	0	0	0	22	0	0	0	0
18	51	54	10	1	33	0	0	0	0	22	0	0	0	0
19	54	57	20	2	33	0	0	0	0	22	0	0	0	0
20	57	60	30	3	33	0	0	0	0	22	0	0	0	0
21	60	63	40	1	8	0	0	0	0	22	0	0	0	0
(t_{L_2}) 22	63	66	60	9	50	1	0	1	1	50	1	1	1	0
23	66	69	80	24	100	1	0	1	1	100	1	1	1	1
24	69	72	90	45	167	1	1	1	1	167	1	1	1	1
25	72	75	100	76	253	1	1	1	1	153	1	1	1	1

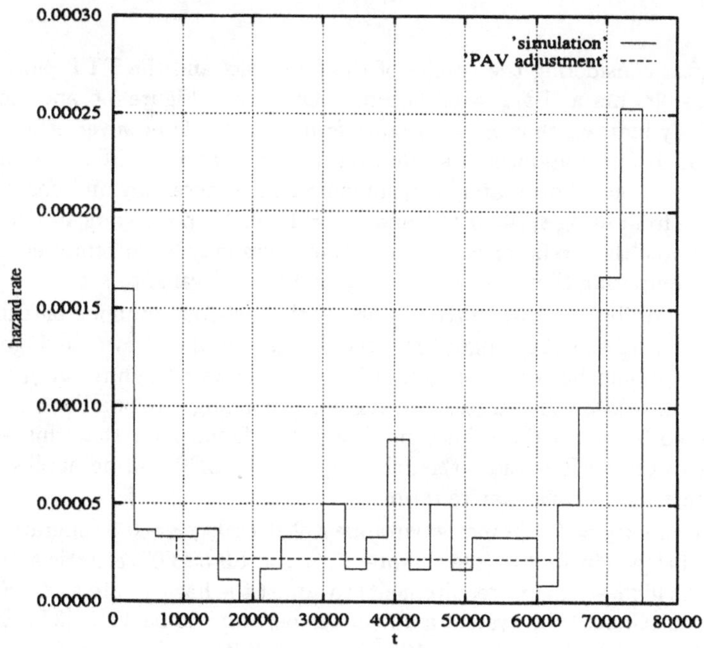

Figure 9.5: Simulated and modified system hazard rates.

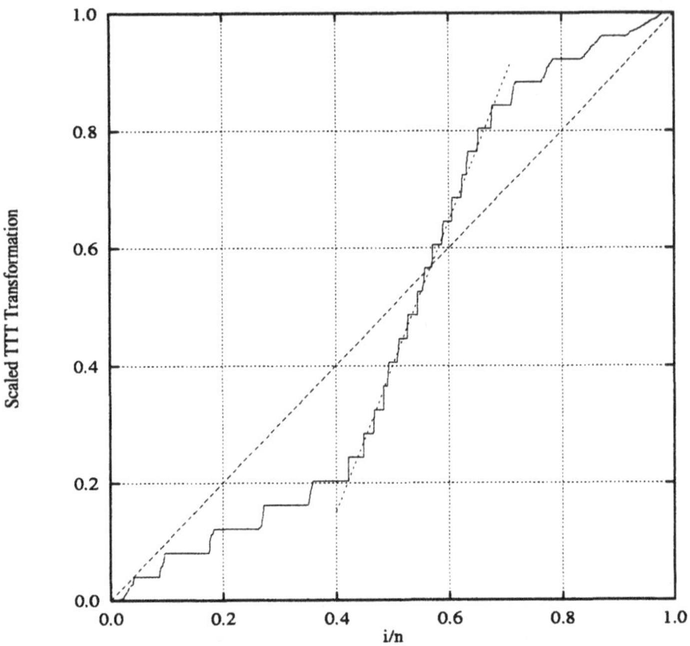

Figure 9.6: TTT plot of the failure times.

After considering the results of the four tests and the TTT plot, we set $t_{L_1} = 15,000$ hrs and $t_{L_2} = 63,000$ hrs. Of course, Figure 9.6 and Eq. (9.7) do strongly indicate that t_{L_1} may start from segment 6; however, since we use the MRL to find the optimal system burn-in time and the MRLs of different burn-in times are shown later to be monotonically increasing and concave, the difference from using $t_{L_1} = 9,000$ and $t_{L_1} = 15,000$ is not significant. It should also be noted that in other cases, large differences may be encountered, and we suggest comparing these two test results and the R^2 value to set t_{L_1}.

The Weibull testing outcomes are in the last four columns of Table 9.5. After applying the PAV algorithm, the $h_s^{'}(t)$s are also drawn in Figure 9.5. Attention should be paid to segment 7, which shows no failure. If $f_k = 0$, θ_k is approximated by the largest θ in the corresponding section. For instance, if there is no failure in the kth segment and it is found later that this segment belongs to the CFR section, then the largest θ in CFR will be applied to the kth segment for a conservative estimate.

We use $\pi_E = 1.0$ (minor environmental stress), $\pi_T = 0.6$ (operated under 30°C or below; Curve # 7 (see Chapter 7), $\pi_S = 1.3$ ($\approx 60\%$ electric stress), π_Q = 1.5 (the highest quality requirement) to adjust the hazard rate as described in Chapter 6. Choose the system ambient temperature to be 40°C [37]. Suppose the system can be burned-in at 125°C (T_b = 125°C) and it follows Curve # 7 in Chapter 7; that is, the accelerated factor is equal to 24. Thus, if we burn-in the system for 125, 250, 375, 500, and 625 hrs, it is equivalent to burning-in the system at 40°C for 3,000, 6,000, 9,000, 12,000, and 15,000 hrs, respectively. The MRLs under different system burn-in times are calculated from Eq. (9.5) and shown in Table 9.6 and Figure 9.7.

Table 9.6: MRL, total cost and $(C_B(t; t_{b,T_b}) - g\mu(t_b))$ under different burn-in times.

Burn-in time (t_b) (at 40°C)	Burn-in time (t_{b,T_b}) (at T_b = 125°C)	MRL ($\mu(t_b)$) (hrs)	Total cost ($C_{b(t;t_{b,T_b})}$)	$(C_{B(t;t_{b,T_b})})$- $g\ \mu(t_b)$
0	0	41,304.8	582.96	479.70
3,000	125	56,444.5	553.90	412.79
6,000	250	57,771.6	553.13	408.70
9,000	375	58,483.1	553.16	406.95* (min)

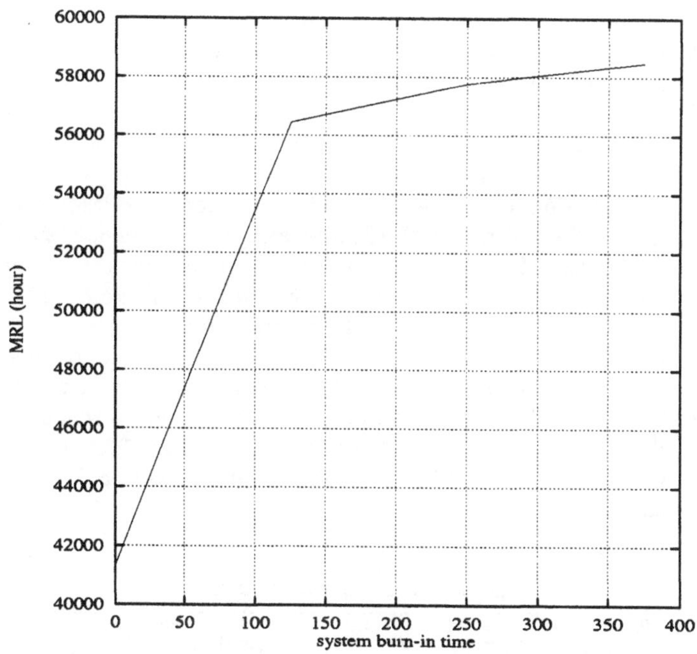

Figure 9.7: MRL under different burn-in times.

Let $C_D = 500$, $B = 10$, $c_v = 0.02$, $c_s = 1$, $c_f = 10$, $c_w = 50$, $g = 0.0025$, $l = 3$ (high penalty), t = 30,000 hrs ($\approx$ 3.4 years), η=0.75, and the warranty duration be 10,000 hrs ($\approx$ 1.1 years). The total costs under different burn-in times are derived by Eq. (9.4), and they are shown in the fourth column of Table 9.6. The outcomes of using Eq. (9.6) appear in the last column of Table 9.6. The optimal burn-in time is set to be 375 hrs (at T_b = 125°C), and the total cost is \$406.95, whose calculation, for example, is:

$$500 + 10 + 375 \times 0.02 + (0.00016 + 0.0000376 + 0.0000333)3000 \times 1 \\ + 40(0.0000222 \times 10)3000 + 37.5(0.0000222 \times \tfrac{10}{3})3000 \\ - 0.0025 \times 58483.1 = 406.95. \quad \square$$

□ Example 9.3

The overall acceleration factor (including both temperature and voltage acceleration), A_a, depends on the product characteristics. The usual A_a varies from 100 to 2,000. Major computer makers (e.g., IBM, Compaq, Dell, and HP) may ask their IC suppliers to specify their own activation energy (E_a) and voltage acceleration multiplier (β) or simply indicate the E_a and β for the suppliers to follow.

The most commonly required quality indices are the HTOL failure rates. The most popular HTOL test conditions, for the 5V products are burning-in ICs at 7V and 125°C for 1,000 hrs; the read points are x, 48, 96, 168, 500, and 1,000 hrs, where x depends on the A_a. Suppose a customer requires two kinds of HTOL failure rates: the failure percentage of the first year (in PPM) and the long-term failure rate (in FIT) in ten years when the products are used in the field. Let us call the failure percentage of the first year and the long-term failure rate FPM and LTFR, respectively. Assume A_a to be 750 with x set to 12. That is, the first reading point is after burning-in the ICs for 12 hrs.

Consider a major account whose requirements for the FPM and LTFR is less than 2,000 PPM and below 40 FIT, respectively; the sample size for the HTOL test is 20,000. To satisfy the customers' requirements, the number of failures at the first reading point should be less than 40. As to the LTFR, the test results at the 168, 500, and 1,000 hr reading points are of concern. The PAV algorithm can be applied if the DFR is believed to be valid for this HTOL test before 1,000 hrs, which is a reasonable assumption for most modern ICs. Consider the test results shown in Table 9.7. The second row indicates the equivalent duration (in years) if used at normal conditions. The sixth and the seventh row show the calculated and adjusted failure rates in FIT through the PAV algorithm, respectively.

Table 9.7: HTB test results.

Reading Point (hrs)	12	48	96	168	500	1,000
T_{normal} (year)	1.03	4.11	8.22	14.38	42.81	85.62
Starting Sample Size	20,000	19,968	19,956	19,943	19,941	19,939
number of failures	32	12	13	2	2	4
Failure % (PPM)	1,600	601	651	100	100	201
Failure Rate (FIT)	177.8	22.3*	18.1	1.86	0.40	0.536
Adj. Failure Rate (FIT)	177.8	22.3	18.1	1.86	0.482*	0.482

The failure rate with the mark $*$ is derived by

$$601 \times 10^{-6}/(48 - 12)/750 \times 10^{9} = 22.3.$$

The failure rate at the last two reading points do not satisfy the DFR definition and need to be adjusted. The adjusted failure rates with the mark $\star$ are obtained by

$$[0.40 \times (500 - 168) + 0.536 \times (1000 - 500)]/(1000 - 168) = 0.482.$$

That is, we treat the failure mechanism between every two adjacent test points to be exponentially distributed because the exponential distribution has a CFR. The LTFR, which is the 10-year average failure rate, is derived by

$$\begin{aligned}&[177.8 \times 1.03 + 22.3 \times (4.11 - 1.03)\\&\quad + 18.1 \times (8.22 - 4.11) + 1.86 \times (10 - 8.22)]/10 = 32.95 < 40.\end{aligned}$$

The IC maker can meet the two requirements according to the above calculations.

Customers may also require the failure rate after a 168-hr or 500-hr HTB test to be less than certain limits at a 60% or 90% upper confidence level (UCL). This requirement calls for use of the relationship between the exponential and the χ^2 distribution. It is well known that, for two random variables X and Y, if $X \sim \text{EXP}(\lambda)$ and $Y = 2\lambda X$, then $Y \sim \chi^2(2)$. Suppose we have m samples from an exponential distribution with parameter λ put at a time-termination (Type-I) test. It can be shown that, if n failures are found, $2n\lambda$ has the χ^2 distribution with degree of freedom $2(n+1)$, $\chi^2(2n+2)$. That is, the $100(1-\alpha)\%$ UCL of λ is derived by multiplying the point estimation of λ to $\chi^2_{1-\alpha}(2n+2)/(2n)$ (also refer to Example 6.5). A Chi-square table under different α and n is available in most statistical texts. Figure 6.3 shows critical values for given α and degree of freedom.

In the example above, suppose the customer requires the 60% UCL after 500-hr HTB test to be less than 10 FITs. The calculation follows:

$$0.482 \times \chi^2_{0.6}(2 \times 4 + 2)/(2 \times 4) = 0.482 \times 1.31 = 0.631 < 10.$$

Similarly, the 90% UCL can be derived by

$$0.482 \times \chi^2_{0.9}(2 \times 4 + 2)/(2 \times 4) = 0.482 \times 2 = 0.964 < 10.$$

The IC maker can meet the customer requirement at both 60% and 90% UCL.

To compensate for a case where no failure is encountered during an HTB test, the 60% or 90% confidence level is usually specified. Suppose the resultant acceleration factor for an IC is 950, then the 96-hr burn-in is equal to 10.4 years field use. If a major account requires the 60% long-term failure rate to be less than 20 FITs, then the minimum sample size required for a 96-hr burn-in is

$$\left[\frac{1}{(20\text{FITs})(950)(96)/0.92} + 1\right] = 505,$$

where 0.92 is equal to $\chi^2_{0.6}(2)/2$. In the same way, the minimum sample size for the HTOL test can be determined based on different failure rates required by customers. □

9.4. Conclusions

This chapter presents a nonparametric approach for the optimal burn-in strategy. This approach used at the system level further enhances the burn-in capability that considers the incompatibility factors existing at component, subsystem, and system levels. Specially,

1. The choice of Δt is based on experience or on the comparison of the system with similar products whose failure behavior is known. Few failures will be found if the λs are too small and Δt is not large enough. Analysts may make an estimation beforehand, based on field reports and the component properties about how large the total simulation time, T_{SIM}, should be. As for the sample size in each segment, we strongly recommend that more samples, over 50, should be put on the first 10,000 and last 10,000 hrs (if a trend of wearing-out is detected). Since it is widely conceived that many integrated circuits have a year-long infant mortality period under normal operating conditions [225], more samples can assure that more reliable t_{L_1} and t_{L_2} be obtained. To compensate for limited field data, Δt may be different for each segment; that is, Δt can be replaced by Δt_i; choose larger Δt_i to contain more samples in a segment.

2. Sometimes it is hard to find t_{L_1} and t_{L_2} if there is no significant trend indicating the start and/or the end of the CFR. One may then select a higher confidence level to find an observable trend. A second method is to increase the sample size, which will not be beneficial if the system being tested is expensive. The previous example is tested under the 95% confidence level and the decisions for t_{L_1} and t_{L_2} are made only when three consecutive nonexponential segments are inferred. Table 9.1 shows that the CFR region should begin at 15,000 hrs ($\max_i T_i^D$). However, t_{L_1} is estimated by the Anderson-Darling test to be 9,000 hrs. A similar situation can be found for t_{L_2}. The reason for these phenomena is that the changing rate of a Weibull failure rate is very small when t is comparatively small for $\beta > 1$ and when t is large for $\beta < 1$; the corresponding λ can be used to estimate the size of t to make the hazard rate increase insignificantly. In this case, t_{L_1} and t_{L_2} will be more accurate if the number of components increases or if many different components are in the system, which means the parameters (both λ and β) of the component differ greatly from one another.

3. The constant hazard rate after the PAV adjustment (0.0000222) is fairly close to the actual failure rate (0.0000270) derived directly from the given parameters of each component by Eq. (9.1). The weighted factors can be used to compensate for this difference.

4. It is possible that some f_ks are zero. Increasing the sample size is the easiest, but not a profitable way, to deal with this. Under the U-shaped hazard rate assumption, $f_k = 0$ can be neglected if a trend is detected that k is not equal to t_{L_1} or t_{L_2}. One should pay attention to the case when $f_k = 0$ occurs in the neighborhood of t_{L_1} and t_{L_2}.

5. Weibull testing is not significant in the example. One possible explanation is that the sample size in each segment is not large enough. Large numbers of failure times are expected to give a better estimation of the $\hat{\lambda}$ and $\hat{\beta}$ in Eq. (9.3) because they are derived by the numerical method.

Bayes or empirical Bayes methods can be applied to model the system behavior, but a suitable system failure and prior distributions which are often unknown, have to be assumed. Furthermore, different system configurations may have different failure time distributions. That is the major reason that the nonparametric approach introduced in this chapter is better since no assumption about distributions is needed.

The nonparametric approach can be combined with other techniques to construct a powerful system monitoring program. The TTT transformation can be first applied on a set of failure times, $\{T\}$, to detect if a system has a U-shaped hazard rate [1, 170]. Depending on $\{T\}$, Δt can be chosen and the changing point t_{L_1} can be found. The test by Park [321] can be applied to see if a failure rate changes its trend based on t_{L_1}, which is a prerequisite of his test. This test might be applied to reduce the sample size if the system's early failure can be modeled by a 2-parameter Weibull distribution which can be checked by a computer package like S-plus.

10. Nonparametric Bayesian Approach for Optimal Burn-in

A nonparametric method, which is simple and can effectively handle the system burn-in time, was introduced in Chapter 9 in situation where only a few failure times are available. The nonparametric approach can avoid the difficulties of

- assuming any parametric form for the component's failure mechanism,
- estimating the parameters of the assumed parametric distributions,
- calculating the overall system reliability, which is especially difficult when the system structure is complicated, and
- modeling the incompatibility factor described in Chapter 7.

As manufacturing technologies become more advanced, the number of layers for many ICs becomes larger than those which were produced 10, even 5, years ago. In other words, the costs of some ICs are high; for example the Pentium CPUs can be as high as $1,800. Thus, it is not economically feasible to test a large number of samples. Hence, when only limited data are available, the Bayesian concept should be absorbed into the burn-in model to address the following three critical issues:

1. High testing costs of ICs
   Sometimes, a product has to undergo a destructive test to make its properties transparent. It is too expensive to do life testing and too time consuming to wait for a significant amount of data for analysis. Thus reliability engineers often have limited data available to analyze new ICs. The classical way of analyzing burn-in data has been shown to take too much lead time and is inefficient.

2. Applying the experts' opinions
   Designers may not have complete knowledge of the manufacturing processes for producing a final product. Hence, the consensus of all parties is important; management has to gather information and opinions from not only designers but also system operators. The Bayes posterior information obtained by blending the prior knowledge of the parameters and the sample information can be more accurate after each revision or observation. In other words, the model is improved after every iteration, which is equivalent to smaller variance at a later burn-in stage. New ICs are manufactured by using knowledge about other ICs; therefore the experts'

opinions on the existing ICs are valuable and should be incorporated into the analysis.

3. Flexibly reflecting the degree of belief
In order to take advantage of prior knowledge, analysts can assign a proper weighting factor. For instance, if we strongly believe that the previous information about the product is correct, we assign more weight to the prior knowledge and less to the sampling scheme (test result). On the contrary, if there is not enough information on hand, it is reasonable that we will rely mostly on the test outcomes.

Based on the aforementioned reasons, a nonparametric Bayesian method is applied in this chapter to decide the system burn-in time by using the Dirichlet distribution. Several assumptions are made for the proposed model in this chapter.

- Assumption 1
Only system burn-in is conducted because it can remove more incompatibility than the lower level burn-ins. Besides, most components are burned-in either by the suppliers or before they are used in forming the systems and thus will not be considered here. The "systems" here may refer to modules which are defined to be subsystems elsewhere.

- Assumption 2
The system is at its infant mortality stage; i.e., it has a DFR. However, it should be noted that our model can also be applied to systems with IFR without any change. In such a case, a similar cost model in later sections can be used to decide when to replace or overhaul a system.

- Assumption 3
Different discrete times are prespecified, as shown in Figure 10.1, where k different burn-in times are illustrated for some test i. This assumption is both useful and practical because, in reality, engineers set discrete burn-in times like 50, 100, 150, 168, 200 ... hrs. At the end of each stage, samples are tested and failures are removed from the burn-in oven. Without loss of generality, we assume that all stage durations are the same (one week; i.e., 168 hrs) for simplicity.

The MRL will be used to determine the optimal burn-in time. If the MRL is not attainable, an alternative decision rule under the cost restriction and reliability requirement will be applied.

10.1. The Dirichlet Distribution

The Dirichlet distribution is used to study spacings, proportions, or the random division of an interval [90, p 199]. Consider $k+1$ mutually stochastically independent random variables, X_is, which follow GAM(β_i, λ). The pdf of X_i, $g_i(x_i)$, is given by

$$g_i(x_i) = \frac{\lambda^{\beta_i}}{\Gamma(\beta_i)} x_i^{\beta_i - 1} e^{-\lambda x_i}, x_i > 0,\ \beta_i > 0,\ \lambda > 0.$$

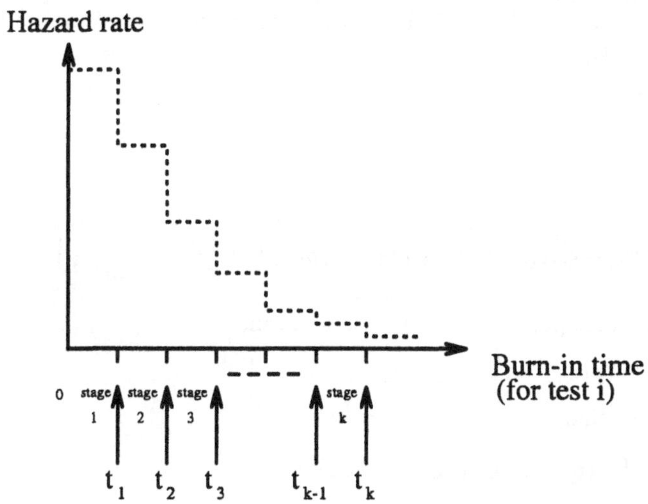

Figure 10.1: Using the Dirichlet distribution to decide system burn-in times.

When λ=1, the joint pdf of $X_1,\ldots,X_{k+1}$ becomes

$$g_{\tilde{X}}(x_1,\ldots,x_{k+1}) = \prod_{i=1}^{k+1} \frac{1}{\Gamma(\beta_i)} x_i^{\beta_i - 1} e^{-x_i}, x_i > 0.$$

Define

$$Y_i = \frac{X_i}{\sum_{j=1}^{k+1} X_j}, \; i = 1, \; 2, \; \cdots, \; k$$

then the joint pdf of $Y_1,\ldots,Y_k$ becomes

$$g_{\tilde{Y}}(y_1,\ldots,y_k) = \frac{\Gamma(\sum_{i=1}^{k+1} \beta_i)}{\prod_{i=1}^{k+1} \Gamma(\beta_i)} \prod_{i=1}^{k} y_i^{\beta_i - 1}(1 - \sum_{i=1}^{k} y_i)^{\beta_{k+1}-1}, \; y_i > 0 \qquad (10.1)$$

where $\sum_{i=1}^{k} y_i < 1$. Any random variables $Y_1,\ldots,Y_k$ that have a joint pdf in Eq. (10.1) are said to have a Dirichlet distribution with parameters $\beta_1,\ldots,\beta_{k+1}$, and the $g_{\tilde{Y}}(y_1,\ldots,y_k)$ is called a Dirichlet density [192, pp 148–149]. It can be easily seen that when $k = 1$, the Dirichlet distribution becomes a beta distribution, BETA(β_1,β_2).

Johnson and Kotz [206, p 65] use the differences of successive order statistics to define the Dirichlet distribution. A more practical definition than the aforementioned one is given in Ross [356, pp 120–121]. Suppose there are n trials; each trial will result in $k+1$ different types of outcome with probabilities $p_1, \ldots, p_{k+1}$, and $\sum_{i=1}^{k+1} p_i = 1$. Let X_i denote the number of the ith outcome in these n trials,

$$n = \sum_{i=1}^{k+1} x_i. \tag{10.2}$$

Then

$$\Pr\{X_1 = x_1, \cdots, X_{k+1} = x_{k+1} \mid p_1, \ldots, p_{k+1}\} = \frac{n!}{x_1! \cdots x_{k+1}!} \prod_{i=1}^{k+1} p_i^{x_i}. \tag{10.3}$$

Suppose the joint pdf of $p_1, \ldots, p_{k+1}$ is uniform,

$$f(p_1, \ldots, p_{k+1}) = c, \quad 0 \leq p_i \leq 1.$$

From the fact that

$$\int \cdots \int f(p_1, \ldots, p_{k+1}) dp_1 \cdots dp_{k+1} = 1,$$

we have $c = k!$. Thus,

$$f(p_1, \ldots, p_{k+1}) = k!, \quad 0 \leq p_i \leq 1 \tag{10.4}$$

where $\sum_{i=1}^{k+1} p_i = 1$. Since

$$\Pr\{p_1, \ldots, p_{k+1} \mid x_1, \ldots, x_{k+1}\}$$

$$= \frac{f(p_1, \ldots, p_{k+1}) \Pr\{X_1 = x_1, \cdots, X_{k+1} = x_{k+1} \mid p_1, \ldots, p_{k+1}\}}{\Pr\{X_1 = x_1, \cdots, X_{k+1} = x_{k+1}\}}, \tag{10.5}$$

$\Pr\{X_1{=}x_1, \cdots, X_{k+1}{=}x_{k+1}\}$ must be derived if we want to evaluate $\Pr\{p_1, \ldots, p_{k+1} \mid X_1{=}x_1, \cdots, X_{k+1}{=}x_{k+1}\}$. As explained in [356, pp 118–119], to evaluate $\Pr\{X_1{=}x_1, \cdots, X_{k+1}{=}x_{k+1}\}$, recall Eq. (10.2) and that $x_1, \ldots, x_{k+1}$ are non-negative integers. In other words, we want to compute the number of non-negative integer valued solutions of Eq. (10.2), and the answer is $\binom{n+k}{k}$. Therefore,

$$\Pr\{X_1 = x_1, \cdots, X_{k+1} = x_{k+1}\} = \frac{1}{\binom{n+k}{k}}. \tag{10.6}$$

From Eqs. (10.3), (10.4), (10.5), and (10.6), we have

$$\Pr\{p_1, \ldots, p_{k+1} \mid X_1 = x_1, \cdots, X_{k+1} = x_{k+1}\}$$

$$= \frac{(n+k)!}{\prod_{i=1}^{k+1} x_i!} \prod_{i=1}^{k+1} p_i^{x_i} = \frac{\Gamma(\sum_{i=1}^{k+1} x_i + k + 1)}{\prod_{i=1}^{k+1} \Gamma(x_i + 1)} \prod_{i=1}^{k+1} p_i^{x_i}.$$

which is the same as Eq. (10.1).

The Dirichlet distribution family is a natural conjugate prior for a multinomial likelihood and is a multivariate generalization of the beta distribution [171, p. 584]. The Dirichlet distribution is applied in reliability growth models during development testing as in [210, 274, 390]. A strength-stress model is developed using the Dirichlet distribution in Papadopoulos and Tiwari [317], in which the effect of using incorrect priors is also discussed.

Define [90]

$$\begin{aligned} S_j &= \sum_{i=1}^{j} P_i, \quad S_0 = 0, \; i = 1, 2, \cdots, k+1 \\ Z_i &= P_i/(1 - S_{i-1}), \quad Z_1 = P_1, \quad Z_{k+1} = 1 \\ \sum_{i=1}^{k+1} P_i &= 1 \end{aligned}$$

where $P_1, \ldots, P_{k+1}$ are non-negative continuous random variables. If Z_i has a beta distribution with parameters a_i and b_i (a_i, $b_i > 0$), then the pdf of the Ps is

$$f_{\vec{p}}(p_1, \ldots, p_{k+1}) = \frac{p_{k+1}^{b_k - 1}}{\prod_{i=1}^{k} \mathcal{B}(a_i, b_i)} \prod_{i=1}^{k} [p_i^{a_i - 1} (\sum_{j=i}^{k+1} p_j)^{b_{i-1} - (a_i + b_i)}], \tag{10.7}$$

which is called the generalized Dirichlet distribution and is applied for reliability analyses in [258, 271]. The moments of a generalized Dirichlet distribution are derived in Connor and Mosimann [90]. The relationship between the isotonic regression and the generalized Dirichlet distribution is given in Ramsey [340]. An excellent introduction to the properties of the generalized Dirichlet distribution can be found in Lochner's paper [258], which also indicates its major advantage—that is "*it takes the relationship between two variables in the distribution into account.* " Lochner [258] also provides sensitivity analyses if one (or more) of the variates increases or decreases. In 1973, Ferguson introduced the Dirichlet process [140, 141, 142, 143, 144], which will be studied and applied to burn-in in Chapter 11.

10.2. The Model Formulation

We define the conditional reliability given that the samples all survive t_{i-1}, which is the starting time of the ith test stage (or the ending time of the $(i-1)^{st}$ test stage; $i = 1, 2, \ldots, k$, and $t_0 = 0$):

$$u_i = \Pr(X > t_i | X > t_{i-1}) = \frac{1 - F(t_i)}{1 - F(t_{i-1})} \tag{10.8}$$

and let $u_0 = 1$ and $u_{k+1} = 0$. Mazzuchi and Singpurwalla [271] record the likelihood of $\mathbf{u} = (u_1, \ldots, u_k)$ as

$$\begin{aligned} l(\mathbf{u}|\mathbf{r},\mathbf{s}) &\propto \prod_{i=1}^{k}(1-u_i)^{s_i} u_i^{v_i} \\ v_i &= \sum_{j=i}^{k} r_j + \sum_{j=i+1}^{k} s_j \equiv (\text{removal } \#) + (\text{failed } \#) \end{aligned} \tag{10.9}$$

where r_i, $i = 1, 2, \ldots, k$, is the number of unfailed items removed at t_i, and s_i and v_i are the numbers of samples which fail and survive in $(t_{i-1}, t_i]$, respectively. Assuming that the unknown quantities u_1, $(u_2 - u_1)$, $(u_3 - u_2)$, ..., $(u_k - u_{k-1})$ and $(1 - u_k)$ follow a Dirichlet density with parameters α_i ($i = 1$, ..., k, $k+1$) and β, we obtain

$$\pi(\mathbf{u}) = \frac{\Gamma(\beta)}{\prod_{i=1}^{k+1}\Gamma(\beta\alpha_i)} \prod_{i=1}^{k+1}(u_i - u_{i-1})^{\beta\alpha_i - 1} \tag{10.10}$$

where $\alpha_i \geq 0$, i=1, 2, $\cdots$, $k+1$, $\sum_{i=1}^{k+1}\alpha_i$=1, and $\beta > 0$. The mean, variance, and covariance of a Dirichlet density in Eq. (10.10) are:

$$E[u_i] = 1 - \sum_{j=i+1}^{k+1} \alpha_j, \tag{10.11}$$

$$Var[u_i] = E[u_i^2] - (E[u_i])^2 = \frac{E[u_i](1 - E[u_i])}{\beta + 1}, \tag{10.12}$$

$$Cov[u_i, u_j] = E[u_i, u_j] - E[u_i]E[u_j] = \frac{E[u_i](1 - E[u_j])}{\beta + 1}, \quad \text{for } i > j,$$

respectively, where

$$E[u_j^m] == \frac{\Gamma(\beta)}{\Gamma(\beta + m)} \frac{\Gamma(\beta\alpha_{j+1} + \cdots + \beta\alpha_{k+1} + m)}{\Gamma(\beta\alpha_{j+1} + \cdots + \beta\alpha_{k+1})}.$$

The Dirichlet distribution in Eq. (10.10) is the generalized one according to Connor's derivation [90]. Using the likelihood in Eq. (10.9) and the prior in Eq. (10.10), we find the posterior distribution to be

$$f(\mathbf{u}|\mathbf{r},\mathbf{s}) = \frac{1}{I_o}\left(\prod_{i=1}^{k} u_i^{v_i}(1 - u_i)^{s_i}(u_i - u_{i-1})^{\beta\alpha_i - 1}\right)(1 - u_k)^{\beta\alpha_{k+1} - 1} \tag{10.13}$$

where

$$I_o = \sum_{l_k=0}^{s_k} \cdots \sum_{l_1=0}^{s_1} (-1)^{\sum_{i=1}^{k} l_i} \prod_{i=1}^{k} \binom{s_i}{l_i} B\left(\beta\alpha_{i+1}, \sum_{j=1}^{i}(v_j + l_j + \beta\alpha_j)\right). \tag{10.14}$$

To sum up, the main idea is to expand the term $(1-u_i)^{s_i}$ to the corresponding binomial summation [274],

$$(1-u_i)^{s_i} = \sum_{l_i=0}^{s_i} \binom{s_i}{l_i} (-1)^{l_i} (u_i)^{l_i}$$

and by the relation

$$\int_a^b (x-a)^m (b-x)^n dx = (b-a)^{m+n+1} \mathcal{B}(m+1, n+1).$$

Let $\hat{u}_i$ be the posterior mean of u_i. Mazzuchi and Singpurwalla [271] use (1-$\hat{u}_i$) as an estimator of h_i, the hazard rate of the ith test stage, for i = 1, 2, ..., k. However, since

$$\begin{aligned} F(t) &= 1 - e^{-H(t)} \\ H(t) &= \int_0^t h(s)ds \approx \sum_{j=1}^{i} h_j \Delta t \end{aligned}$$

we have

$$u_i = e^{-\left(H(t_i) - H(t_{i-1})\right)} \approx e^{-h_i \Delta t}. \tag{10.15}$$

If $h_i \Delta t$ is small, $u_i \approx 1 - h_i \Delta t$. That is, the validity of using (1-$\hat{u}_i$) for h_i holds only if $h_i \Delta t$ is small, which is not true for large Δt. We will release this restriction and derive a better estimation for h_i.

10.2.1. Adding Extra Samples

It is possible for items which survive stage i of burn-in tests to be put under test at the beginning of stage $i+1$. This step can prevent a shortage of samples remaining at the later test stages. If additional samples are added as described above, the v_i in Eq. (10.9) is modified by

$$v_i^{'} = \begin{cases} v_i + \sum_{j=2}^{i} a_j, & \forall\, i \geq 2 \\ v_1, & i = 1 \end{cases} \tag{10.16}$$

where a_i is the number of extra samples put on test at the beginning of the ith test stage; these extra samples must survive the previous $(i-1)^{st}$ test stage (i = 2, ..., k). All the v_is used in the rest of the chapter are the $v_i^{'}$ in Eq. (10.16). Removal of sample(s) at the end of test stages is also allowed. Adding and withdrawing samples are very common in practice.

10.2.2. Setting αs

By comparing similar systems' failure records or from experts' opinions, we set

$$h_i = \frac{1}{2}(h_{e,i-1} + h_{e,i}), \; i = 1, 2, \; \cdots, \; k \tag{10.17}$$

where $h_{e,i-1}$ and $h_{e,i}$ are the empirical (with the subscript e) hazard rates of the beginning and the ending point hazard rate of the ith test stage, respectively. It is easy to see that if $h_{e,i}$s are decreasing, then so are the h_is.

By the assumption that a system has a DFR,

$$\begin{aligned} & h_1 \geq h_2 \geq \cdots \geq h_k \\ \Rightarrow\ & h_1 \Delta t \geq h_2 \Delta t \geq \cdots \geq h_k \Delta t \\ \Rightarrow\ & \ln(u_1^*) \leq \ln(u_2^*) \leq \cdots \leq \ln(u_k^*), \qquad \text{(by Eq. (10.15))} \\ \Rightarrow\ & u_1^* \leq u_2^* \leq \cdots \leq u_k^* \end{aligned} \tag{10.18}$$

where u_i^*s are the best prior guesses (knowledge) for $\mathbf{u}=(u_1, \ldots, u_k)$. That is, $\mathbf{u}^*$ is an increasing sequence. From Eq. (10.1) and the $E(u_i)$ in Eq. (10.11), the α_is can be set as

$$\alpha_i = u_i^* - u_{i-1}^*,\ i = 1, 2, \cdots, k+1. \tag{10.19}$$

From Eq. (10.19), it is obvious that $\sum_{\forall i} \alpha_i = 1$, and, if a convex DFR is considered, we also have $\alpha_2 \geq \alpha_3 \geq \cdots \geq \alpha_k$.

10.2.3. The Posterior Hazard Rate

The Bayes estimator of u_i under square error loss is the posterior mean, which can be expressed as

$$\hat{u}_i = \frac{I_{i,1}}{I_o}, \tag{10.20}$$

where $I_{i,j}$ is obtained by replacing v_i with $(v_i + j)$ in I_o, which is very similar to the expression in Mazzuchi and Singpurwalla [271]. From Eq. (10.15),

$$\hat{h}_i = -\frac{\ln(\hat{u}_i)}{\Delta t}$$

Let X be a random variable with mean μ and $H(X)$ be a function of X. Expand $H(X)$ on a Taylor series about $x = \mu$; we have

$$\begin{aligned} H(x) &\approx H(\mu) + H'(\mu)(x-\mu) + \tfrac{1}{2}H''(\mu)(x-\mu)^2 \\ \Rightarrow Var(H(X)) &\approx (H'(\mu))^2 Var(X). \end{aligned}$$

Hence, for i=1,2,$\cdots$,k,

$$Var(\hat{h}_i|\mathbf{r},\mathbf{s}) = \frac{1}{(\hat{u}_i \Delta t)^2} Var(\hat{u}_i|\mathbf{r},\mathbf{s}) = \frac{1}{(\hat{u}_i \Delta t)^2}\left[\frac{I_{i,2}}{I_0} - (\frac{I_{i,1}}{I_0})^2\right]. \tag{10.21}$$

The positivity of $Var(\hat{h}_i|\mathbf{r},\mathbf{s})$ is guaranteed due to the relationship, $I_{i,2} I_o > I_{i,1}^2$, which is proved using the property of

$$\mathcal{B}(a, b+2)\mathcal{B}(a,b) > \mathcal{B}(a,b+1)^2,\ \forall\ a, b > 0.$$

The posterior covariance is

$$Cov(\hat{h}_i, \hat{h}_j|\mathbf{r},\mathbf{s}) = \frac{1}{(\hat{u}_i \Delta t)^2}\left[\frac{I_{(i,j),1}}{I_0} - (\frac{I_{i,1}}{I_0})(\frac{I_{j,1}}{I_0})\right] \tag{10.22}$$

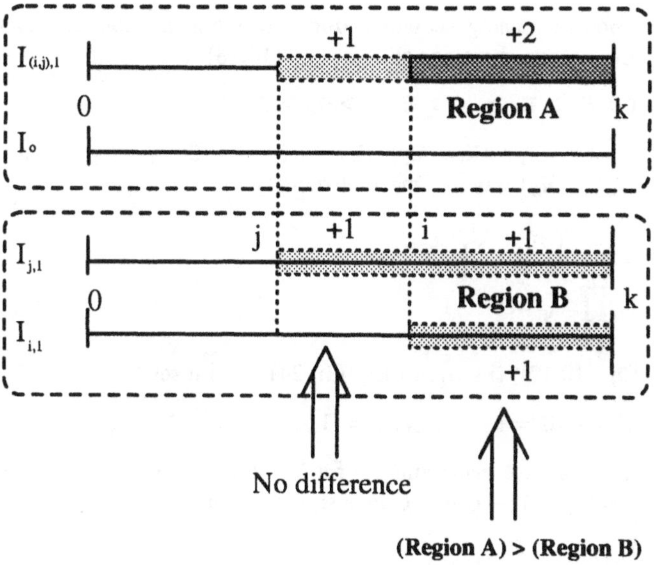

Figure 10.2: The terms in the covariance.

where $i, j = 1, 2, \cdots, k$; $i \neq j$; and $I_{(i,j),1} = E[u_i u_j]$ and is derived by replacing v_i and v_j in Eq. (10.13) by $v_i + 1$ and $v_j + 1$, respectively. Similar to $I_{i,2} I_o > I_{i,1}^2$, we can see $I_{(i,j),1} I_o > I_{i,1} I_{j,1}$ which makes Eq. (10.22) a legitimate expression. Figure 10.2 helps to clarify the idea.

It is easy to see that h_i and h_j are positively correlated because of the properties of the generalized Dirichlet distribution [90]. Besides, this observation is intuitive: any increase of the hazard rate in any stage implies the increase in the hazard rate in all stages.

The decrease or increase of the $\hat{h}_i$s depend on v_is and s_is. If the observed s_is are roughly proportional to α_is, it is reasonable to expect a decreasing posterior hazard rate because the proportionality implies that our previous information about the system is consistent with the experimental results.

10.2.4. When s_is Are Large

If many failures are observed, it would be time consuming to derive $\hat{u}_i$ in Eq. (10.20) by using Eq. (10.13). And since there are many terms in Eq. (10.13), serious round-off errors could occur. One solution for this is using the $\overline{u}_i$, which is defined as

$$\overline{u}_i = 1 - u_i = \Pr(X \leq t_i | X > t_{i-1}) = \frac{F(t_i) - F(t_{i-1})}{1 - F(t_{i-1})}. \quad (10.23)$$

The likelihood of $\overline{u}_i$ is denoted by

$$L(\overline{\mathbf{u}}; \mathbf{r}, \mathbf{s}) \propto \prod_{i=1}^{k} \overline{u}_i^{s_i} (1 - \overline{u}_i)^{v_i}.$$

The $\overline{u}$ is a non-increasing sequence under the Assumption 2, which can be verified in an analogous way as that in Eq. (10.18):

$$\overline{u}_0 = 1 \geq \overline{u}_1 \geq \overline{u}_2 \geq \cdots \geq \overline{u}_{k-1} \geq \overline{u}_k \geq 0 = \overline{u}_{k+1}.$$

Let $(\overline{u}_0 - \overline{u}_1 = 1 - \overline{u}_1)$, $(\overline{u}_1 - \overline{u}_2)$, $(\overline{u}_2 - \overline{u}_3)$, $\cdots$, $(\overline{u}_{k-2} - \overline{u}_{k-1})$, $(\overline{u}_{k-1} - \overline{u}_k)$, and $(\overline{u}_k - \overline{u}_{k+1} = \overline{u}_k)$ have a Dirichlet density,

$$\pi(\overline{\mathbf{u}}) = \frac{\Gamma(\beta)}{\prod_{i=1}^{k+1} \Gamma(\beta\alpha_i)} \prod_{i=1}^{k+1} (\overline{u}_{i-1} - \overline{u}_i)^{\beta\alpha_i - 1}. \tag{10.24}$$

Similar to Eq. (10.19), the α_is in Eq. (10.24) can be set by

$$\alpha_i = \overline{u}_{i-1}^* - \overline{u}_i^* = u_i^* - u_{i-1}^*, \; i = 1, 2, \cdots, k+1$$

where $\overline{u}_i^*$s are the best prior guesses for $\overline{\mathbf{u}}$. The expected value of $\overline{u}_i$ can be obtained by $\mathrm{E}[\overline{u}_i] = 1 - \mathrm{E}[u_i]$. Under square error loss, we have

$$\hat{\overline{u}}_i = \frac{\overline{I}_{i,1}}{\overline{I}_o} \tag{10.25}$$

where

$$\overline{I}_o = \sum_{l_k=0}^{v_k} \cdots \sum_{l_1=0}^{v_1} (-1)^{\sum_{i=1}^{k} l_i} \prod_{i=1}^{k} \binom{v_i}{l_i} B\left(\beta\alpha_i, \sum_{j=i}^{k} (s_j + l_j + \beta\alpha_{j+1})\right) \tag{10.26}$$

and $\overline{I}_{i,1}$ (for $i = 1, 2, \cdots, k$) is obtained by substituting s_i as $(s_i + 1)$ in $\overline{I}_0$ by the similar procedures in Eq. (10.14). Therefore,

$$\hat{h}_i(t) = -\frac{\ln(1 - \hat{\overline{u}}_i)}{\Delta t}, \; i = 1, 2, \cdots, k. \tag{10.27}$$

Eqs. (10.26) and (10.27) may be used when s_is are large to expedite program execution and to minimize computer round-off errors. The variance of $\hat{h}_i(t)$ in Eq. (10.27) can be derived by replacing the I_o, $I_{i,1}$, and $I_{i,2}$ with $\overline{I}_o$, $\overline{I}_{i,1}$, and $\overline{I}_{i,2}$, respectively, in Eq. (10.21).

10.2.5. Setting β

The posterior hazard rate estimator, $\hat{h}_i(t)$, can be used as a prior guess if a subsequent burn-in test on the same population is performed. If systems are modified, an improvement of reliability is expected. That is, we can use

$$u_i^{**} \Leftarrow \hat{u}_i, \; i = 1, 2, \cdots, k$$

as the prior guess along with a larger β for the new test.

When $\beta = 0$, the mode of the joint density is the isotonic regression; whereas for $\beta = \infty$, the posterior distribution is dominated by the prior distribution [340]. In intermediate cases ($0 < \beta < \infty$), the joint posterior densities are convex and unimodal [340, p. 846]. Most researchers assign different values of β according to experience. Rule 2 in [107] is developed with respect to the one that is termed Rule 1 in [107] and [340, p. 853]. It states: .... $\alpha = prior\ \alpha + \min_i(D_i)$, where $D_i = n_i +$ (the number of successes at dose levels less than x_i) + (the number of failures at dose levels greater than x_i); and the α suggested in Rule 1 is $\alpha = n_o$, where n_o is the number of observations. Disch [107] claims that the choice of α used in Rule 1 is too conservative in its use of the prior information.

We propose sequences for β under the following situations.

Case I We know the product pretty well (e.g., the system under test is simply a revised or advanced model) and no modification or re-design is done.

Case II We know the product pretty well and some modifications or re-designs are performed.

Case III We have limited knowledge about the product and no modification or re-design is done.

Case IV We have limited knowledge about the product and some modifications or re-designs are performed.

The β sequences for the four cases above are shown in Table 10.1 and Figure 10.3 where $(K = i)$ represents the ith test.

Table 10.1: The β sequence for the four cases.

	$K = 1$	$K = 2$	$K = 3$	$K = 4$	$K = 5$
Case I	5	15	30	50	70
Case II	5	10	20	30	40
Case III	1	7	14	21	28
Case IV	1	5	10	15	20

This is from the engineering viewpoint and is very similar to the weighted averages used in [164]. One can use any value of β according to one's knowledge about the product. If an engineer has reliable prior knowledge about the sample to test, he can simply start from $K = 2$ or even $K = 3$ in Table 10.1. Larger β should be used for successive burn-in tests to reflect the variance reduction of the analyses as shown in Eqs. (10.12) and (10.21).

10.3. Other Considerations

10.3.1. Determining the Optimal System Burn-in Time

The MRL is applied here again to find the optimal burn-in time. The optimal burn-in time is determined by

$$\min_{t_b} \ (C_B(t_o;\ t_b) - g\mu(t_b)). \tag{10.28}$$

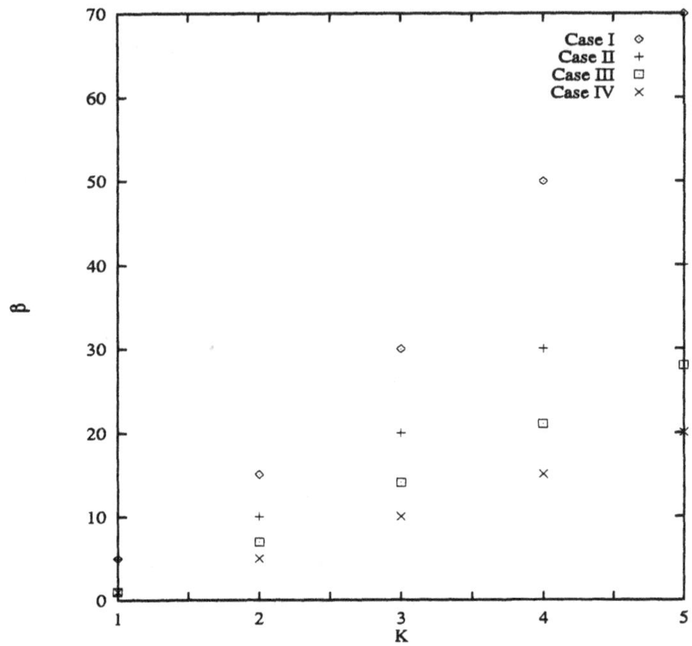

Figure 10.3: The proposed β sequence.

The MRL is calculated by assuming the system has the hazard rate $\hat{h}_b(t)$, where b is the stage number of the optimal system burn-in of 10 years, which is longer than the duration of most real world systems' operating conditions. An alternative method is proposed here to avoid going through complex calculations in Eq. (10.28).

Two factors can be considered to decide the appropriate burn-in time, under a pre-specified observation time (t_o):

1. A minimum system reliability, $R_{\min}(t_o)$, is required. This will remove some smaller t_bs which result in $R(t_o) < R_{\min}(t_o)$. A lower bound of the burn-in stage, i_L, can be found. If the longest t_b produces an inferior reliability at t_o, we might extend the burn-in test to the stage that achieves $R_{\min}(t_o)$;
2. A maximum burn-in cost, C_{total}, may be set by management. A larger cost will be inferred for a longer burn-in time. This consideration will help us set an upper bound of the stage, i_U.

We should pick the i such that $i_L \leq i \leq i_U$. If $i_U < i_L$; we can either (1) use smaller $R_{\min}(t_o)$, or (2) increase C_{total}. More test stages are recommended to ensure that enough potential defects are removed from the system, if there are many failures at the last test stage.

10.3.2. Determining Sample Size

The system reliability for an experiment whose failures are not replaced can be viewed as a binomial distribution with parameters n (number of samples to test) and d (number of failures). It is known that the binomial parameter has confidence bounds expressed by the F distribution [168]. Grosh [168, p. 177] shows the reliability bound for such nonreplacement tests is

$$\Pr\{R_L(t_b^*) \leq R(t)\} = 1-\gamma, \tag{10.29}$$

where $R_L(t_b^*)= (1+\frac{d+1}{n-d}F)^{-1}$ and $F{=}F_\gamma(2d+2, 2n-2d)$. From Eq. (10.29), n can be set as the one with smallest

$$\left| \frac{\frac{1}{R_L(t_b^*)}-1}{d+1} - \frac{F_\gamma(2d+2,2n-2d)}{n-d} \right|. \tag{10.30}$$

□ Example 10.1

The Arrhenius model is used for time transformation. The system burn-in time is set at 70~80°C as shown in [76] to avoid causing any damage to temperature vulnerable components, such as jump wires and plastic-cover capacitors. From [37, p.87], we know that the reading from Curve #7 for 80°C is 5.4 (i.e., η = 5.4) (supposing the system ambient temperature is 40°C at normal condition); we will thus use the accelerated factor 5.4 as an approximation.

Consider the $h_{e,i}$s in Figure 10.4, and Table 10.2 where the system burn-in temperature is 80°C and each stage continues for one week. The H_2 and F_2, for example, are obtained:

$$\begin{aligned} H_2 &= \sum_{i=1}^{2} h_i \eta \Delta t = 5.4 \times 168(2250+625) \times 10^{-6} = 2.6082, \\ F_2 &= 1-\exp(-H_2) = 1-0.07367 = 0.9263. \end{aligned} \tag{10.31}$$

Table 10.2: Set the u_is (iteration 1).

i	$h_{e,i}(\times 10^{-6})$	$h_i(\times 10^{-6})$	$H_i(\times 10^{-6})$	F_i	u_i^*	α_i
0	3,650	–	–	0.0000	0.0000	–
1	850	2,250	2.0412	0.8701	0.1299	0.1299
2	400	625	2.6082	0.9263	0.5674	0.4375
3	250	325	2.9030	0.9451	0.7449	0.1775
4	200	225	3.1072	0.9553	0.8142	0.0693
5	150	175	3.2659	0.9618	0.8546	0.0404
6	–	–	–	–	1.0000	0.1454

After comparing the samples they want to test with several previous products with known failure mechanisms, the engineers believe the hazard rates listed in Table 10.2 are not suitable and instead choose one with the failure history as shown in Table 10.3, where the $h_{e,i}$s, h_is, and H_is are in 10^{-6}. The hazard rate curve for the second iteration is given in Figure 10.5. It should be

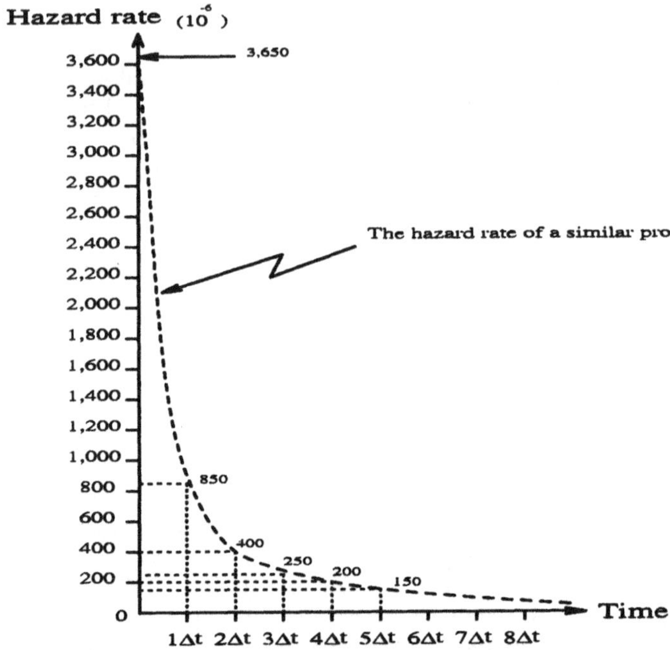

Figure 10.4: Hazard rate for choosing the u^*s (iteration 1).

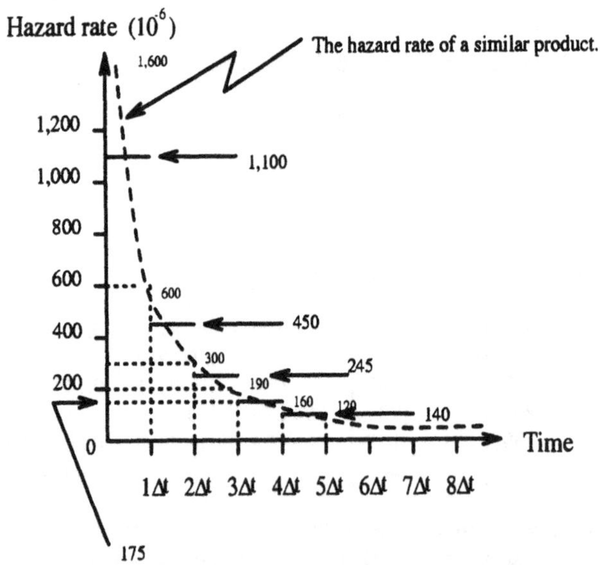

Figure 10.5: Hazard rate for choosing the u^*s (iteration 2).

Table 10.3: Set the u_is (iteration 2).

i	$h_{e,i}(\times 10^{-6})$	$h_i(\times 10^{-6})$	$H_i(\times 10^{-6})$	F_i	u_i^*	α_i
0	1,600	–	–	0.0000	0.0000	–
1	600	1,100	0.9979	0.6314	0.3686	0.3686
2	300	450	1.4062	0.7549	0.6649	0.2963
3	190	245	1.6284	0.8038	0.8005	0.1356
4	160	175	1.7872	0.8326	0.8532	0.0527
5	120	140	1.9142	0.8525	0.8811	0.0279
6	–	–	–	–	1.0000	0.1189

noted that, in some cases, the similar products used to construct the prior information may not have smooth hazard rate functions as shown in Figures 10.4 and 10.5. Instead, they may exhibit empirical hazard plots like the one shown in Figure 10.6.

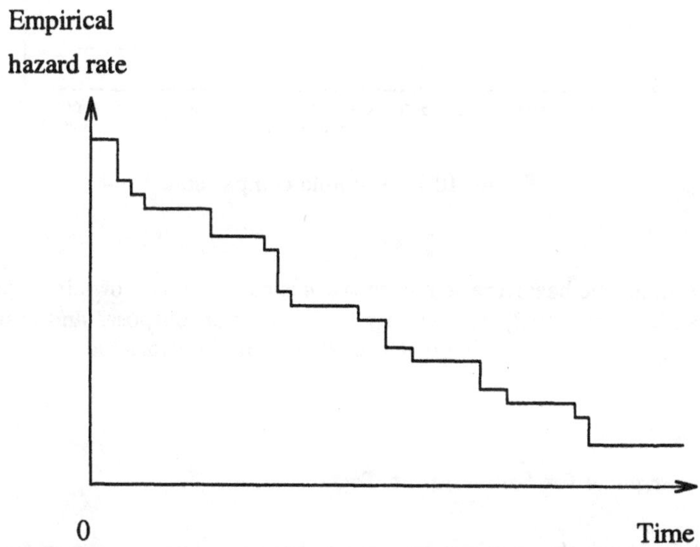

Figure 10.6: A possible hazard rate function of a comparison product.

A Comparison

Let $\beta = 5$, $r_1 = r_2 = r_3 = 1$, $r_1 = 0$, $a_2 = 3$, $a_3 = 2$, $a_4 = 1$, $a_5 = 2$, $s_1 = 6$, $s_2 = 4$, $s_3 = 2$, $s_4 = s_5 = 0$, the initial sample size is 9, and u_i^*s are from Table 10.2, which indicates the h_is obtained by Eq. (10.17). Figure 10.7 illustrates the difference between the approximated hazard rate estimation presented in [271] ($h_i = (1-u_i)/\Delta t$) and the exact estimation introduced here. It is obvious that, when Δt is not small (like the one used in this chapter), the approximations tend to give falsely optimistic results at earlier stages. It should be pointed out

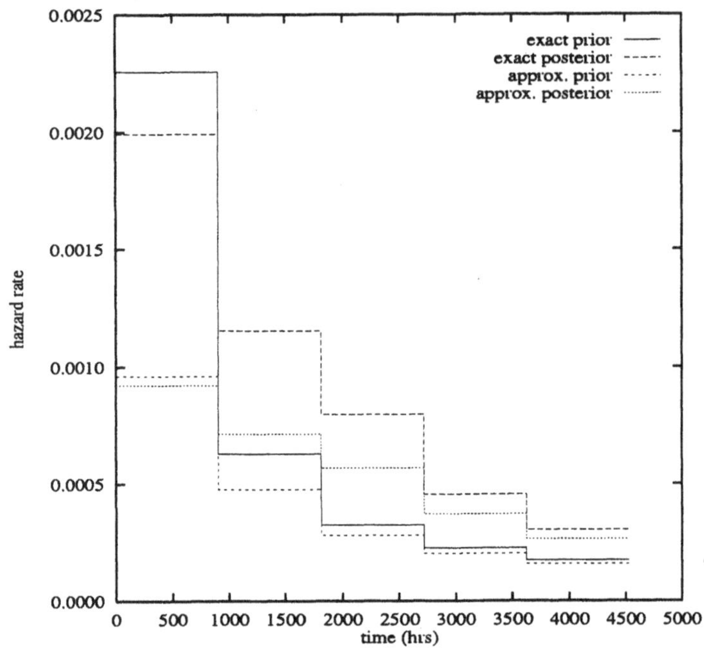

Figure 10.7: A simple comparison.

that, although the hazard rate in each stage is constant as shown in Figure 10.7 (and also in Figure 10.8), this is only for illustration purposes and is not true because, as one can see from Eq. 10.20, h_i is a random variable.

Using Larger β for Subsequent Tests

To continue from the previous section, we suppose three 5-stage system burn-in tests are performed with samples from the same population. The results are tabulated in Table 10.4. The β sequence in Case II (Table 10.1) is chosen and starts from $K = 1$ because, according to the analysts' knowledge and experience, the systems are selected from a common population without re-design or modification. Using the u_i^*s in Table 10.2, which is also the first row in Table 10.5, to begin the analysis, the $\hat{u}_i$s are then used as the best prior guess for the second test (as shown in the second row of Table 10.5). A similar procedure is used for the third test (the standard deviation is in unit 10^{-6} in Table 10.5). As can be seen from Table 10.5, a larger β results in smaller variance. The estimated system hazard rate after each test for stages 1 trough 5 is listed in Table 10.5. Figure 10.8 depicts the hazard rates of the prior and the three test samples.

Table 10.4: Sample size for the three tests and failure number at each stage.

	Beginning Sample size	Stage 1 failure (removal) [add]	Stage 2 failure (removal) [add]	Stage 3 failure (removal) [add]	Stage 4 failure (removal) [add]	Stage 5 failure
Test 1	10	5 (0) [1]	2 (1) [0]	1 (0) [0]	0 (1) [0]	0 — —
Test 2	9	4 (1) [1]	1 (2) [1]	1 (0) [0]	0 (1) [0]	0 — —
Test 3	10	5 (1) [2]	2 (1) [2]	2 (1) [2]	1 (2) [1]	0 — —

Table 10.5: Results for the three tests in Example 10.1.

	β	Stage 1	Stage 2	Stage 3	Stage 4	Stage 5
Test 1 (u^*)	5	0.3686	0.6649	0.8005	0.8532	0.8811
$\hat{h}(10^{-6})$		875.61	423.36	254.19	153.39	118.88
std. dev. of $\hat{h}$		284.00	183.06	158.22	124.23	112.23
Test 2 (u^*)	10	0.4528	0.6817	0.7945	0.8704	0.8980
$\hat{h}(10^{-6})$		741.77	356.51	226.70	124.17	94.79
std. dev. of $\hat{h}$		230.27	145.85	115.44	90.01	79.63
Test 3 (u^*)	20	0.5111	0.7243	0.8145	0.8937	0.9178
$\hat{h}(10^{-6})$		800.32	426.66	298.04	170.98	125.65
std. dev. of $\hat{h}$		189.80	126.00	104.20	83.18	75.65
$\hat{u}$	–	0.4847	0.6797	0.7636	0.8567	0.8925

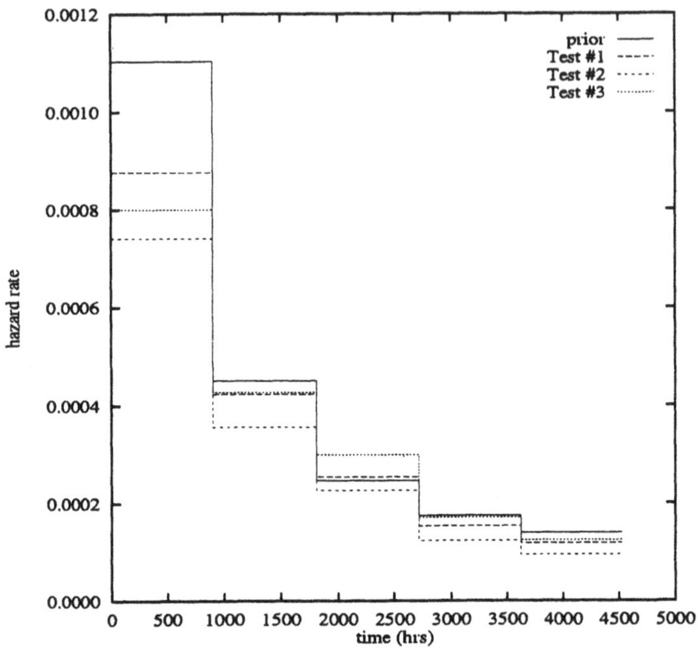

Figure 10.8: Hazard rates of the 3-test samples, 5-stage example.

Optimal System Burn-in Time

By Eq. (10.30), it is found that at least 8 samples are needed to have one failure under 90% ($1 - \gamma = 0.90$) confidence.

Using MRL

Let B=2, c_v=0.002, c_s=4, c_f=15, g=0.001, l=1 (low penalty), and t_o=10,000 hrs (slightly longer than a year), use the posterior hazard rate after the third test; Table 10.6 illustrates different burn-in times, MRLs, and the corresponding costs. After burning-in for $t_{b,80^\circ C,i}$ hrs (i = 0, 1, ..., 5), the minimum cost occurs when burn-in time is 840 hrs as illustrated in Table 10.6 where the estimated hazard rates, $\hat{h}_i(t)$s (in the fourth column), are in 10^{-6}/h. If the reward of MRL is not considered, burning-in the system for 672 hrs can give us a minimum total cost of $59.97.

Using Alternative Method

The reliability at t_o=10,000 hrs is listed in the last column of Table 10.6. If C_{total}=$60, then i_U=4. So we set the system burn-in time to be 672 hrs. This example does not really represent a high reliability system. The best system reliability is only 0.40. Thus, if we want to assure the system reliability to be better than 0.40, we have to burn-in the system for at least 672 hrs.

Table 10.6: Burn-in times, MRLs, and corresponding costs.

Stage	$t_{b,40^oC}$ (hrs)	$t_{b,80^oC}$ (hrs)	$\hat{h}_i(t)$ (10^{-6})	MRL $(\mu(t_b))$ (hrs)	$C_B(t_o;t_b)$ (\$)	$C_B(t_o;t_b)$ $-g\ \mu(t_b)$ (\$)	$R(t_o)$
0	0	0	–	2,772.10	89.20	86.42	0.0964
1	907	168	800.32	4,822.46	89.02	84.19	0.1778
2	1,814	336	426.66	10,868.35	70.15	59.28	0.2336
3	2,722	504	298.04	18,943.60	63.15	44.21	0.3431
4	3,629	672	170.98	27,549.96	59.95	32.42	0.4007
5	4,536	840	125.65	34,997.70	60.67	25.67	0.4007

The hazard rate after i=5, i.e., $\hat{h}_5(t)$=125.65 × 10^{-6}, is used as the system hazard rate if the system is burned-in for 840 hrs (i.e., if $t_b^* = 1{,}680$), and this is why the reliabilities of burning-in the system for 672 and 840 hrs are the same. That is a conservative estimate because in practice a system usually has an infant mortality longer than 10,000 ($= t_o$) hrs. In other words, the hazard rate for the system after t_o should be less than or equal to 125.65 × 10^{-6}. Thus, a larger (or at least the same) $R(t_o)$ can be expected in reality. □

10.4. Conclusions

To have better estimates for high reliability products, we can (1) put more items (systems, or samples) to test, (2) lengthen the stage duration–that is, use larger Δt (e.g., use 3 or 4 weeks instead of one week)–or (3) use higher burn-in temperature if possible. Engineers can use temperature-durable components instead of the prototype components which are more vulnerable to high temperature (e.g., they will be damaged at over 80°C). Of course, it might increase the testing cost. One can use more test stages to get better cost estimation.

One important future work is a model for blending experts' opinions on the $h_{e,i}$s in Eq. (10.17).

11. The Dirichlet Process for Reliability Analysis

The Dirichlet process is among the most common models used for nonparametric Bayesian analysis. Ferguson [140] described this process as follows. Let X be a space and A a σ-field of a subset, and let a be a finite non-null measure on (X,A). Then a stochastic process P, indexed by elements of A, is said to be a Dirichlet process $D(a)$ on (X,A) with parameter a if for any measurable partition $(A_1,\ldots,A_k)$ of X, the random vector $(P(A_1),\ldots,P(A_k))$ has a Dirichlet distribution with parameter $(a(A_1),\ldots,a(A_k))$.

Kuo's simulation procedure [223] for the Dirichlet process under a variance reduction technique introduced in Section 11.2 is applied for a Weibull-distributed system [80]. Optimal burn-in time is determined given the cost parameters.

11.1. Method

The Dirichlet distribution is chosen by many authors for nonparametric Bayesian analysis [210, 221]. Some important properties of the Dirichlet process are discussed in Ferguson [140, 143]. A random environment model is developed by Kumar and Tiwari [221] for an exponential-distributed system. We modify their model to fit a system whose failure mechanism can be described by a Weibull distribution. A weighting function suggested in [80] for variance reduction is used.

Consider a system whose failure behavior can be described by a Weibull distribution, WEI(θ,γ). From Eq. (3.17),

$$h(t) = \frac{\gamma}{\theta}\left(\frac{t}{\theta}\right)^{\gamma-1}. \tag{11.1}$$

If $\gamma < 1$, then the system has a DFR, and if $\gamma=1$, it has a CFR. Otherwise, it has an IFR. The scale parameter, θ, may be adjusted by the stress factors [76].

Let τ be the intensity of the environment. The system is used in a random environment whose strength has the gamma distribution, GAM(α,β). Thus, the system hazard rate can be expressed as

$$h'(t) = \tau h(t).$$

The $h'(t)$ is used in the plots of this section. The uncertainty of τ is described by a Dirichlet distribution, $D(MG_0)$ where G_0 is the prior guess of an unknown CDF which is a random variable and is expressed by a prior distribution.

Let $R_0(t)$ and $R_n(t)$ be the prior estimation of the system reliability at time t and the system reliability after n observations, respectively. The Lo's representation [257] is used for $\hat{R}_n(t|\theta)$,

$$\hat{R}_n(t|\theta) = \frac{g(t, t_1, \ldots, t_n)}{g(t_1, \ldots, t_n)} \tag{11.2}$$

where

$$g(t_1, \ldots, t_n) = \frac{1}{M^{(n)}} \int \cdots \int \Big\{ \prod_{i=1}^{n} \exp[-\tau_i (\frac{t_i}{\theta})^\gamma] \Big\} \prod_{i=1}^{n} D(MG_0 + \sum_{j=1}^{i-1} \delta_{\tau_j})(\tau_i).$$

$M^{(n)}$ is the Pochhammer's symbol [257],

$$M^{(n)} = M(M+1) \cdots (M+n-1).$$

The Dirac delta function $\delta(\cdot)$ is defined via [419]

$$\int_{-\infty}^{\infty} \delta(x)dx = \begin{cases} 1 & , x = 0 \\ 0 & , x \neq 0 \ (\delta(x) = 0 \text{ for } x \neq 0). \end{cases}$$

We need more definitions to represent the system reliability:

n	number of observations
$\lvert K_i \rvert$	number of elements in set i
L	number of simulations
M	measure of the strength of belief in the prior guess; a parameter of the Dirichlet distribution
$\lvert Q \rvert$	number of non-empty set Q
$\{T_i\}$	i^{th} data set
t_i	observing time of the ith sample
v	parameter used in variance reduction

Following the expressions in [221], we have the system reliability at time t:

$$R_n(t|\theta) = \frac{M}{M+n} R_o(t|\theta) + \frac{n}{M+n} \hat{R}_n(t|\theta) \tag{11.3}$$

where

$$R_o(t|\theta) = \left[\beta/(\beta + (\frac{t}{\theta})^\gamma)\right]^\alpha$$

$$\hat{R}_n(t|\theta) = \sum_{i=1}^{L} Y(Q_i) / \sum_{i=1}^{L} Z(Q_i)$$

$$Z(Q_i) = \prod_{K \in Q_i} \left[\beta/(\beta + \frac{1}{\theta^\gamma} \sum_{t_i \in K} t_i^\gamma) \right]^\alpha$$

$$Y(Q_i) = \frac{Z(Q_i)}{n} \sum_{K \in Q_i} |K| \left[\frac{\beta + \frac{1}{\theta^\gamma} \sum_{t_i \in K} t_i^\gamma}{\beta + \frac{1}{\theta^\gamma}(\sum_{t_i \in K} t_i^\gamma + t^\gamma)} \right]^\alpha .$$

The variance of system reliability at time t after n observations is calculated by

$$\text{Var}(\hat{R}_n(t|\theta)) = \left(\sigma_y^2 - 2\sigma_{yz}\frac{\mu_y}{\mu_z} + \sigma_z^2 \frac{\mu_y^2}{\mu_z^2}\right)/(L\mu_z^2) \tag{11.4}$$

where

$$\mu_z = \sum_{i=1}^{L} Z(Q_i)/L$$

$$\mu_y = \sum_{i=1}^{L} Y(Q_i)/L$$

$$\sigma_z^2 = \sum_{i=1}^{L} [Z(Q_i) - \mu_z]^2/[L(L-1)]$$

$$\sigma_y^2 = \sum_{i=1}^{L} [Y(Q_i) - \mu_y]^2/[L(L-1)]$$

$$\sigma_{yz} = [\sum_{i=1}^{L} Y(Q_i)Z(Q_i) - L\mu_y\mu_z]/[L(L-1)],$$

which can be calculated by the recurrent relations below:

$$\mu_{z,0} = 0$$

$$\mu_{y,0} = 0$$

$$\mu_{z,j+1} = \mu_{z,j} + \frac{Z(Q_{j+1}) - \mu_{z,j}}{j+1}$$

$$\sigma_{z,j+1}^2 = (1 - \tfrac{1}{j})\sigma_{z,j}^2 + (j+1)(\mu_{z,j+1} - \mu_{z,j})^2$$

$$\sigma_{yz,j+1} = \frac{j-1}{j+1}\sigma_{yz,j} + \frac{1}{(j+1)^2}\Big[\mu_{z,j}\mu_{y,j} - \mu_{z,j}Y(Q_{j+1}) - \mu_{y,j}Z(Q_{j+1}) - \tfrac{1}{j}Z(Q_{j+1})Y(Q_{j+1})\Big].$$

The μ_y and σ_y^2 can be derived in the same manner. Eq (11.4) is used in the variance reduction technique.

11.2. Variance Reduction in the Dirichlet Process

The flow chart of Kuo's simulation method [223] along with Ferguson's variance reduction technique [143] is shown in Figure 11.1.

Ferguson's procedures for variance reduction [143], which are based on importance sampling, are:

Step 1 Order the t_is such that $t_1 < t_2 < \ldots < t_n$.

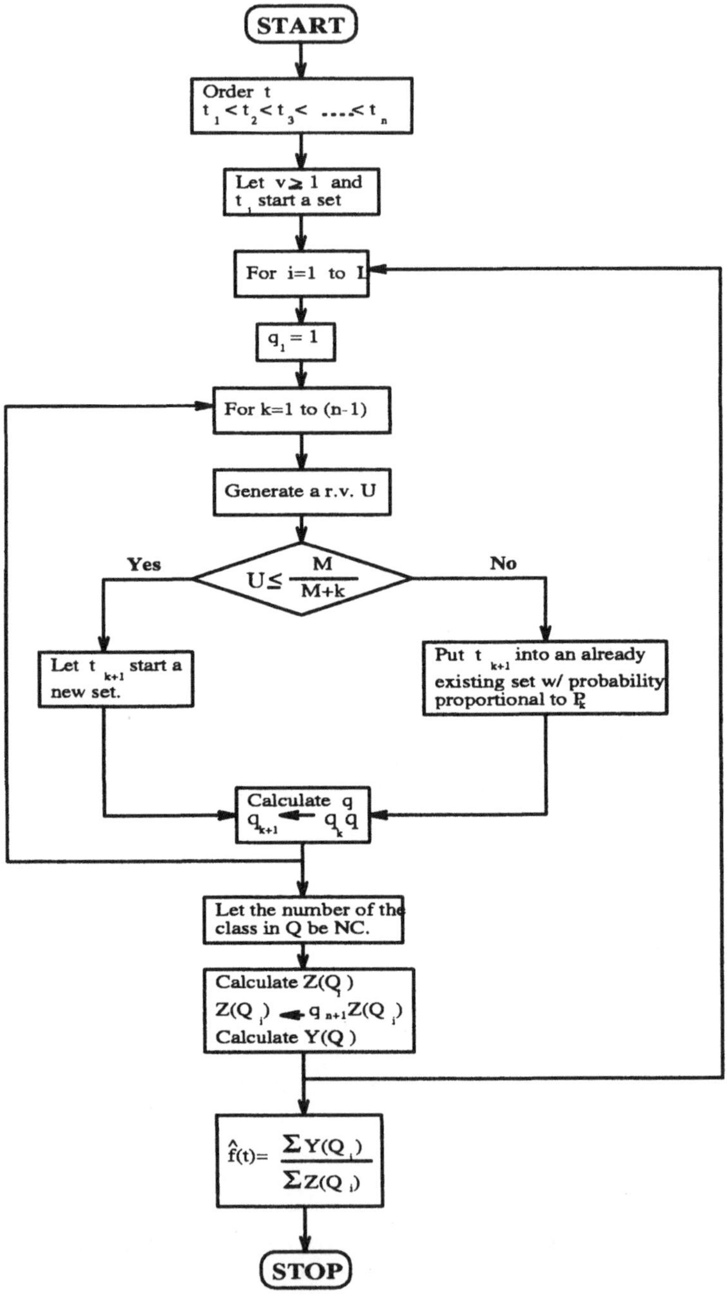

Figure 11.1: Simulation procedures with variance reduction for series systems.

Step 2 Choose $v \geq 1$ and let t_1 start a set of the partition.

Step 3 For $k = 1, 2, \ldots, (n-1)$, repeat the operation:
Let t_{k+1} start a new set of the partition with probability $\frac{M}{M+k}$. Otherwise, put t_{k+1} into the already-created set K with probability proportional to $\sum_{i \in K} f(v)$. And, let

$$\text{ratio} : q = \frac{f(v, i = 1, \cdots, k)}{k} / \frac{f(v, i \in K)}{|K|}. \tag{11.5}$$

Step 4 Keep a running product of the qs as one goes along and multiply $Z(Q_i)$ by this product when finished.

Two weighting functions for $f(v, i = 1, \cdots, k)$ are compared as suggested by Ferguson [143]:

$$\begin{aligned} f_1(v, i = 1, \cdots, k) &= \sum_{i=1}^{k} v^i \\ f_2(v, i = 1, \cdots, k) &= \sum_{i=1}^{k} e^{-v|t_i - t_{k+1}|}. \end{aligned} \tag{11.6}$$

The form of the corresponding $f(v, i \in K)$ is the same except that i denotes the number ("ID") of the t_i in set K. When $v = 1$ and $v = 0$, for f_1 and f_2, respectively, the model degenerates to the case without variance reduction.

To make the concept clear, see Figure 11.2, where $|Q_1| = 4$, $|Q_2| = 3$, and $|Q_3| = 4$ when $L = 3$. Furthermore, if $n = 5$, $\sum_{i=1}^{4} |K_i| = 5$ for Q_1, Q_3 and $\sum_{i=1}^{3} |K_i| = 5$ for Q_2. Suppose we are at $L = 2$, $k = 4$, the present $\{Q\}$ is $\{(t_1, t_3), (t_2), (t_4)\}$; i.e., the last observed data is going to be assigned to either K_1, K_2, or K_3. Assume that K_1 is chosen, then q should be $(\sum_{i=1}^{4} v^i/4)/(\frac{v+v^3}{2})$ if f_1 is used.

We suggest another weighting function, f_3, which is both order- and value-dependent.

$$f_3(v, i = 1, \cdots, k) = \sum_{i=1}^{k} v^i e^{-(v-1)|t_i - t_{k+1}|}. \tag{11.7}$$

f_3 degenerates to the case without variance reduction when $v = 1.0$.

□ **Example 11.1**

Assume that the reliability of time t, $t_{i-1} < t < t_i$, can be estimated by connecting $R_n(t_{i-1}|\lambda)$ and $R_n(t_i|\lambda)$, for $i = 1, \cdots, n$. The curve of reliability versus time will consist of "stairs" without the above assumption. Let γ=1,

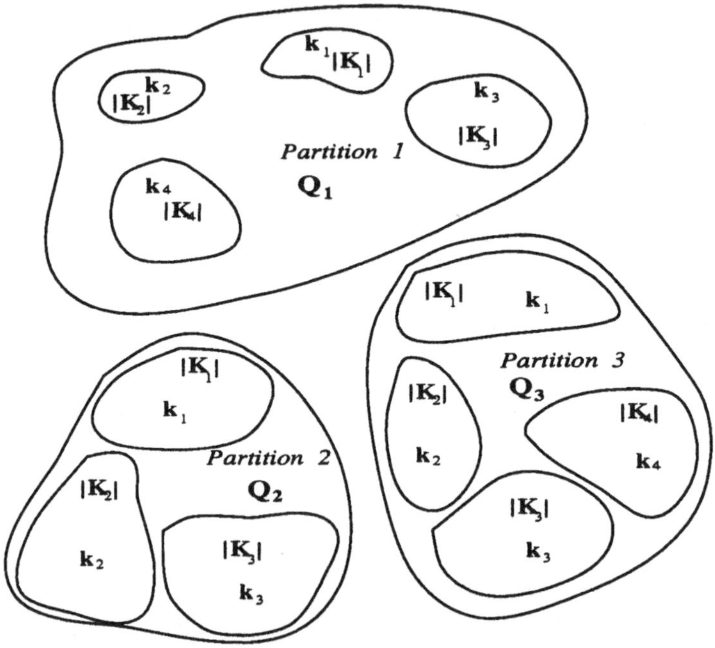

Figure 11.2: Partitioning the data.

$\theta = 1.0$, $\alpha = 1.0$, $\beta = 1.0$, $M = 10$, $L = 1000$, and the fourteen observations, $\{T_1\}$:

$$\begin{array}{llll} t_1 = 0.11, & t_2 = 0.13, & t_3 = 0.15, & t_4 = 0.19, \\ t_5 = 0.22, & t_6 = 0.26, & t_7 = 0.30, & t_8 = 0.38, \\ t_9 = 0.50, & t_{10} = 0.68, & t_{11} = 0.89, & t_{12} = 1.31, \\ t_{13} = 19.52, & t_{14} = 90.78. & & \end{array}$$

Suppose we want to know the system reliability at times 0.1, 0.3, 0.5, 0.7, 1.0, 1.5, 2.0, 2.5, 3.0, 3.5, 4.0, 4.5, and 5.0.

After several trials using f_1, we find that when $v = 1.09$, we have the largest variance reduction (around 19%). Table 11.1 depicts the posterior reliability, variance (in 10^{-3}) reduced percentage (in %; $v = 1.09$ over $v = 1.00$) for every t. The prior guessed, the posterior, and the estimated system reliabilities are drawn in Figure 11.3.

The Effect of Different Ms

The effect of using different Ms is shown in Figure 11.4 ($v = 1.09$).

The Difference among Data

Suppose we have another data set $\{T_2\}$: $t_{13} = 6.52$ and $t_{14} = 10.78$; all other t_is remain the same. It is generally found that smaller variance can be obtained

Table 11.1: Results of data set $\{T_1\}$.

t	0.1	0.3	0.5	0.7	1.0	1.5	2.0	2.5	3.0	3.5	4.0	4.5	5.0
prior	0.91	0.77	0.67	0.59	0.50	0.40	0.33	0.29	0.25	0.22	0.20	0.18	0.17
$\underline{v = 1.00}$													
mean	0.96	0.90	0.85	0.80	0.76	0.69	0.64	0.60	0.58	0.57	0.53	0.52	0.50
Var.	10.6	9.6	10.9	6.7	10.5	5.8	6.4	4.2	4.3	6.1	3.8	4.6	4.5
$\underline{v = 1.09}$													
mean	0.96	0.90	0.85	0.80	0.76	0.69	0.64	0.60	0.57	0.56	0.52	0.51	0.50
Var.	10.5	8.8	10.0	5.4	7.7	6.3	4.4	3.8	3.3	4.9	2.7	3.1	3.5
red. %	1.0	8.2	8.7	8.8	26.1	-9.6	30.9	10.4	22.6	19.2	27.8	32.0	22.6

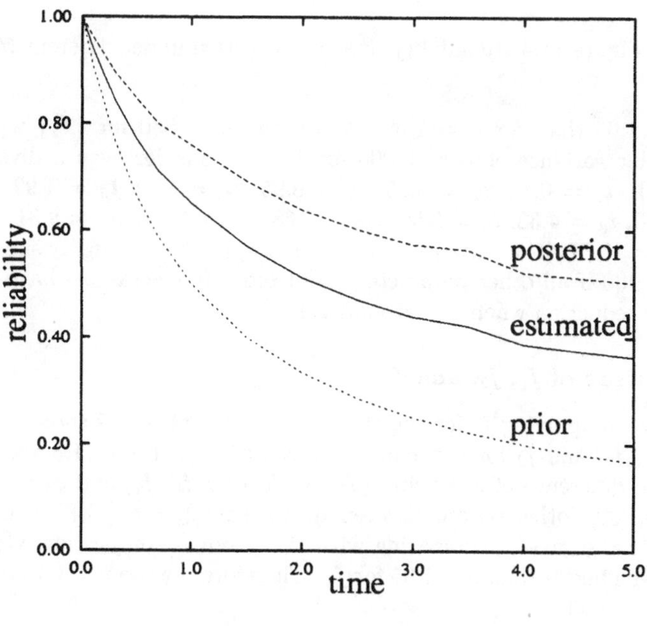

Figure 11.3: Reliability of a series system given observations $\{T_1\}$.

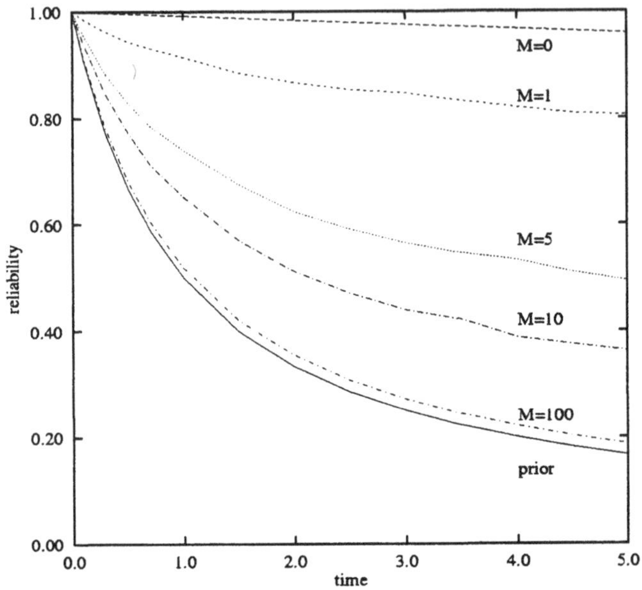

Figure 11.4: Reliability of a series system under different Ms.

for $v = 1.05$ than for $v = 1.09$. Moreover, $L = 830$ for $\{T_2\}$ will result in the similar variance of $L = 1{,}000$ for $\{T_1\}$. Consider even a divergent data set, $\{T_3\}$: $t_1 = 0.11$, $t_2 = 0.33$, $t_3 = 0.95$, $t_4 = 1.19$, $t_5 = 1.92$, $t_6 = 2.26$, $t_7 = 3.30$, $t_8 = 4.38$, $t_9 = 5.50$, $t_{10} = 6.68$, $t_{11} = 7.89$, $t_{12} = 8.31$, $t_{13} = 9.52$, and $t_{14} = 10.57$. The variance of $\{T_3\}$ for $L = 10{,}000$ is roughly similar to $\{T_1\}$ for $L = 400$ if all other parameters are fixed. In general, we have the largest variance reduction when v is around 1.3.

Comparison of f_1, f_2, and f_3

We focus on applying f_1 for the above analyses. Table 11.2 shows the effects of using f_1, f_2, and f_3 for reducing variance of $\{T_1\}$. Table 11.3 and Table 11.4 show the difference of analyzing $\{T_2\}$ ($\{T_3\}$) by f_1, f_2, and f_3 for $L = 1{,}000$ ($L = 10{,}000$) (other parameters remain the same). For $\{T_2\}$, we use $v = 1.05$ for f_3 for the sake of comparing with the goodness of f_2; however, $v = 1.10$ works even better than $v = 1.05$ for f_3. Therefore, we use $v = 1.10$ in Table 11.4 for f_3. $\square$

Notes on Variance Reduction

From the tables shown in Example 11.1, the simulation is proved to be a valid one because the posterior reliabilities, with and without variance reduction, are very close. A large L should be used when n becomes large and large M should

Table 11.2: Reducing variance of data set $\{T_1\}$ by f_1 and f_2.

t	0.1	0.3	0.5	0.7	1.0	1.5	2.0	2.5	3.0	3.5	4.0	4.5	5.0
prior	0.91	0.77	0.67	0.59	0.50	0.40	0.33	0.29	0.25	0.22	0.20	0.18	0.17
$f_1, v = 1.09$													
mean	0.96	0.90	0.85	0.80	0.76	0.69	0.64	0.60	0.57	0.56	0.52	0.51	0.50
Var.	10.5	8.8	10.0	5.4	7.7	6.3	4.4	3.8	3.3	4.9	2.7	3.1	3.5
red. %	1.0	8.2	8.7	8.8	26.1	-9.6	30.9	10.4	22.6	19.2	27.8	32.0	22.6
$f_2, v = 1.0$													
mean	0.96	0.90	0.85	0.80	0.76	0.69	0.63	0.60	0.57	0.56	0.52	0.51	0.49
Var.	6.3	6.6	7.2	4.6	6.4	5.1	2.6	2.9	3.3	4.9	2.0	2.7	2.3
red. %	40.7	31.4	33.6	31.2	38.6	11.7	59.4	31.1	23.8	20.0	47.9	41.0	48.3
$f_3, v = 1.09$													
mean	0.96	0.90	0.85	0.80	0.76	0.69	0.64	0.60	0.57	0.57	0.52	0.52	0.50
Var.	7.7	5.1	6.6	3.6	5.1	4.2	2.8	2.5	2.0	4.3	1.9	2.1	2.4
red. %	27.8	46.5	39.8	46.7	51.6	27.9	56.3	41.5	54.1	30.5	51.2	53.6	47.3

Table 11.3: Reducing variance of data set $\{T_2\}$ by f_1 and f_2.

t	0.1	0.3	0.5	0.7	1.0	1.5	2.0	2.5	3.0	3.5	4.0	4.5	5.0
$f_1, v = 1.0$													
mean	0.96	0.89	0.83	0.79	0.74	0.66	0.60	0.56	0.52	0.50	0.46	0.44	0.42
Var.	2.8	2.3	2.6	1.5	2.4	1.3	1.6	1.0	0.9	1.3	0.8	0.9	0.8
$f_1, v = 1.05$													
mean	0.96	0.89	0.83	0.79	0.73	0.66	0.60	0.56	0.52	0.50	0.46	0.44	0.42
Var.	2.4	2.1	2.2	1.4	2.0	1.3	1.2	0.9	0.9	1.5	0.6	0.8	0.7
$f_2, v = 1.0$													
mean	0.96	0.89	0.84	0.79	0.73	0.66	0.60	0.56	0.53	0.49	0.46	0.45	0.41
Var.	27.9	24.1	21.8	21.1	14.0	15.4	12.5	12.3	11.9	6.9	6.8	11.5	5.1
$f_3, v = 1.05$													
mean	0.96	0.89	0.83	0.79	0.73	0.66	0.60	0.56	0.52	0.50	0.46	0.44	0.42
Var.	2.2	2.0	2.0	1.3	1.8	1.3	1.1	0.9	0.9	1.5	0.6	0.7	0.6
red. %	21.2	14.7	23.5	17.9	24.8	1.5	32.4	14.5	6.7	-13.1	24.7	22.5	23.3

Table 11.4: Reducing variance of data set $\{T_3\}$ by f_1 and f_2.

t	0.1	0.3	0.5	0.7	1.0	1.5	2.0	2.5	3.0	3.5	4.0	4.5	5.0
$f_1, v = 1.0$													
mean	0.99	0.96	0.94	0.93	0.90	0.86	0.84	0.82	0.79	0.77	0.76	0.73	0.73
Var.	20.3	17.5	31.3	17.9	7.3	9.5	8.3	22.9	17.7	13.2	16.2	8.6	13.8
$f_1, v = 1.3$													
mean	0.99	0.96	0.95	0.93	0.90	0.86	0.84	0.82	0.79	0.78	0.76	0.74	0.72
Var.	12.0	17.1	29.1	12.0	20.9	6.2	13.1	12.8	13.3	13.2	18.5	7.9	6.7
$f_2, v = 1.0$													
mean	0.98	0.96	0.95	0.91	0.90	0.88	0.84	0.82	0.78	0.77	0.76	0.74	0.72
Var.	147.6	35.8	133.4	170.5	367.1	692.4	139.1	90.2	20.1	16.9	26.3	81.9	21.2
$f_3, v = 1.1$													
mean	0.99	0.96	0.95	0.93	0.90	0.86	0.84	0.82	0.79	0.77	0.76	0.74	0.73
Var.	13.2	19.2	20.8	14.0	11.5	4.2	11.1	12.9	12.4	11.5	14.3	8.1	7.6
red. %	35.2	-9.5	33.6	21.6	-56.2	55.9	-33.3	43.7	30.0	13.3	11.8	6.7	45.3

be used if we have a strong belief that the prior guess is correct. When M is small, the estimate is close to the MLE. The v remains the same for different M. It is not easy to find the v with the largest variance reduction in a technical way; however, a small v can be expected if the difference among data is not significant.

Generally speaking, difficulties exist for expressing $Z(Q_i)$ and $Y(Q_i)$. The model considered in this chapter is specially designed to have a neat $Z(Q_i)$ and $Y(Q_i)$. It is worth mentioning that usually systems have more complicated configurations; that is, the series assumption can be treated only as an approximation.

Smaller variance and smaller v are expected if the data are closer together in value. Data can be scaled by using different units. For example, hours can be used instead of minutes to make the value of the data smaller and, therefore, closer to one another. If data are more evasive (like $\{T_3\}$), a larger v will result in larger variance reduction. Moreover, the reduced percentage is larger than the closer data set (like $\{T_1\}$) if the variance reduction works.

From Table 11.1 and $\{T_1\}$, one can see that the prior belief about the system reliability is too pessimistic. A bigger β ($\beta > 1$) should be used if the analyst feels confident of the observations. With $\{T_1\}$ and the parameters used to create Table 11.1 (except β), it is equivalent to use $\alpha = 1$ and $\beta = 5$ for our prior guess. From Table 11.2, it is clear that the value-dependent f_2 works much better than the order-dependent f_1 for $\{T_1\}$. Also, from our experiments, the variance reduction of f_2 is usually better than that of f_1 under the same L for most $v > 0$. However, f_2 cannot work for $\{T_3\}$ and $\{T_2\}$ as shown in Figure 11.4.

As indicated in Tables 11.2, 11.3, and 11.4, f_3 works better than f_1 and f_2 in the sense that

- it works for all data sets ($\{T_1\}$, $\{T_2\}$, and $\{T_3\}$),
- in general, it produces larger variance reductions than f_1 and f_2, and
- it is easy to choose the v with large reductions in the variance; we suggest using $v \approx 1.1$.

□ Example 11.2

The weighting function f_3 in Eq. (11.7) is used. Let $\theta = 2{,}000.0$, $\gamma = 0.35$, $\alpha = 1.5$, $\beta = 1.0$, $M = 3$, $L = 1{,}000$, $v = 1.1$, and the twelve observations (in hrs), $\{T_4\}$:

$$
\begin{array}{llll}
t_1 = 1.2, & t_2 = 12.5, & t_3 = 53.9, & t_4 = 76.1, \\
t_5 = 115.0, & t_6 = 269.9, & t_7 = 402.1, & t_8 = 668.5, \\
t_9 = 1{,}198.8, & t_{10} = 1{,}826.7, & t_{11} = 2{,}950.6, & t_{12} = 5{,}302.8.
\end{array}
$$

After deriving the estimated system reliability from Eqs. (11.3) and (11.4), the estimated hazard rate can be obtained from Eq. (11.1). Figures 11.5 and 11.6 illustrate, respectively, the resultant reliability and the corresponding hazard rate of the estimated reliability.

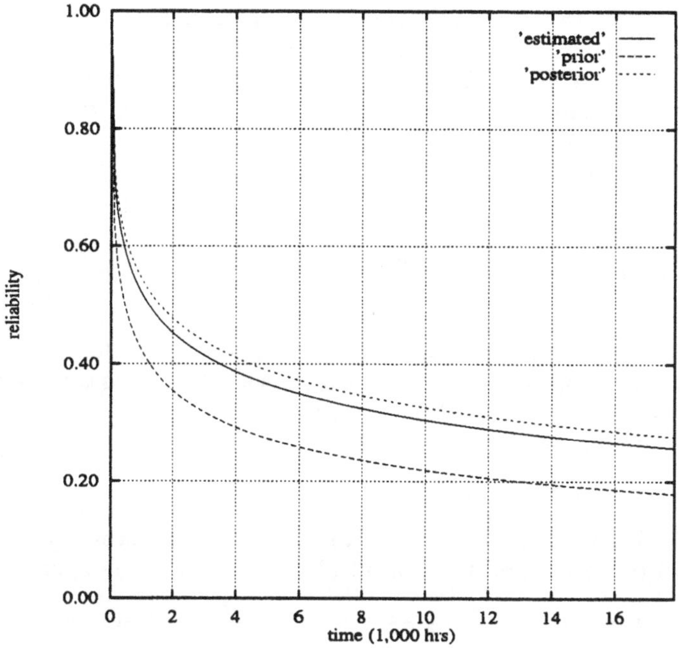

Figure 11.5: The prior, the posterior, and the estimated reliability.

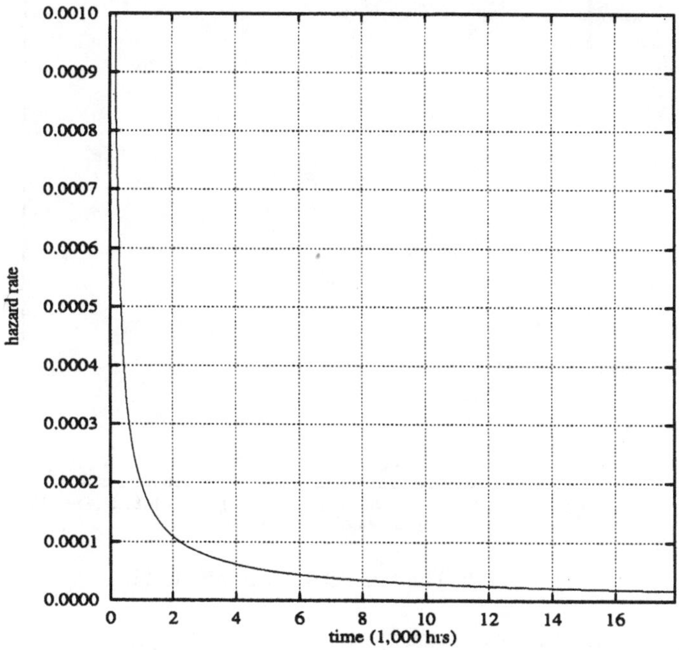

Figure 11.6: The estimated hazard rate.

Let TC_j be the total cost if burn-in time is $j\Delta t$. TC_j is a function of c_b (burn-in cost/hr), c_s (shop repair cost/failure), and c_f (field repair cost/failure). The s-expected number of failures during burn-in and under the field repair is defined as e_s and e_f, respectively. The total number of items in the ith inspection is defined as n_i and the number of good items as s_i.

Let the observation time $t_{ob} = 10,000$hrs, $c_b = 0.05$, $c_s = 0.5$, and $c_f = 100.0$. The e_s and e_f can be obtained by

$$e_{s,j} = \Delta t \sum_{i=1}^{j} h_i, \quad e_{f,j} = \Delta t \sum_{i=j+1}^{k} h_i.$$

Assume the burn-in time is $j\Delta t/\eta$, where η is the time transformation factor, and $t_{ob} = k\Delta t$. Suppose the Arrhenius equation is adequate to transform time, and the system is burned-in at 80°C; then, η is found to be 5.4. Hence, the total cost with burn-in time $t_b = j\Delta t$ is

$$TC_j = c_b j\Delta t/\eta + c_s e_{s,j} + c_f e_{f,j}.$$

The TC_j is drawn in Figure 11.7, from which the optimal burn-in time is set to be 2,350/5.4 = 435.19 hrs. Using smaller Δt will give a better approximation;

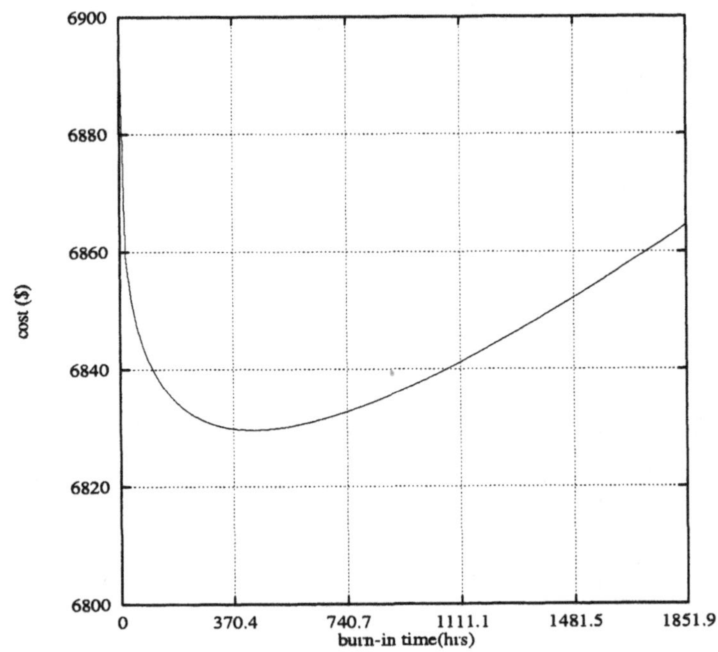

Figure 11.7: The total cost in Example 11.2.

however, it has a longer execution time.

The random environment assumed in the previous example, call it environment 1, has mean 1.5 and variance 1.5 (an exponential distribution). If environment 2 (GAM(6,4)) and environment 3 (GAM(0.3,0.2)) are used for comparison, the hazard rates are drawn in Figure 11.8. All 3 environments

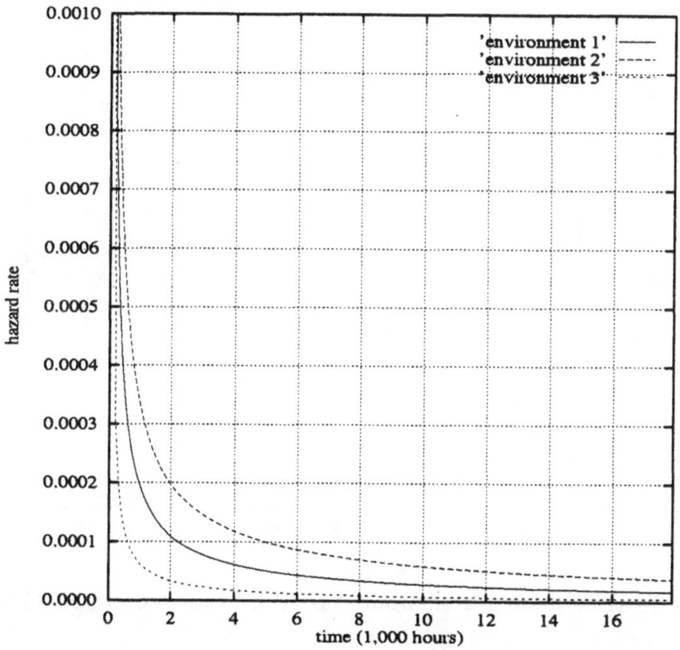

Figure 11.8: Hazard rates at different environments in Example 11.2.

have the same mean but different variance (1.5, 0.375, and 7.5, for environments 1, 2, and 3, respectively). As indicated in Figure 11.8, the system used in environment 3 has the lowest hazard rate. The 12 observations result in aposterior reliability which is higher than our prior guess (from Figure 11.5). If a more unsure environment is assumed, the influence of the observations becomes more significant, and thus a lower hazard rate can be expected.

For a harsher environment 4 (GAM(3,1)), we have higher a hazard rate, as shown in Figure 11.9. One way to use Eq. (11.3) for a non-random environment is to assign large α and β; for example, let $\alpha = 10,000m$ and $\beta = 10,000$ where m is the estimated intensity of the environment. To avoid overflow during computation, the $R_o(t|\theta)$ in Eq. (11.3), for example, can be calculated by $R_o(t|\theta) = \exp\left[\alpha \ln(\frac{\beta}{\beta+t/\theta})^{\gamma}\right]$. □

11.2.1. An Alternative Approach

We will use the following model to illustrate the case where the Dirichlet process is not a good choice. Suppose there are s suppliers of a raw material for a company; multiple suppliers reduce the risk derived from the loss of any one supplier. The ratio of the material from each supplier is known as p_i. The quality of the incoming raw material is not fully known. However, it is believed

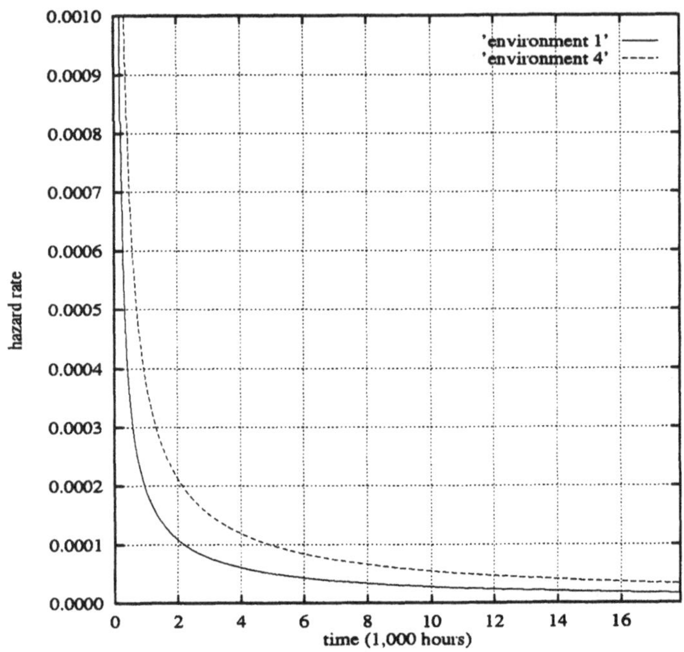

Figure 11.9: A harsher environment hazard rate in Example 11.2.

that the percentage of good items, x_i, in a shipment from each supplier can be treated as a beta distribution, BETA(a_i,b_i). The engineers are able to estimate a_i and b_i based on their prior knowledge and confidence about each supplier. Suppose we are furnished with the data in Table 11.5.

Table 11.5: Parameters of alternative approach.

Supplier (i)	1	2	3	4	5
p_i	0.05	0.15	0.45	0.25	0.10
a_i	38	33	27	22	14
b_i	2	2	3	3	2
mean	0.950	0.943	0.900	0.880	0.875
Var.($\times 10^{-3}$)	1.159	1.497	2.903	4.062	6.434

The density of each x_i is drawn in Figure 11.10. Under this model, the

$Z(Q_i)$ and $Y(Q_i)$ can be derived as:

$$Z(Q_i) = \prod_{K \in Q_i} \sum_{j=1}^{s} p_j \left[\mathcal{B}(a_j, b_j)\right]^{|K|} \left[\prod_{x_i \in K} x_i\right]^{a_j} \left[\prod_{x_i \in K} (1 - x_i)\right]^{b_j}$$

$$Y(Q_i) = \frac{Z(Q_i)}{L} \times \sum_{K \in Q_i} |K| \frac{\sum_{j=1}^{s} p_j \left[\mathcal{B}(a_j, b_j)\right]^{|K|+1} \left[x \prod_{x_i \in K} x_i\right]^{a_j} \left[(1-x) \prod_{x_i \in K} (1 - x_i)\right]^{b_j}}{\sum_{j=1}^{s} p_j \left[\mathcal{B}(a_j, b_j)\right]^{|K|} \left[\prod_{x_i \in K} x_i\right]^{a_j} \left[\prod_{x_i \in K} (1 - x_i)\right]^{b_j}}.$$

The prior density is

$$R_0(x) = \sum_{j=1}^{s} p_j \left[\mathcal{B}(a_j, b_j)\right] x^{a_j} (1 - x)^{b_j}.$$

The raw material is stored in a common area after receipt from the suppliers,

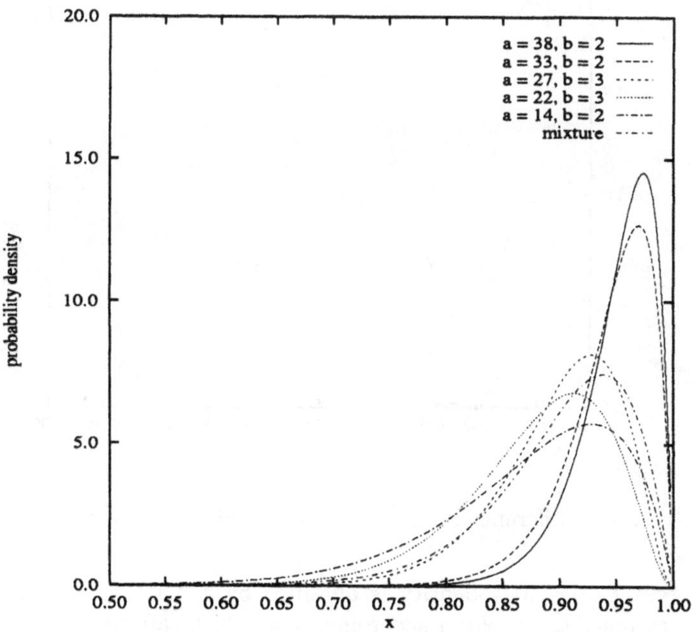

Figure 11.10: The density of x_i.

and ten inspections are recorded. The x_i, s_i, and n_i are shown in Table 11.6.

As before, for the Dirichlet process scheme, the weighting function Eq. (11.7) is used and $v = 1.005$ and $L = 10,000$ are chosen. However, it is not valid to apply the Dirichlet process to this model because the observations indicate the prior knowledge is improper; to have the Dirichlet process work, we have to

Table 11.6: Data of the inspection.

Inspection (i)	1	2	3	4	5	6	7	8	9	10
n_i	21	12	13	15	17	12	15	8	10	11
s_i	11	9	10	12	14	10	13	7	9	10
$x_i(= s_i/n_i)$	0.52	0.75	0.77	0.80	0.82	0.83	0.87	0.88	0.90	0.91

be confident that the prior information is quite accurate. A similar conclusion is described in [313]. Figure 11.11 shows the difference between the density estimated by the Dirichlet process scheme and the "exact" posterior density.

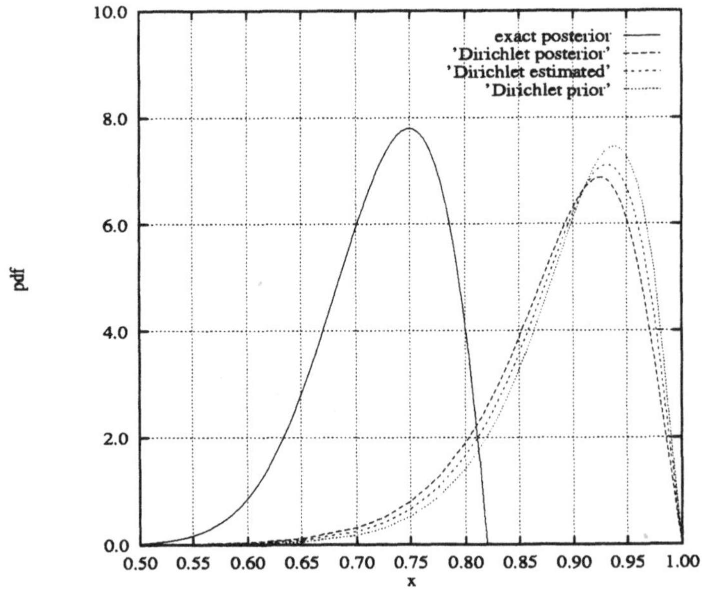

Figure 11.11: The difference between the estimated and the exact density.

The "exact" posterior density is obtained by simulation in which the sample x is assigned to supplier i according to q_i, which satisfies

$$q_i = \frac{p_i f_i(x|a_i, b_i)}{\sum\limits_i p_i f_i(x|a_i, b_i)}$$

where $f_i(x|a_i, b_i)$ is the density of the corresponding Beta distribution for supplier i. Repeat the simulation for L times, the average quantity of the total samples ($\overline{N}_{n,i}$) and the successful number ($\overline{N}_{s,i}$) assigned to the supplier

$$\overline{N}_{n,i} = (\sum_{j}^{L} \sum_{k \in i} n_k)/L$$
$$\overline{N}_{s,i} = (\sum_{j}^{L} \sum_{k \in i} s_k)/L.$$

Under binomial sampling, the posterior distribution of quality from the ith supplier becomes

$$f_i(x|a_i', b_i') = B(a_i', b_i')x^{a_i'}(1-x)^{b_i'}$$
$$a_i' = a_i + \overline{N}_{s,i}$$
$$b_i' = b_i + \overline{N}_{n,i} - \overline{N}_{s,i}.$$

After simulation, the quality distribution of the ith supplier is tabulated in Table 11.7. From the previous tables, it is clear that the prior information about

Table 11.7: Parameters of the supply quality.

Supplier (i)	1	2	3	4	5
a_i'	38.949	37.236	70.086	56.659	36.070
b_i'	2.133	2.642	11.433	11.624	13.168
mean	0.948	0.934	0.860	0.830	0.733
Var.($\times 10^{-3}$)	1.170	1.513	1.461	2.039	3.900

the suppliers is not accurate; the inspection outcome indicates that the quality level is not as high as expected. The simulation results in putting more weight on the fifth supplier, whose quality level is the lowest and, from Tables 11.5 and 11.7, whose distribution has the largest change. The prior information is modified to form the posterior distribution for each supplier, which can be used as the prior belief for later analysis.

When engineers believe they have reliable and proper prior information, the Dirichlet process seems to be a good candidate for Bayesian analysis. Consider the other experiment containing five inspections as shown in Table 11.8.

Table 11.8: Data of alternative inspection.

Inspection (i)	1	2	3	4	5
n_i	20	18	15	13	10
s_i	19	17	14	12	9
$x_i(= s_i/n_i)$	0.950	0.944	0.933	0.923	0.900

The prior, posterior, estimated, and "exact" probability density are plotted in Figure 11.12. In this case, after 10,000 simulations, only supplier #1's parameters are updated (38 $\rightarrow$ 109 and 2 $\rightarrow$ 7). Figure 11.12 indicates that the estimated and the "exact" densities are almost the same when the prior information and the samples match.

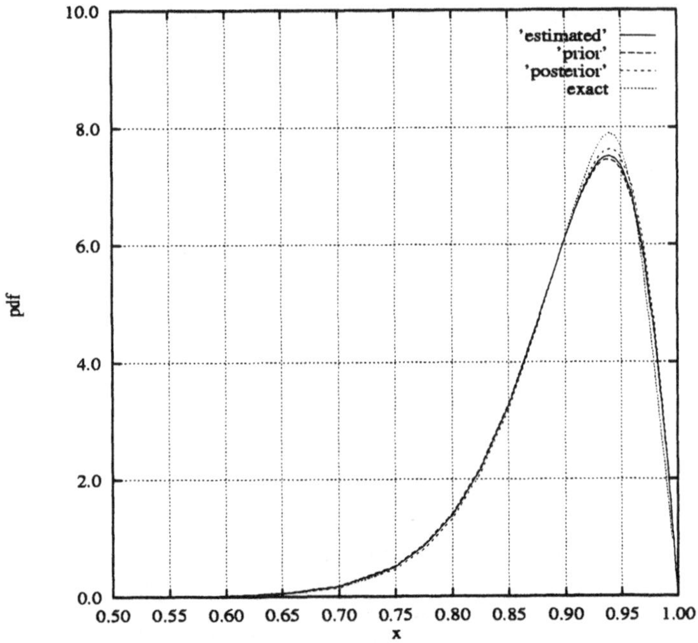

Figure 11.12: Four densities of inspection #2.

11.3. Determining Optimal Burn-in Time Using the Dirichlet Process

A tree representation for the Dirichlet process is introduced in [141] where the dyadic intervals are suggested to clarify the concept. When the cost parameters are given, we apply the same notion by modifying the tree representation for the Dirichlet process to find the optimal burn-in time of a complex system whose failure mechanism is not fully known but will be approximated by comparison with other similar products. This approach can be considered a nonparametric one since we do not assume any probability distribution for the system of interest.

11.3.1. Method

We define some notation used in this section.

$h(t)$	s-expected hazard rate without adjustment by the PAV algorithm (in 10^{-6})
$h_a(t)$	s-expected hazard rate with adjustment by the PAV algorithm (in 10^{-6})
p_i	s-expected number of failure in the ith interval

t_b, $t_{b,i}$	optimal burn-in time and the burn-in time after the ith interval, respectively (in hrs)
t_i	end point of the ith interval (in hrs); define $t_0 = 0$
t_q	terminating time of the experiment (in hrs)
x_i	failure time of the ith sample
α, m	parameters of the Dirichlet process

The dyadic intervals used in [141] is shown in Figure 11.13. The condition

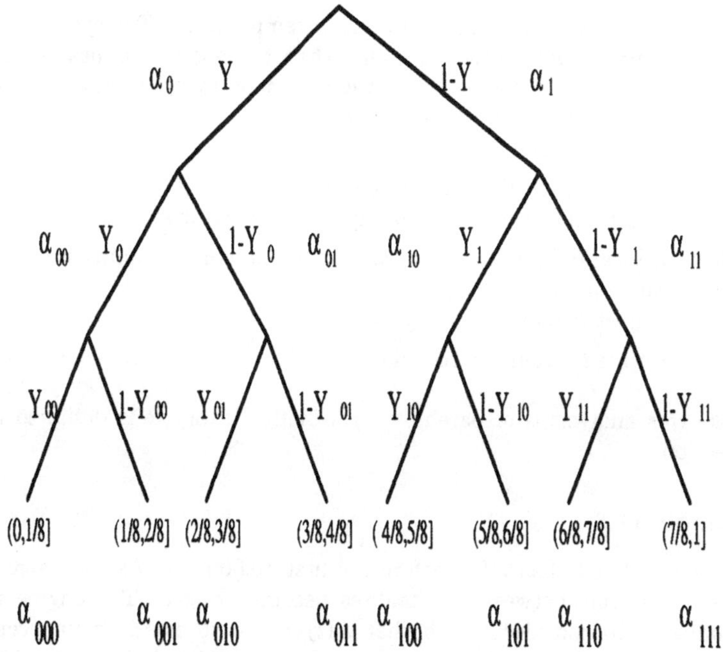

Figure 11.13: The Dyadic intervals used by Ferguson.

for this tree to be a Dirichlet process is

$$\alpha_{\epsilon_1\epsilon_2\cdots\epsilon_m} = \alpha_{\epsilon_1\epsilon_2\cdots\epsilon_m^0} + \alpha_{\epsilon_1\epsilon_2\cdots\epsilon_m^1} \tag{11.8}$$

where $\epsilon_1\epsilon_2\cdots\epsilon_m$ denotes the binary expansion of the dyadic rational $\sum_{j=1}^{m}\epsilon_j 2^{-j}$ and ϵ_j is either 1 or 0.
The non-negative random variables Y are independent between rows [141]; for example,

$$Y_{00} \sim \text{BETA}(\alpha_{000}, \alpha_{001}). \tag{11.9}$$

Suppose a complicated, newly-developed system, Q, is under investigation. The analysts are interested in estimating its failure rate so that a system burn-in time can be determined. Since the system under test Q is composed of many components that are not necessarily arranged in a series, it is difficult to express the Qs failure rate in a technical manner, even when Qs component failure mechanisms are known. Thus, a nonparametric approach is needed, and its procedure is as follows.

Step 1 Look for a system (call it a comparison system) whose structure, components, and other factors are similar to Q and whose failure mechanisms (hazard rate, reliability, and others) are well-known; the prior knowledge of Q can be established after this comparison.

Step 2 Find the t_is and form a tree, which might be a truncated one.

Step 3 Decide the safest system burn-in temperature. The system burn-in temperature should be higher than the highest endurable temperature of the components, which can be found in many electronic handbooks, used in the system.
Decide t_{ob} and t_q.

Step 4 Proceed with the accelerated life testing.
Record the failure time if any failure is detected.

Step 5 Allocate the number of failures in the corresponding interval and calculate p_i and $h(t_i)$.
Adjust $h(t_i)$ if necessary.

Step 6 Derive t_b^* from the cost data.

To make this analysis transparent, a hypothetical example is designed in the next section.

□ Example 11.3

It is believed that Q has DFR before the first 10,000 hrs. As the hazard rate decreases, the time between two failures becomes longer. The engineers feel it is sufficient to consider t such that $F(t) < 0.5$ to avoid an unacceptably long testing time (i.e., avoid using large t_q). A truncated tree as depicted in Figure 11.14 is formed. The values of α_0, α_{00}, α_{000}, $\cdots$, α_{0110}, and α_{0011} can be obtained by Eq. (11.8) once $m_1 \sim m_{17}$ are given. That is,

$$\begin{aligned}
\alpha_{0000} &= m_1 + m_2, \alpha_{0001} = m_3 + m_4, \alpha_{0010} = m_5 + m_6, \alpha_{0011} = m_7 + m_8, \\
\alpha_{000} &= \sum_{i=1}^{4} m_i, \alpha_{001} = \sum_{i=5}^{8} m_i, \\
\alpha_{00} &= \sum_{i=1}^{8} m_i, \alpha_0 = \sum_{i=1}^{16} m_i, \quad \text{and others.}
\end{aligned}$$

m_1 through m_{17} are chosen to reflect the possibility that a failure will fall in the corresponding interval based on the prior information; the larger the value, the higher the probability of a failure.

The system burn-in temperature is set to 80°C. The Arrhenius equation is assumed for time transformation; that is, the re-scaling factor is 5.4 [76], which means the effect of having systems on the stress test for 10 hrs, say, is

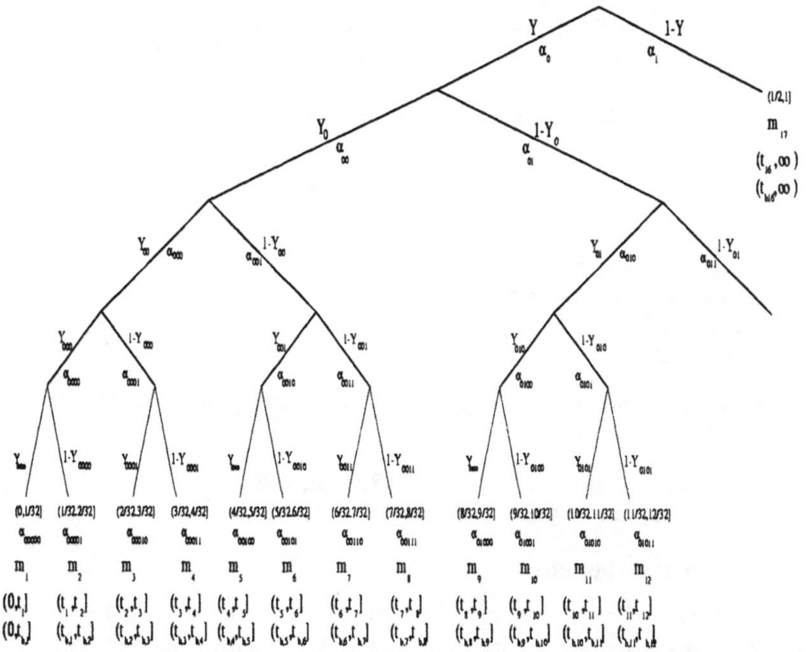

Figure 11.14: A truncated tree for Example 11.3.

Table 11.9: Information on the similar system.

i	$R(t_i)$	$F(t_i)$	$H(t_i)$	t_i (hrs)	$t_{b,i}$ (hrs)
0	1	0	0	0	0
1	31/32	1/32	0.03175	100	18.5
2	30/32	2/32	0.06454	400	74.1
3	29/32	3/32	0.09844	950	175.9
4	28/32	4/32	0.13353	1,750	324.1
5	27/32	5/32	0.16990	2,900	537.0
6	26/32	6/32	0.20764	4,300	796.3
7	25/32	7/32	0.24686	6,100	1,129.6
8	24/32	8/32	0.28768	8,250	1,527.8
9	23/32	9/32	0.33024	10,900	2,018.5
10	22/32	10/32	0.37469	13,850	2,564.8
11	21/32	11/32	0.42121	16,900	3,129.6
12	20/32	12/32	0.47000	20,100	3,722.2
13	19/32	13/32	0.52130	23,500	4,351.9
14	18/32	14/32	0.57536	27,100	5,018.5
15	17/32	15/32	0.63252	30,850	5,713.0
16	1/2	1/2	0.69315	34,850	6,453.7

equivalent to using the systems for 54 hrs in normal (unstressed) conditions. The $R(t_i)$ and $H(t_i)$ of the comparing system is tabulated in Table 11.9.

If 64 records of the comparing system are collected, we let $\sum_j m_j = 64$; this is based on the Rule 2 in [107]. Thus, we set

$$m_1 = m_2 = \cdots = m_{16} = 2, \text{ and } m_{17} = 32.$$

That implies

$$\begin{aligned}
&\alpha_{0000} = \alpha_{0001} = \cdots = \alpha_{0111} = 4,\\
&\alpha_{000} = \cdots = \alpha_{011} = 8,\\
&\alpha_{00} = 16,\\
&\alpha_0 = 32.
\end{aligned}$$

Suppose twenty samples of Q are put under stress test for $t_q = 6{,}453.7$ hrs, and the eight failure times are

$$\begin{aligned}
&x_1 = 1.3, \quad && x_2 = 15.0, \quad && x_3 = 30.8, \quad && x_4 = 81.1,\\
&x_5 = 201.3, && x_6 = 448.7, && x_7 = 920.1, && x_8 = 2051.1.
\end{aligned}$$

The new m' and α' become

$$\begin{aligned}
&m'_1 = m_1 + 2 = 4, && m'_2 = m_2 + 1 = 3, && m'_3 = m_3 + 1 = 3,\\
&m'_4 = m_4 + 1 = 3, && m'_5 = m_5 + 1 = 3, && m'_6 = m_6 = 2,\\
&m'_7 = m_7 + 1 = 3, && m'_8 = m'_9 = 2, && m'_{10} = 3,\\
&m'_{11} = \cdots = m'_{16} = 2, && m'_{17} = m_{17} + 12 = 44,
\end{aligned}$$

$$\begin{aligned}
&\alpha'_{0000} = 7, && \alpha'_{0001} = 6, && \alpha'_{0010} = 5, && \alpha'_{0011} = 5, && \alpha'_{0100} = 5,\\
&\alpha'_{0101} = 4, && \alpha'_{0110} = 4, && \alpha'_{0111} = 4, && \alpha'_{000} = 13, && \alpha'_{001} = 10,\\
&\alpha'_{010} = 9, && \alpha'_{011} = 8, && \alpha'_{00} = 23, && \alpha'_{01} = 17, && \alpha'_0 = 40.
\end{aligned}$$

Let $P(B)$ be the product of all the variables associated with the path in the tree from $(0, 1]$ to B, then

$$P(B) = \Big(\prod_{j=1;\epsilon_j=0}^{n} Y_{\epsilon_1\cdots\epsilon_{j-1}} \Big) \Big(\prod_{j=1;\epsilon_j=1}^{n} (1 - Y_{\epsilon_1\cdots\epsilon_{j-1}}) \Big). \tag{11.10}$$

From Eqs. (11.9) and (11.10), we have, with p_1 as an example,

$$p_1 = \frac{40}{84} \times \frac{23}{40} \times \frac{13}{23} \times \frac{7}{13} \times \frac{4}{7} = \frac{4}{84}.$$

The p_i in each interval, denoted by I_1 through I_{17}, is shown in Table 11.10. The expected hazard rate in I_i is obtained by

$$h(t_i) = p_i/(t_i - t_{i-1}).$$

As indicated in Table 11.10, $h(t_i)$ may not be strictly decreasing. To solve this situation, if we believe Q has a DFR, the PAV algorithm can be applied to adjust the $h(t_i)$. The values for $h(t_i)$ and the $h_a(t_i)$ are also listed in Table 11.10.

Table 11.10: The expected failure number in each interval.

Interval	I_1 $(0,t_1]$	I_2 $(t_1,t_2]$	I_3 $(t_2,t_3]$	I_4 $(t_3,t_4]$	I_5 $(t_4,t_5]$	I_6 $(t_5,t_6]$	I_7 $(t_6,t_7]$	I_8 $(t_7,t_8]$	
$t_i - t_{i-1}$	100	300	550	800	1,150	1,400	1,800	2,150	
p_i	4/84	3/84	3/84	3/84	3/84	2/84	3/84	2/84	
$h(t_i)$	476.2	119.0	64.9	44.6	31.1	17.0	19.8	11.1	
$h_a(t_i)$	476.2	119.0	64.9	44.6	31.1	18.6	18.6	11.1	
Interval	I_9 $(t_8,t_9]$	I_{10} $(t_9,t_{10}]$	I_{11} $(t_{10},t_{11}]$	I_{12} $(t_{11},t_{12}]$	I_{13} $(t_{12},t_{13}]$	I_{14} $(t_{13},t_{14}]$	I_{15} $(t_{14},t_{15}]$	I_{16} $(t_{15},t_{16}]$	I_{17} (t_{16},∞)
$t_i - t_{i-1}$	2,650	2,950	3,050	3,200	3,400	3,600	3,750	4,000	—
p_i	2/84	3/84	2/84	2/84	2/84	2/84	2/84	2/84	44/84
$h(t_i)$	8.98	12.1	7.81	7.44	7.00	6.61	6.35	5.95	—
$h_a(t_i)$	10.6	10.6	7.81	7.44	7.00	6.61	6.35	5.95	—

Suppose we want to evaluate the system's performance at t_{ob} = 8,760 hrs (= 1 year), then TC_i can be expressed by

$$\begin{aligned} TC_i = & \; c_b t_{b,i} + c_s \sum_{j=1}^{i} h_a(t_j)\Delta t_j + c_f \sum_{j=i+1}^{8} h_a(t_j)\Delta t_j + c_f h_a(t_9)(t_{ob} - t_8) \\ = & \; c_b t_{b,i} + c_s \sum_{j=1}^{i} p_j + c_f \sum_{j=i+1}^{8} p_j + c_f p_9 \frac{t_{ob} - t_8}{t_9 - t_8} \end{aligned}$$

where $\Delta t_j = t_j - t_{j-1}$. Let c_b=0.05, c_s=40, and c_f=320 ($t_0 = 0$), we have the TC_i summarized in Table 11.11. Hence, t_b^* is set to be 324.1 hrs (= $t_{b,4}$) which

Table 11.11: The expected total cost.

Burn-in time	$t_{b,1}$	$t_{b,2}$	$t_{b,3}$	$t_{b,4}$	$t_{b,5}$	$t_{b,6}$	$t_{b,7}$	$t_{b,8}$
TC_i	76.68	69.46	64.55	61.96	62.60	68.90	75.57	88.81

results in the smallest expected total cost. After burning-in Q for $t_{b,4}$ hrs, we have

$$\begin{aligned} m_{\text{new},i} &= m_{\text{old},i} + m_t(1/32), \quad i = 1, \cdots, 12 \\ m_{\text{new},13} + m_{\text{new},14} &= m_{\text{old},17} + m_t(20/32) \\ m_t &= \sum_{i=1}^{4} m_{\text{old},i}. \end{aligned}$$

In other words, the m_js, where $j \le i$ and $t_{b,i} = t^*_{b,i}$, of the previous tree are distributed to the new tree according to the length of $\left(F(t_k) - F(t_{k-1})\right)$, $k = 1$, $\cdots$, 14 by defining $t_{14} = \infty$. The knowledge of the comparison system and field failures, whether from tests or customer reports, is used to estimate $m_{\text{new},13}$, $m_{\text{new},14}$, and $t_{\text{new},13}$. The collected failure records can also be used to make the tree reliable so that it can be used as the comparison system for revised models or similar systems. □

11.4. Conclusions

If the available data for a previous system is large, e.g., over 100, it implies that

- our prior information is more accurate because more precise analysis can be expected if more data is provided,
- the analysts have more confidence in the prior information, and
- unless the sample size from Q is large enough, the posterior estimate will show little change.

That is, the available information about the comparison system plays a role similar to the α used in [107] and [141], the degree of confidence of the prior information. A modification for assigning the m_is is to adjust the available sample size of the comparing system by a positive weighting factor, w; for $w < 1$, we reduce the importance of the prior information. On the contrary, any $w > 1$ implies that more weight is put on the comparison system.

12. Software Reliability and Infant Mortality Period of the Bathtub Curve

In the past 30 years, numerous papers have been published in the areas of software reliability modeling, model validation, measurement, and practice [226]. Software has become an essential part of many industrial, military and commercial systems. In today's large systems, software life-cycle cost (LCC) typically exceeds that of hardware, with 80 $\sim$ 90% of these costs going into software maintenance to fix, adapt, and expand the delivered program to meet the changing and growing needs of the users [589, p 305]. Compiling cost expenditures from all industries, in 1960 about 20% of the system's cost was spent on software. That percentage has risen to 80% in 1985 and 90% in 1996 [385]. The investment in software increased even more dramatically for the military industry.

The current trend in systems is approaching software domination rather than hardware domination [176, p 216]. Unfortunately, the relative frequency of failure could be as high as 100:1 more software failures than hardware failures [145, p 332]. This ratio may be higher for extremely complicated chips. In January 1986, Reliability Plus Inc. measures showed the MTBF for IBM central processors (3080 class machines) to be between 20,000 and 30,000 hrs; by the fall of 1986, these same machines had demonstrated MTBFs of over 80,000 hrs, a very substantial reliability growth (in hardware). However, newly developed software has an average of only 160 to 200 hrs between faults, although the vast majority (more than 90%) of these software faults were not considered fatal [219, p 190]. In 1994, some Pentium processors were reported to be defective, which forced the Intel Corporation to take enormous effort to replace the bad ones.

Although there have been many software reliability models suggested and studied, none are valid for all situations. The reason is probably due to the fact that the assumptions made for each model are only approximations or are correct for only some, but not all, situations. Analysts have to choose models whose assumptions match their systems. This chapter presents a review of some well-known reliability models, both stochastic and non-stochastic (static) models as a basis for paving the way to the future development and evaluation of highly reliable software and of systems involving software and hardware.

12.1. Basic Concept and Definitions

Before specific models can be addressed, we need to first understand the basic terms and concepts of software reliability [295].

12.1.1. Failures and Faults

A software failure is defined as a departure of the external results of program operation from program requirements on a run. A software fault is defined as a defective, missing, or extra instruction or set of related instructions that is the cause of one or more actual or potential failure types. An entire set of defective instructions which is causing a failure is considered to be a fault.

12.1.2. Environment and Operational Profile

The operational condition of software refers to the environment under which the software is designed to be used. The software environment is usually described by the operational profile. We need to illustrate the concept of the operational profile with several terms: input variable, input state, and run type. The input variable is a variable that exists external to the program and is used by the program in executing its functions. Every program generally has a large number of associated input variables. Each set of values for different input variables characterizes an input state. Every individual input state identifies a particular run type of the program. Therefore, runs can always be classified by their input states. Each run type required of the program by the environment can be viewed as being randomly selected. Operational profile is the frequency distribution of a set of run types that the program executes. It shows the relative frequency of occurrence as a function of different input states. A realistic operational profile will be a fuzzy function of possible input states. Thus, we can define the operational profile as a set of run types that the program can execute along with the probabilities with which they will occur.

12.1.3. Software Reliability

Software reliability is defined as the probability that a software will function without failure for a specified period of time in a specified environment. There are four general ways of characterizing failure occurrences in time: time of failure, time interval between failures, cumulative failures experienced up to a given time, and failures experienced in a time interval. From the definition, reliability quantities have usually been defined with respect to time, although it is possible to define them with respect to other variables; e.g., number of failures in 100 printed pages.

Three kinds of time are considered here: execution time, calendar time, and clock time. The execution time for a program is the time that is actually spent by a processor in executing the instructions of the program. Though the execution time is very important, the calendar time is more meaningful to the users or managers. The clock time represents the elapsed time from start to end of a program executed on a running computer. It includes the waiting time and the execution time of other programs. Execution time, calendar time, and clock time are exchangeable if we are given the usage information of personnel and computer [295].

12.1.4. Characteristics of Software Reliability

The hardware and software contributions to reliability are not independent [145, p 334]. Changes in either the hardware or the software can (and usually do) affect the other [453, p 428]. In hardware testing, the statistical emphasis is often on estimating the failure rate of an item; whereas in software testing, the main statistical emphasis is on estimating the number of errors remaining in the system [220, p 338]. Unlike software reliability models, where no universally accepted one has been found, hardware reliability models, especially the ones for electronic components and systems, are generally accepted to describe the failure mechanisms of components. For example, the Weibull distribution is widely used for the failure behavior of many semiconductor components as we have demonstrated in previous chapters.

Software reliability and hardware reliability have distinct characteristics. Some characteristics of software are:

- **Imperfect debugging**
  When attempting to remove an error, new errors may be introduced into the program during the error correction process [260, p 253].

- **Not all errors are created equal**
  Different errors have different implications and thus need different handling [260, p 253]. Different faults may differ greatly in contribution to the total failure rate [448, p 718].

- **Fast growing complexity**
  The basic problem in software is that the complexity of the tasks which software must perform has grown faster than the technology for designing, testing, and managing software development [383, p 48].

- **No universally applicable model**
  It is shown in [2, p 965] that a good "model" is not sufficient to produce good predictions, and that there is no universal "best buy" among the competing methods. Prediction systems perform with varying adequacy on different data sources. In other words, every software project is different; the specific process models, tools, or even management techniques that have improved the productivity of one site will not necessarily have the same effect on all sites [54, p 288]. Users need to be able to select, from a plethora of prediction systems, one (or more) that gives trustworthy predictions for the particular software being studied.

- **Data aging**
  It is not necessarily the case that all the failure data should be used to estimate model parameters and to predict failures [366, p 1095]. The reason is that old data may not be as representative of the current and future failure process as recent data.

- **Fatal side-effects from a hardware-software system**
  Software is not intrinsically hazardous; however, when software is integrated with hardware, it has the potential for catastrophic effect [176].

- **Critical early testing**
  The earlier in the development cycle we can test, the earlier we can predict the operational reliability, and the more time we have to take action to improve the software in response to a gloomy prediction. However, the trade-off is: the earlier we go in the test cycle, the less realistic, the less documented, and the more scattered the test results are [383, p 50].

- **Changeable and unpredictable user environments**
  In general, software error detection during the operational phase is different from that during testing [306, p 596].

12.1.5. Unified Theory for Software and Hardware Reliability

Imperfect debugging commonly exists in software development. However, in the long-term, most debugging efforts are hopefully successful, and consequently the failure chance becomes smaller as debugging activities continue. Differing from the hardware failure, which follows the bathtub curve, software should not be aging and is fully expected to stay in the decreasing failure rate mode. Also once a software bug is removed, the population that contains bugs changes. Therefore, reliability should be used in software with special attention.

While hardware burn-in is "the pre-conditioning of assemblies and the accelerated power-on tests performed on equipment subjected to temperature, vibration, voltage, and humidity cycling," we can consider software debugging as a software burn-in process described as "an intensive effort for error debugging, system updating, modeling validation, and others by system analysts." Thus, the software failure mode stays in the infant mortality stage of the bathtub curve. Namely, software reliability improvement is essentially the burn-in process applied in electronic systems. This observation will be clear after reading the following sections.

12.2. Stochastic Software Reliability Models

The stochastic software reliability models (SWRM) include the reliability growth models and the reliability prediction models. To have good predictive results by using the reliability growth models, we have to make sure that the experimental environment is represented by the operational profile.

12.2.1. Types of Reliability Models

A stochastic process is usually incorporated in the description of the failure phenomenon by the Markov assumption: given the process is at a specific state, its future development does not depend on its past history and the transition probabilities among these states depend only on the present state. Markov models are very useful in studying software fault-removal processes, especially during the testing phase. Here the state of the process at time t is the number of remaining faults at that time.

The fault-removal process can usually be described by a pure death process, since the number of remaining faults is a decreasing function of time, provided that no new fault is introduced. If we also consider the introduction of new faults due to incorrect debugging, the birth-death process can be applied. Some Bayesian models deal mainly with inference problems based on failure data.

Selected stochastic models are discussed with their characteristics, merits, and shortcomings. The static models, such as the input-domain-based models, which do not take the failure process into consideration, are included in the next section.

The notation used in Sections 12.2.2 ~ 12.2.8 are as follows:

- N number of faults present in the software at the beginning of the test phase
- T time at which we want to estimate the software reliability measures after testing has begun (observation time), assuming that the test begins at time 0, t is a realization of T
- X_i time between the $(i-1)^{st}$ and the ith failures i.e., the debugging interval however, the time interval is used when data is collected with respect to constant time intervals, such as weeks or months, x_i is a realization of X_i
- T_i the ith software failure time; $T_0 = 0$ and $X_i = T_i - T_{i-1}$, t_i is a realization of T_i
- m number of software failures observed; set $T = T_m$ if we terminate the test immediately after the mth failure and employ the software reliability model; otherwise, $T > T_m$
- m_i number of software faults removed during the ith debugging interval
- n_i cumulative number of faults removed from the software during the first i debugging or time intervals; i.e., $n_i = \sum_{j=1}^{i} m_j$
- n total number of faults removed from the software during $[0, T]$; i.e., $n = n_m$
- N_t number of faults remaining in the software at time t; $N_0 = N$ and $N_T = N - n$
- λ_i hazard rate between the $(i-1)^{st}$ and the ith software failures
- ϕ the size of a fault.

12.2.2. The Jelinski-Moranda Model

The Jelinski-Moranda (JM) model [202], one of the earliest software reliability models, assumes:

1. the number of initial software faults is an unknown but fixed constant,
2. a detected fault is removed immediately and no new fault is introduced,

3. times between failures are independent, exponentially distributed random quantities; i.e., $T_1, T_2, \ldots$ are independent r.v. with exponential pdf:

$$\begin{aligned} \Pr\{t_i \mid \lambda_i\} &= \lambda_i e^{-\lambda_i t_i}, \quad t_i > 0 \\ \lambda_i &= (N - n_{i-1})\phi, \end{aligned} \tag{12.1}$$

and

4. all remaining software faults contribute the same amount to the software failure intensity.

The failure rate vs time diagram for the JM model is shown in Figure 12.1. The parameters of the JM model are N and ϕ, which can be estimated by

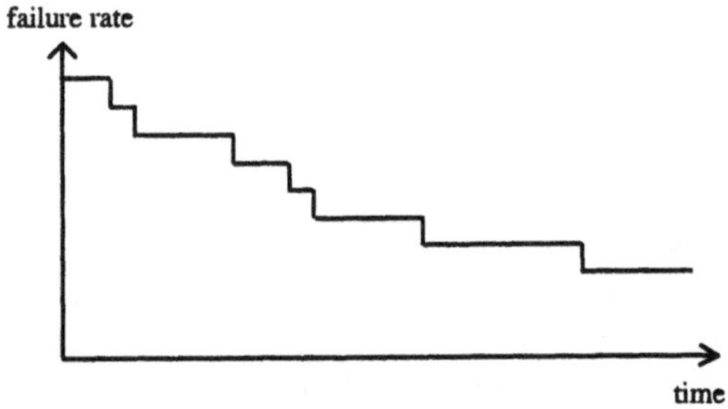

Figure 12.1: The failure rate of the JM model.

the maximum likelihood method given x_i, m, m_i. The maximum likelihood estimators, $\hat{N}$ and $\hat{\phi}$, satisfy the following equations

$$\begin{aligned} \sum_{i=1}^{m} \frac{1}{\hat{N} - n_{i-1}} - \sum_{i=1}^{m} \hat{\phi} x_i &= 0 \\ n/\hat{\phi} - \sum_{i=1}^{m} (\hat{N} - n_{i-1}) x_i &= 0. \end{aligned}$$

Note that $\hat{N}$ is an integer and $\hat{\phi}$ is a real number. Some performance measures can then be derived:

- The number of remaining errors
  $N_T = N - n$
- The mean time to failure
  $\text{MTTF} = \frac{1}{(N-n)\phi}$
- The reliability
  $R(t) = \exp[-(N - n)\phi t], \quad t \geq 0$
- The mean and variance of a perfect debugging system

$$\mathrm{E}[T_n] = \sum_{i=1}^{N-n} \frac{1}{i\phi}, \quad \mathrm{Var}[T_n] = \sum_{i=1}^{N-n} \frac{1}{(i\phi)^2}.$$

The disadvantages of the JM model are that it assumes that all faults contribute equally to the unreliability of a program and in certain cases, one might have $\hat{N}=\infty$, $\hat{\phi}=0$ [606, p 31].

12.2.3. The Schick and Wolverton Model

Schick and Wolverton (SW) [363] modified the JM model by assuming that

$$\lambda_i = \phi(N - n_{i-1})x_i.$$

The failure rate of the SW model is given in Figure 12.2. The failure rate

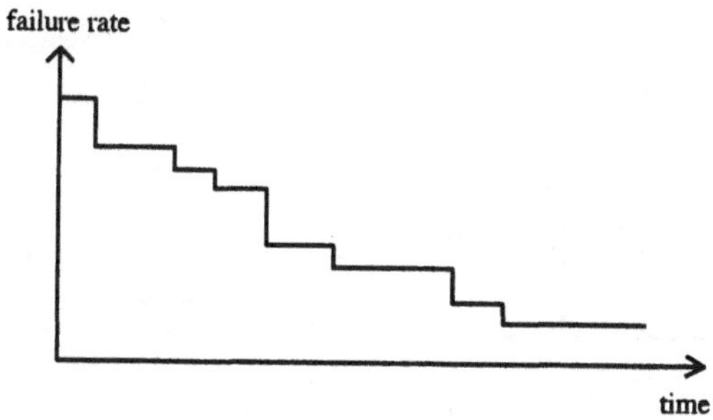

Figure 12.2: The failure rate of the SW model.

diagram for this model is similar to that in Figure 12.1 except that the failure rate on each "stair" is dependent on x_i.

12.2.4. The Littlewood Model

By assuming that the faults contribute different amounts to the unreliability of the software, Littlewood (L) rewrites the first equation in Eq. (12.1) [254, p 953] as:

$$\Pr\{x_i \mid \Lambda_i = \lambda_i\} = \lambda_i e^{-\lambda_i x_i}$$

where $\{\Lambda_i\}$ represents the successive current rate of occurrence of failures' arising from the gradual elimination of faults. Here,

$$\Lambda_i = \Phi_1, \ldots, \Phi_{N-i+1}$$

where Φ_j represents the random rate associated with fault j. When the program is executed for a total time τ, the Bayes approach shows that the remaining rates are i.i.d. GAM(α, ($\beta+\tau$)) random variables. Furthermore, by first conditioning on $\Phi_i = \phi_i$, we have

$$\Pr\{x_i \mid \Phi_i = \phi_i\} = \phi_i e^{-\phi_i x_i} \tag{12.2}$$

if Φ_i has a GAM(α,β), the unconditional probability of X_i (from Eq. (12.2)) is a Pareto distribution

$$\Pr\{x_i\} = \frac{\alpha\beta^\alpha}{(\beta + x_i)^{\alpha+1}}.$$

The maximum likelihood method is applied for estimating the model parameters α, β, and N if m and $x_1, \ldots, x_{i-1}$ are known. The estimated current reliability based on data $x_1, \ldots, x_{i-1}$ is [2, p 954]

$$\hat{R}_i(t) = \left(\frac{\hat{\beta} + \tau}{\hat{\beta} + \tau + t}\right)^{(\hat{N}-i+1)\hat{\alpha}}, \quad \tau = \sum_{j=1}^{i-1} x_j.$$

The failure rate of this model is shown in Figure 12.3.

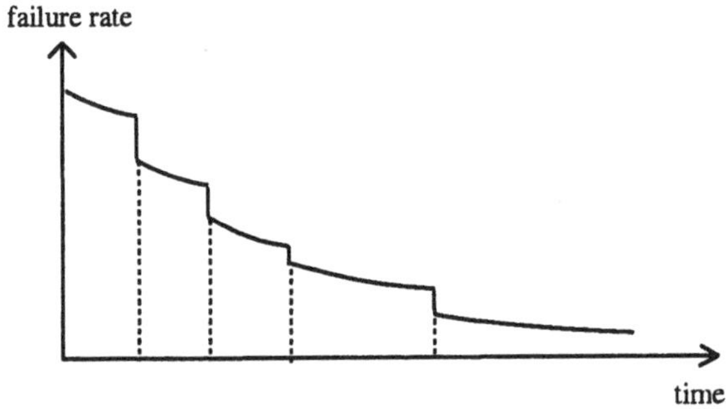

Figure 12.3: The failure rate of the L model.

12.2.5. The Weibull Order Statistics Model

The JM and L models can be treated as special cases of a general class of stochastic processes based on order statistics [2, p 954]. These processes exhibit intervening times, which are exponential and Pareto for JM and L, respectively. The intervening times are the spacings between order statistics from a random sample of N observation with pdf $f(x)$. For the Weibull order statistics (W) model, $f(x)$ has the Weibull density. The estimated parameters can also be obtained via the maximum likelihood method.

12.2.6. The Bayesian JM Models

For software systems, there is usually information available about the software's development. Useful information can also be obtained from similar software products through, for example, some measurements of the software complexity, the fault-removal history during the design phase, and others. The information can be used in combination with the collected test data to make a more accurate estimation and prediction of the software reliability. Bayesian analysis is a common technique for incorporating previous information.

The most important feature of a Bayesian model is that prior information can be incorporated into the estimation procedure. Bayesian methods are especially useful in reliability analysis because an increase in reliability is usually achieved by improving a previously developed similar product. Bayesian methods require fewer test data to achieve a high accuracy of estimation.

To avoid the use of MLE for making inferences in the JM model, Littlewood and Sofer [606] propose a Bayesian approach (the BJM model, also refer to [2]) by assuming that λ and ϕ are described by a GAM(b, c) and a GAM(f, g), respectively [606, p 32]. Thus,

$$\begin{aligned} \hat{R}_i(t) &= \int\int R_i(t \mid \lambda, \phi) p(\lambda, \phi \mid t_1, \cdots, t_{i-1}) d\lambda d\phi \\ R_i(t \mid \lambda, \phi) &= \exp[-(\lambda - (i-1)\phi)t]. \end{aligned}$$

An earlier model of Littlewood and Verall (LV) [256] assumes that only λ_i has a gamma prior with parameters α and $\psi(i)$,

$$\begin{aligned} \Pr\{x_i \mid \Lambda_i = \lambda_i\} &= \lambda_i e^{-\lambda_i x_i}, \quad x_i > 0 \\ \Pr\{\lambda_i\} &= \tfrac{1}{\Gamma(\alpha)} \psi(i)(\psi(i)\lambda)^{\alpha-1} \exp[-\psi(i)\lambda_i] \\ \psi(i) &\equiv \psi(i, \beta) = \beta_1 + \beta_2 i \end{aligned}$$

where $\psi(i)$ is chosen to be an increasing function so that $\{\Lambda_i\}$ will be stochastically decreasing, which is the condition for reliability growth [2, p 954]. Using the maximum likelihood method to find parameters α and β, we have

$$R_i(t) = \left[\frac{\psi(i, \hat{\beta})}{t + \psi(i, \hat{\beta})}\right]^{\hat{\alpha}}.$$

One can also assume that the scale parameter, α, is known and, instead, induce reliability growth via the shape parameter; i.e.,

$$\begin{aligned} \Pr\{\lambda_i\} &= \beta(\beta\lambda)^{\psi(i)-1} e^{-\beta\lambda_i} / \Gamma(\psi(i)) \\ \psi(i) &\equiv \psi(i, \alpha) = 1/(\alpha_1 + \alpha_2 i) \end{aligned}$$

where $\psi(i)$ is again chosen to be an increasing function to have reliability growth. Given $x_1, \ldots, x_{i-1}$ and deriving MLE (β and α), we have

$$R_i(t) = \left(\frac{\hat{\beta}}{t + \hat{\beta}}\right)^{\psi(i, \hat{\alpha})}.$$

The drawback of the JM and BJM models is that they treat debugging as a deterministic process where each fix is effective with certainty, and all fixes have the same effect on the reliability.

12.2.7. Empirical Bayes

Since there exists a great deal of uncertainties in software failure data and the software engineer's experience on the previous software development is highly non-quantifiable, we suggest that Bayesian approach does not have to be used in a formal way. Specifically, the prior information used in the L model or the Bayesian JM models can be modeled nonparametrically. Simulation can be conducted for displaying the prior and the posterior distributions. No specific well-behaved distribution function needs to be introduced.

12.2.8. Nonhomogeneous Poisson Process Models

A random process whose probability distribution varies with time is known as a nonhomogeneous process. The class of nonhomogeneous Poisson process (NHPP) models is widely used since most failure processes during tests fit this situation. The main characteristic of this type of model is that there is a mean value function, which is defined as the expected number of failures up to a given time. The failure intensity function is the rate of change of the mean value function or the number of failures per unit time (see Kuo [226]). The typical debugging experience of the NHPP is shown in Figure 12.4 (x indicates a debugging effort occurred).

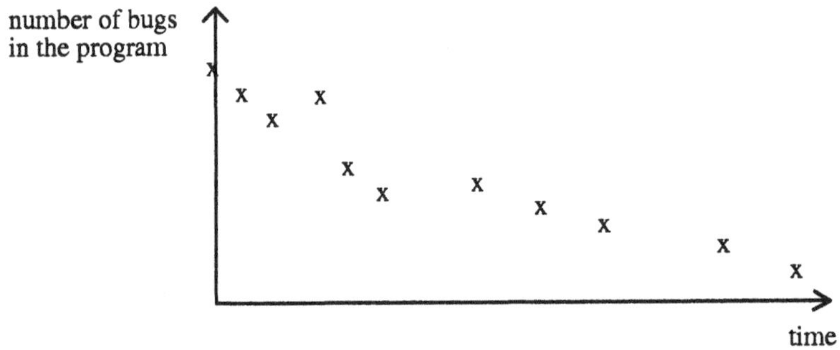

Figure 12.4: The number of bugs remaining in the program at time t.

NHPP models have been widely applied by practitioners. The theory of NHPP is well developed and can be found in most of the established literature on stochastic processes. The calculation of the expected number of failures up to time t is very simple due to the existence of a mean value function. The estimates of the parameters are easily obtained by using either the method of maximum likelihood or the method of least squares. The NHPP models are useful in describing reliability growth. By using different mean value functions, we can fit software failure data to some satisfactory degree.

The Goel-Okumoto (GO) model [166] is similar to the JM model except that N is treated as a Poisson random variable with mean m. The unconditional process is exactly an NHPP with failure intensity

$$m\phi e^{-\phi t}.$$

Goel and Okumoto indicate differences from the JM model in two aspects [166, p 206]:

1. The initial error content of a software system is treated as a random variable in the GO model while in the JM model, it is an unknown, fixed constant.

2. In the JM model, the times between software failures are assumed to be s-independent of each other, while in the GO model, the time between failure $k-1$ and k depends on the time to failure $k-1$.

Musa's basic execution time model [295] assumes that failures occur as an NHPP. The decrement in the failure intensity function remains constant for the basic execution time model irrespective of whether it is the first failure that is being corrected or the last. This model implies a uniform operational profile. Let μ denote the average number of failures experienced at a given point of time. Then the failure intensity λ as a function of μ is

$$\lambda(\mu) = \lambda_0 \left(1 - \frac{\mu}{\nu_0}\right)$$

where λ_0 is the initial failure intensity at the start of execution and ν_0 is the total number of failures that would occur in infinite time. Let the execution time be denoted by τ. Then for this model, we have

$$\mu(\tau) = \nu_0 \left[1 - \exp(-\frac{\lambda_0}{\nu_0}\tau)\right] \tag{12.3}$$

and the failure intensity as a function of the execution time is

$$\lambda(\tau) = \lambda_0 \exp(-\frac{\lambda_0}{\nu_0}\tau).$$

The logarithmic Poisson execution time model [295] also assumes that failures occur as an NHPP. The decrement per failure in the failure intensity function becomes smaller with each failure experienced, i.e., later fixes contribute less to program reliability than earlier ones. In fact, the decrement decreases exponentially. Following the same definitions for μ, λ_0, and τ from the basic execution time model, the failure intensity λ as a function of μ is

$$\lambda(\mu) = \lambda_0 \exp(-\theta\mu)$$

where θ is called the failure intensity decay parameter. It represents the relative change of failure intensity per failure experienced. Similar to Eq. (12.3), we have

$$\mu(\tau) = \frac{1}{\theta} \ln(\lambda_0\theta\tau + 1).$$

Hence, the failure intensity as a function of the execution time is

$$\lambda(\tau) = \frac{\lambda_0}{\lambda_0\theta\tau + 1}.$$

For the two models just described, we need the value of the initial failure intensity λ_0 in both models, the total failures experienced ν_0 for the basic execution time model, and the failure intensity decay parameter θ for the logarithmic Poisson model. Once enough failure data are gathered, we can estimate these parameters by the maximum likelihood method. It should be noted that the NHPP models are capable of coping with nonhomogeneous testing and changing software, and hence they are very useful for software reliability measurements.

The basic execution time model and the logarithmic Poisson execution time model both have certain advantages not possessed by the other. According to [295], the reasons for selecting the basic execution time model are:

1. it has a satisfactory record of prediction,

2. it is simple and easy to understand,

3. it is the model that is most thoroughly developed and has been most widely applied to actual projects,

4. its parameters have a clear physical interpretation and can be related to information (characteristics of the software, such as size, and characteristics of the development environment) existing prior to the program's execution, and

5. it can handle evolving systems (systems that change in size) by adjusting the failure time to estimate results had all the code been present.

However, the logarithmic Poisson execution time model has a high predictive validity which it attains early in the system test phase. Studies by Downs [109] and Trachtenberg [422] demonstrate that for a highly non-uniform operational profile, the logarithmic Poisson execution time model may be a better choice. If early predictive validity is very important or the operational profile is highly non-uniform, it is suggested in [226] that the logarithmic Poisson execution time model is a better choice.

12.3. The Non-stochastic Software Reliability Models

In the model presented in Section 12.2, we normally need to assume that the initial number of software faults is known. In this section, we present some existing software reliability models for which no dynamic assumption about the software failure process has been made. This class of models is mainly useful for the estimation of the number of software faults, and it is usually assumed that faults are not removed immediately after detection. Two models included here are an input-domain-based model and a fault-seeding model.

12.3.1. The Input-domain-based Model

Nelson [644] proposes a method for estimating software reliability by taking representative samples from the input domain and looking at the resultant failure rate when the sample data is input to the system for execution. It is assumed that a number of test cases are chosen from the operational profile of the software. Assume that the input space (which is also called the input domain) E is divided into M subsets, that is

$$E = \{E_1, E_2, \cdots, E_M : M > 0\}.$$

The reliability of a single execution of the software, R_1, is then equal to

$$R_1 = \sum_{i=1}^{M} p_i X_i$$

where p_i is the probability of choosing an input datum from E_i, and

$$X_i = \begin{cases} 1 & \text{if input } E_i \text{ leads to correct output,} \\ 0 & \text{otherwise.} \end{cases}$$

The quantity $\{p_i : i = 1, 2, \cdots, M\}$ is the operational profile of the software, which describes the user condition, and it is assumed to be known completely. If each test run consists of n executions of the software, then the reliability of the test run is

$$R(n) = \prod_{j=1}^{n} R_j = \exp\left(\sum_{j=1}^{n} \ln R_j\right).$$

The weaknesses of the Nelson model and other input-domain-based models include that [357, p 236]

- a large amount of testing is necessary in order to get a highly accurate estimate of the reliability,
- the representative samples may not exist, and
- there may be no correlation between results from one sample to another.

12.3.2. The Fault-seeding Model

A known number of defects can be seeded into software, which is then tested using a process that presumably has an equal probability of finding a seeded or an indigenous defect. The numbers of indigenous and seeded faults found are used to estimate system reliability. Ferrara [145, p 333] uses "fault injection" where known faults are introduced purposely into the design. The best known seeding model is the Mills model [287], in which a statistical sampling technique called "capture-recapture" sampling is used to estimate software reliability.

Let M denote the number of seeded faults. Suppose that during the testing, k faults are detected and m of them are seeded faults. If both inherent faults and seeded faults are detected with equal likelihood, then an estimate of the total number of inherent faults is

$$\hat{N} = \frac{M(k-m)}{m}.$$

Let X_k be the number of seeded faults among the total number of k detected faults. And let p_i denote the probability of detecting i seeded faults, that is $p_i = P(X_k = i \mid M)$. Then X_k follows the hypergeometric distribution utilizing the results of capture-recapture sampling. It can be shown that the above estimate of inherent faults $\hat{N}$ is the maximum likelihood estimate which maximizes p_m with respect to N.

One drawback of this model should be pointed out. The model's primary assumption is that the probability of detecting the seeded faults is the same as the probability of detecting the inherent faults. This is in practice impossible, since seeded faults can hardly be made representative of the inherent faults, and this will make the estimate biased and inaccurate.

12.4. A Proposed Procedure of Testing Data

In this section, a statistic is introduced to test the trend of data sets. *GINO*, a commercial package for optimization problems developed by the University of Pennsylvania, can help us solve minimization programs. The smoothing method is then extended to handle exact failure times and is explained by using a computer package *GAMS*.

The likelihood ratio statistic $\Lambda_n^*(\vec{x})$, Barlow [22]:

$$\Lambda_n^*(\vec{t}) = \left(\frac{n-1}{\sum_{i=i}^{n} t_i} \right)^{n-1} \prod_{i=1}^{n-1} \frac{n-i}{t_{i+1} - t_i}. \tag{12.4}$$

can be used to test

$$H_o : F \in \{IFR\}$$
$$H_a : F \in \{DFR\},$$

and we conclude that for large Λ_n^*, we are in favor of DFR, and thus reject the exponentiality, which has CFR, and for small Λ_n^*, the data has IFR. If a data set is tested to have DFR, then one of the smoothing techniques can be applied for reliability prediction. The general procedure is discussed in the next section.

12.4.1. Completely Monotone

It is mentioned in [282, p 344] that all the JM, GO, and L models satisfy the completely monotone

$$(-1)^n \frac{d^n m(t)}{dt^n} \geq 0, \quad n, \; t \geq 0 \tag{12.5}$$

where $m(t)$ is the failure intensity. The relationship between completely monotone and the doubly stochastic process (especially the exponential order statistic processes) appears in [282, p 344], which indicates that "the class of completely monotone functions is identical to the totality of intensity functions for the family of doubly stochastic exponential order statistic processes."

One more clear explanation is given in [283, p 98], which describes two reasons for requiring complete monotonicity. Virtually, all of the various software reliability growth models in the literature have this property. If the times until each bug manifests itself are (non-identically distributed) exponential random variables, then by definition of the exponential distribution and the additive property of the expectation operator,

$$L(t) = \sum_{i=1}^{\infty} (1 - e^{-\lambda_i t})$$
$$l(t) = \sum_{i=1}^{\infty} \lambda_i e^{-\lambda_i t}.$$

The $l(t)$ in the above equation is the Laplace transform of a measure, which assigns mass to λ_i at $t = \lambda_i$, $i = 1, 2, \cdots$, and therefore must be completely monotone.

Define

$$h(t) = \frac{\text{\# of failures}}{(\text{\# of samples on test})\Delta t} \tag{12.6}$$

for estimating hazard rate and, since there is only one software under test, Eq. (12.6) reduces to

$$\hat{h}_i(t) = \frac{1}{t_{i+1} - t_i} = \hat{h}_i, \quad t_i \le t < t_{i+1}, \quad i = 0, \cdots, n-1 \tag{12.7}$$

for the hazard rate after the ith and before the $(i+1)^{st}$ failures. Then our objective is to find the solution of the following optimization model [283, p 100]:

$$\begin{aligned} \text{Minimize}: \quad & D(\hat{h}, h) = \sum_{i=1}^{k} w_i(\hat{h}_i - h_i^*)^2 & \\ \text{s.t.} \quad & (-1)^d \Delta^d h_i \ge 0, & d+1 \le i \le k+l \\ & (-1)^j \Delta^j h_{k+l} \ge 0, & 0 \le j \le d-1 \end{aligned} \tag{12.8}$$

where w_i is a given weighting factor and $h_i^* = h(t_i^-)$ as the estimator in Eq. (12.7) is left-continuous.

□ **Example 12.1**

Consider the experimental hazard rates in the second column of Table 12.1. *GINO*, which is capable of handling problems with less than or equal to 30 rows (including the objective function), is used to find the optimal solution. The window size for the predicted segment is set to be 4 ($l = 4$) so that we have exactly 30 rows for use in *GINO* (see the sample *GINO* program in Appendix E for $d = 4$). Figure 12.5 shows the results by using $d = 1$, 2, 3, 4, and 5 (we use $w_i = 1$, for all i). The estimated hazard rate for each k and the minimum D are summarized in Table 12.1 (the hazard rates are in 10^{-4}). Different w_is are used and the results are in Figure 12.6 where

`model4c1:` $w_i = i$

`model4c2:` $w_i = i^2$

`model4c3:` $w_i = \sqrt{i}$

The step curves in Figures 12.5 and 12.6 are calculated by Eq. (12.6). For this particular data set, no significant difference can be observed for different w_i. □

12.4.2. Using *GAMS* to Solve the Optimization Model

GAMS can be used for finding optimal solutions if the number of rows exceeds 30. It is reported in the *GAMS* manual [158, p 169] that, for nonlinear programming, the maximum number of rows is approximately 240. An example

Table 12.1: h_is and related information for Example 12.1.

i	h_i	h_i^*				
		$d=1$	$d=2$	$d=3$	$d=4$	$d=5$
1	3.583	3.583	3.583	3.583	3.583	3.582
2	1.667	1.667	1.685	1.695	1.701	1.699
3	1.333	1.333	1.298	1.249	1.227	1.227
4	0.875	0.931	0.911	0.959	0.959	0.968
5	0.667	0.931	0.860	0.824	0.835	0.839
6	1.250	0.931	0.809	0.791	0.792	0.782
7	0.500	0.750	0.758	0.766	0.767	0.758
8	0.250	0.750	0.752	0.751	0.754	0.750
9	1.250	0.750	0.746	0.744	0.746	0.747
10	0.500	0.750	0.741	0.738	0.740	0.742
11	1.000	0.750	0.735	0.732	0.733	0.737
12	1.000	0.750	0.729	0.727	0.727	0.732
13	0.500	0.700	0.724	0.722	0.721	0.727
14	0.750	0.700	0.718	0.717	0.716	0.721
15	0.750	0.700	0.712	0.712	0.711	0.715
16	0.500	0.700	0.707	0.707	0.706	0.710
17	1.000	0.700	0.701	0.703	0.702	0.705
18	0.500	0.679	0.695	0.699	0.698	0.700
19	0.750	0.679	0.689	0.696	0.694	0.695
20	0.750	0.679	0.684	0.692	0.691	0.690
21	0.000	0.679	0.678	0.689	0.688	0.686
22	1.000	0.679	0.678	0.686	0.686	0.683
23	0.750	0.679	0.678	0.684	0.684	0.680
24	1.000	0.679	0.678	0.681	0.682	0.677
25	0.500	0.500	0.678	0.679	0.680	0.675
26		0.500	0.678	0.676	0.679	0.673
27		0.500	0.678	0.673	0.678	0.672
28		0.500	0.678	0.671	0.678	0.672
29		0.453	0.678	0.668	0.678	0.672
D		1.814	1.943	1.971	1.978	1.981

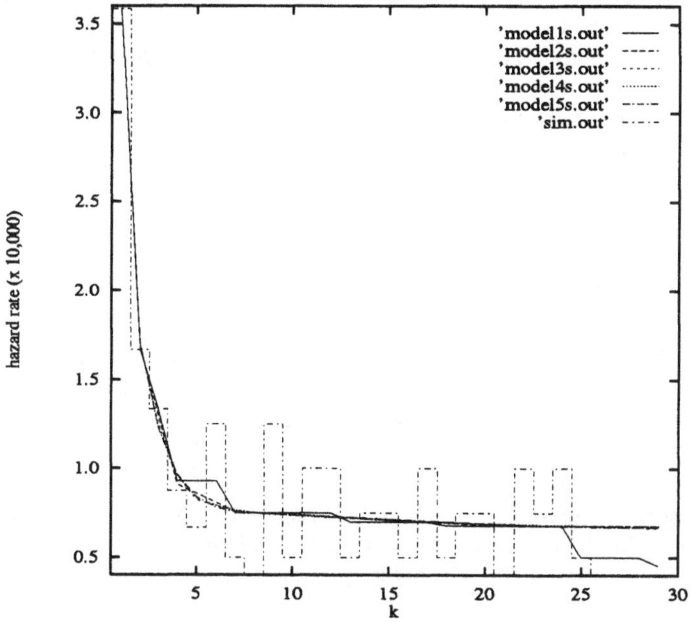

Figure 12.5: Hazard rates from the experiment and for $d = 1, 2, 3, 4$, and 5.

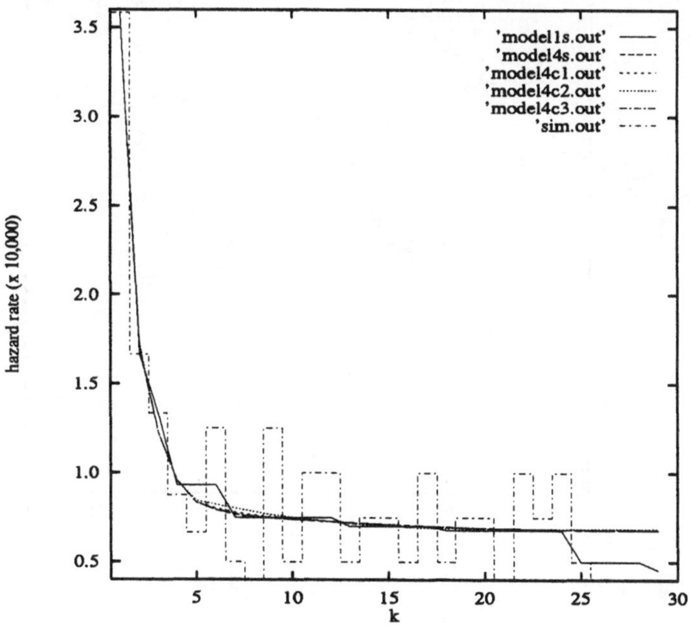

Figure 12.6: Hazard rates under different w_is for $d = 4$.

is displayed here for illustrating the use of *GAMS* for smoothing failure plots, which draw failure time versus the number of failures (see the sample *GAMS* program in Appendix E solving the hazard rates in the second column of Table 12.1).

The mixed exponential distribution is considered here, whose probability density function is

$$\begin{aligned} f(x) &= (1-p)f_1(x) + pf_2(x) \\ f_1(x) &= \lambda_1 e^{-\lambda_1 x} \\ f_2(x) &= \lambda_2 e^{-\lambda_2 x} \end{aligned} \tag{12.9}$$

where

p estimated percentage of the normal parts; this value can be derived based on experiences or from inspection sampling

λ_1 hazard rate of the weaker parts in a population

λ_2 hazard rate of the normal parts in a population; usually $\lambda_2 \gg \lambda_1$.

Data set `mono1` is generated by the mixed exponential distribution in Eq. (12.9) under $p = 0.95$, $\lambda_1 = 0.1$, and $\lambda_2 = 0.02$; and the number of samples in the 4 groups is 20, 30, 15, and 25. The failure plot for data set `mono1` is in Figure 12.7.

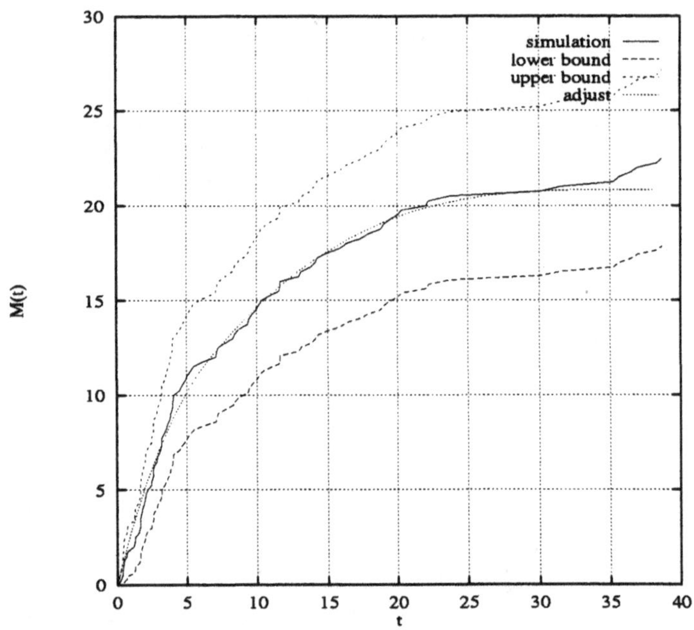

Figure 12.7: The mean value function plot for data set `mono1`.

The upper and lower bounds in Figure 12.7 are derived by [244]. The Eq. (7.18) is used for the asymptotically exact 100(1-α)% confidence interval for $\hat{\Lambda}(t)$ (the expected number of events by time t; we use $\alpha = 0.05$ here). From Eq. (7.18), it can be seen that the more realizations (different data sets from the common NHPP) we have, the smaller the variance becomes. Once the smoothed curve in Figure 12.7 is obtained, the corresponding R, $\ln R$, and the $h(t)$ can be easily prepared; from these curves, we are able to determine if the data sets are CFR, DFR, or IFR by definitions in Section 3.1.3.

12.4.3. Using Different *ds*

For the model in Eq. (12.8), two propositions are given in [283]:

1. For $d = 3$, and a fixed $l > 0$, we can use Eq. (12.8) to extrapolate and make forecasting iff

$$\frac{j(j+1)}{2}\Delta^2 h_k + j\Delta h_k + h_k \geq 0, \quad j = 1, 2, \cdots, l.$$

2. For $d = 4$, and a fixed $l > 0$, we can use Eq. (12.8) to extrapolate and make forecasting iff

$$\begin{aligned} \tfrac{j(j+1)}{3}\Delta^3 h_k + j\Delta^2 h_k + \Delta h_k &\geq 0, & j &= 1, 2, \cdots, l \\ \tfrac{j(j-1)}{6}\Delta^2 h_k + \tfrac{2(j-1)}{3}\Delta h_k + h_k &\geq 0, & j &= 2, 3, \cdots, l \\ \tfrac{l(l+1)}{2}\Delta^2 h_k + k\Delta h_k + h_k &\geq 0. \end{aligned}$$

Let the mean value fuction $M(t)=E[X(t)]$ where $\{X(t),\ t \in T\}$ is a stochastic process. The other way of dealing with hazard mechanisms is to use $M(t)$ instead of $h(t)$:

$$\begin{aligned} \text{Minimize}: \quad & D(\hat{M}, M) = \sum_{i=1}^{k} w_i(\hat{M}_i - M_i^*)^2 & & \\ \text{s.t.} \quad & (-1)^{d+1}\Delta^d M_i \geq 0, & d &\leq i \leq k+l \\ & (-1)^j \Delta^j M_{k+l} \geq 0, & 0 &\leq j \leq d-1 \\ & M_k \equiv n + \delta, & k &> 0 \\ & M_0 = 0, & & \end{aligned} \tag{12.10}$$

since $M(t)$ is a non-decreasing function no matter whether the hazard rate function is decreasing or increasing. Note the differences between Eqs. (12.8) and (12.10). They are

- the power of (-1) in the first constraint,
- the starting value of i in the first constraint, and
- two more constraints in Eq. (12.10).

A concave w_i can be selected so that more weights will be put on the more recent observations.

12.4.4. The Envelopes

Define $gilb(\cdot)$ as the greatest integer lower bound and

$$\begin{aligned}\Delta^0 M_k &= M_k\\ \Delta^1 M_k &= M_k - M_{k-1}\\ \Delta^j M_k &= \Delta^{j-1} M_k - \Delta^{j-1} M_{k-1},\quad j > 1.\end{aligned}$$

The lower and upper envelopes for d =4 are [392, p 334]

$$M_{k+i} = \begin{cases} M_k + i\Delta^1 M_k + \frac{i(i+1)}{2}\Delta^2 M_k \\ \qquad + \frac{i(i+1)(i+2)}{6}\left(\frac{-2(\Delta^1 M_k + u\Delta^2 M_k)}{u(u+1)}\right), & i = 1,\cdots,u \\ M_{k+u}, \quad i = u+1,\cdots,l \end{cases} \tag{12.11}$$

where

$$u = min\Big(l, 1 + gilb(-2\frac{\Delta^1 M_k}{\Delta^2 M_k})\Big),$$

and

$$M_{k+i} = \begin{cases} M_k + i\Delta^1 M_k + \frac{i(i+1)}{2}\Delta^2 M_k \\ \qquad + \frac{i(i+1)(i+2)}{6}\Delta^3 M_k, \quad i = 1,\cdots,q \\ M_{k+q} + (i-q)\alpha, \quad i = q+1,\cdots,l \end{cases} \tag{12.12}$$

where

$$\begin{aligned}\alpha &= \Delta^1 M_k + q\Delta^2 M_k + \tfrac{q(q+1)}{2}\Delta^3 M_k\\ q &= \begin{cases} gilb(-\frac{\Delta^2 M_k}{\Delta^3 M_k}), & \text{if } \Delta^3 M_k > 0\\ l, & \text{if } \Delta^3 M_k = 0,\end{cases}\end{aligned}$$

respectively. Eqs. (12.11), (12.12), and (7.18) can be used to set the confidence intervals.

12.5. Software Reliability Management

Bologna [50] proposes the following design philosophies to help construct an error-free software:

1. Top-down structured design,
2. Structured programming,
3. Formal specification methods,
4. Formal programming conventions using programming standards and naming conventions,
5. High modularization of code and requirements-to-code traceability,
6. Project organization and configuration management techniques,
7. Independent verification process,
8. Use of integrated software engineering environment, and
9. Increasing reusability of the modulars.

Although these concepts are indeed very helpful, they are not sufficient to develop a highly reliable software. Managerial actions are also required. Ryerson [357, p 238] suggests four guidelines to assure cost effective products:

1. Build up historical records of failures
2. Act quickly on customers' complaints
3. Develop or adopt good tools for
    - Visibility of status and trend
    - Projection of expected reliability and cost
    - Summary of data, and
4. Confirm customers' satisfaction.

Informative documents and a user-friendly database are keys for cost savings in software life cycles. Once the predicted system reliability falls within an acceptable region, an error-free and user-friendly document can be designed as indicated in Figure 12.8.

Customer reports and complaints are recorded in a database as references for developing better and more competitive software in the future. The information needed to establish a software reliability database include [383, p 51]:

1. The nature and scope of the project, the organization involved, the time frame, the outcome, field performance, and any preceding and succeeding projects,
2. Requirements and specifications,
3. The program language, compiler, operating system, major tools used, and target machine,
4. The size of the program (source and object, comments and executable instructions), the development approach, key milestones, and the number of person-hrs expended on the program,
5. The number of errors discovered and corrected each week (or month) during integration testing, and if possible during module testing, and
6. The number of tests, level, severity, running time to failure, and running time without failure (for development testing and for simulation testing).

A viable means of building product assurance is the establishment of a field service depot so that early feedback on field problems can be used in engineering and manufacturing to provide the following:

1. The sooner the fault is detected and removed, the cheaper it is to fix [145, p 332],
2. Timely feedback promotes better focus and gives developers the opportunity to make timely corrections [145, p 334],
3. Operational measurements: technical reporting on representative samples and failure analysis [86, p 338],
4. Error avoidance in design is much more cost effective than error resolution during testing [453, p 427], and

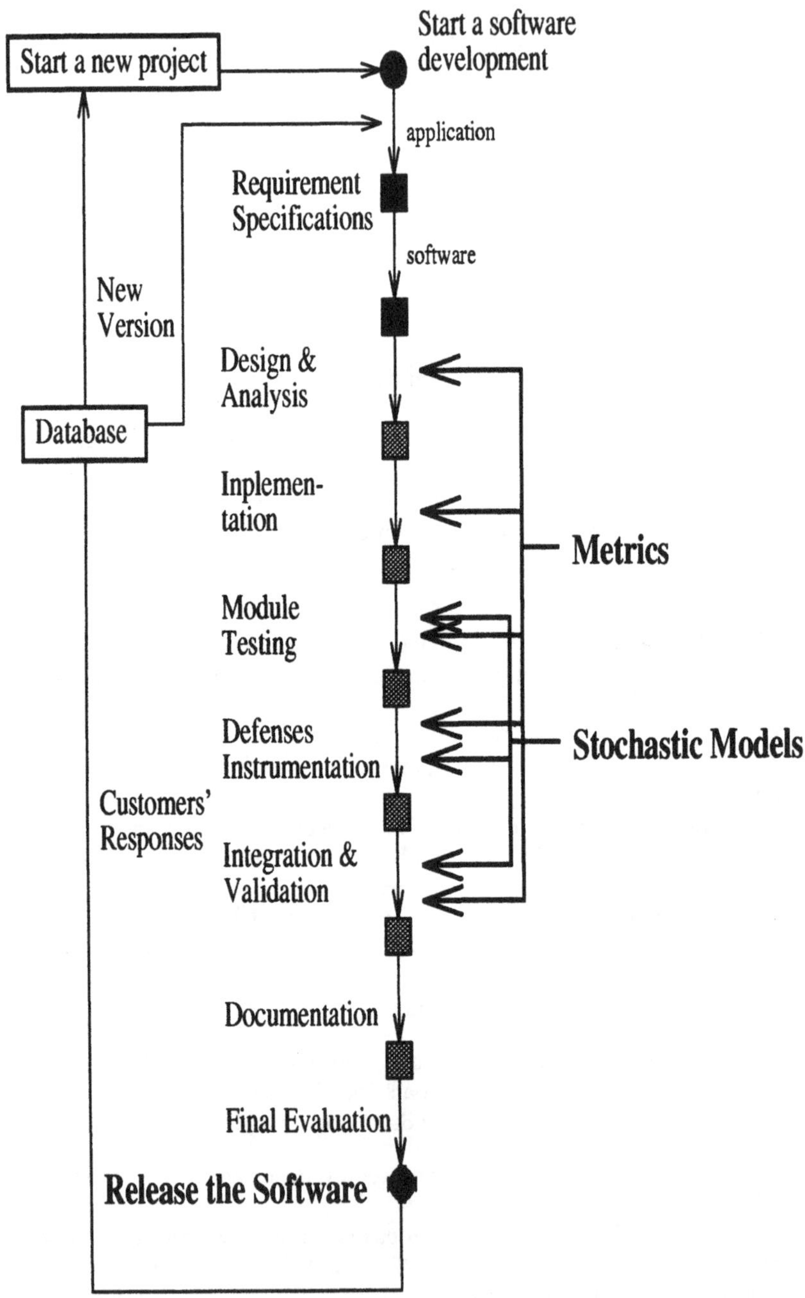

Figure 12.8: A software life cycle.

5. In software, every time an error is corrected during testing, the probability of inducing new error into the program is quite high [453, p 427].

Strict reporting guidelines must be emphasized in order to collect meaningful failure information [145, p 334]. Users need to be educated to benefit from the accurate reporting of software failures [145, p 334]. Detailed analysis of field failure causes can be most cost effective [357, p 238].

Very high software reliability is particularly important for its applications in, for example, the space shuttle, the commercial flights, food and medical industries, and nuclear reactor control.

12.6. Conclusions

At present, software reliability modeling is considered to be a major part of software quality and of software engineering. Software reliability concerns itself with how well the software meets the requirement of the customers. Understanding software reliability measurement and prediction has become a vital skill for both software managers and software engineers.

In this chapter, we present some well-known reliability models from the literature. Many new models also exist, which are either generalizations of older models or special cases of existing models. The biggest difficulty in software measurement is analyzing the context in which measurement is to take place, in order to determine beforehand which model is likely to be most trustworthy. Practitioners have no reliable way of knowing in advance which model is likely to produce the most trustworthy predictions. Usually the following criteria can be employed to help in the model selection [295]:

- Model validity
  Four measures can be used to rate model validity: accuracy, bias, trend, and noise.

- Ease of measuring parameters
  This criterion concerns the number of parameters a model requires and the difficulty in estimating them.

- Quality of assumption
  The assumptions the model is based upon should simulate real project testing and operation as closely as possible.

- Capability
  Capability refers to a model's ability to estimate reliability-related quantities for software systems, such as the software's present reliability, expected date of releasing, and others.

- Applicability
  Applicability refers to the usefulness of the model in different development environments, operational environments, and life-cycle phases.

- Simplicity
  Simplicity refers to the ease of gathering project-specific data.

- Insensitivity to noise
  Insensitivity makes the model appealing if it can make an accurate prediction, even when the failure data contains uncertainties.

However, software development and operation involve many intricate human activities, and software failure patterns are usually uncertain.

A nonparametric statistical test and the smoothing method based on the completely monotone are introduced. We would like to emphasize here that most stochastic models only work well for certain data sets; that is, there is no universally good model for software reliability modeling. Some other smoothing methods, such as the kernel estimator and the penalized maximum likelihood estimation, can also be applied; however, they may be more difficult to solve than the model in Eq. (12.8).

Our ideas described in this chapter are summarized below.

1. Use the statistic in Eq. (12.4) to test the trend of the data set. That is, consider the hypothesis

$$\begin{aligned} &H_o : F \in \{IFR\} \\ &H_a : F \in \{DFR\}. \end{aligned} \tag{12.13}$$

   If the null hypothesis in Eq. (12.13) is rejected under the specified confidence level, the data implies reliability growth. On the contrary, if the hypothesis in Eq. (12.13) cannot be rejected, some possible reasons may be that the experiment is incorrect or we have imperfect debugging.

2. The concept of the completely monotone is applied to smooth and extrapolate the hazard rate or/and the failure plot curve(s). We should have monotonic decreasing hazard rates if the debugging processes are stable and reliable. Thus, there will not be any "surprise" in the hazard rate curve, like an increase pattern at a later time. In other words, if we have a monotonic DFR, the extrapolation would be acceptable, which is a prerequisite of most reliability predictions.

3. It is helpful for software designers to ask themselves the following questions:

   - Does your operation include an organized brain pool approach to make the best use of available specialists' ability?
   - Are your top staff people and technical support specialists buried in a single project or program that does not begin to tax their full mental capacity?
   - Do you regularly call these experts into conference to seek their ideas on problem solving on other programs or just to brainstorm on general subjects?
   - Do you emphasize participation in an organized company Brain Pool that gives them opportunity to stretch their thinking and to get recognition for their contributions to the company?

- Do you sponsor periodic Brain Pool Retreats as you do management retreats, perhaps to some isolated retreat facility?
- How are these people paid?
- Are you personally familiar with these specialists at your level so that you can assist in making use of their unused mental power?
- Do you have an established Brain Pool Journal open to articles from Brain Pool members at all levels of the organization?
- Are your Brain Pool members thoroughly indoctrinated in the least reliability theory and techniques so that they can support your quality and reliability people in carrying out effective measures for trouble prevention?

Finally, similar to the burn-in approach used with microelectronics products, finding the optimal debugging period is an optimization problem given limited resources such as space, time, and computer facilities. We should bear in mind that software reliability is exactly the same type of problem that is encountered in the infant mortality period of hardware burn-in.

EPILOGUE: COST-EFFECTIVE DESIGN FOR STRESS BURN-IN

Burn-in can accelerate the failure realization of manufactured products. Although burn-in is widely used in the microelectronics industry, the concept of, and exercises in, burn-in can be just as applicable in enhancing reliability and yield in

- non-electronics products including mechanical systems, such as rubber tires, copiers, and others,
- software products through software testing and development including software reusage, and
- electronics systems after repairs.

□ **Example**

Example 6.6 is used to explain the incompatibility issue in Section 6.1.5. This example indicates that it is frequently seen that SIMM makers will use the SIMM tester (e.g., the CST tester) to perform initial E-test to find out if there are any failed ICs on the SIMM. The failed ICs may be the escapees from the QC tests as mentioned in Section 2.1.3. The SIMMs failing the E-test can be repaired by replacing the failed IC, which can be found by the SIMM tester. As described in Example 6.6 in Section 6.1.5, several simulation softwares can be used to "burn-in" the SIMMs to check compatibility. However, this kind of burn-in is done on a sampling basis. To guarantee better quality and reliability, we may burn-in all the repaired SIMMs.

The RRT (refer to the Example 5.6 in Section 5.3.1) for SIMM is an important tool for periodic monitoring. Except the usual visual inspections, the SIMMs must pass 1,000-cycle T/C (the two extreme temperatures are -20°C and 90°C with dwelling time 10 minutes) and 500-hour 85°C/ 85%RH (the THB test without bias) tests in the RRT for SIMM. □

Stress burn-in is a 100% screening procedure. One purpose for applying burn-in to the products is to guarantee high reliability of the end products. In addition, we learn early from the failed products which design modifications should be made for future products. Keeping this in mind, we update the design and manufacturing processes in order to enhance both the manufacturing yield and the product reliability. If and when this purpose is achieved, screening products seems unnecessary. However, microelectronics products using the new technology come to the market place almost daily; therefore, information obtained from screening is valuable for a limited number of manufacturing processing updates using the existing technology. Beyond that, once the existing technology becomes obsolete, the products using new technology must be evaluated to meet the quality and reliability standards again. Information obtained from burn-in on current products can serve as prior knowledge for burn-in on the design of the products due to new technology. Unless we can forecast the exact causes of design and manufacturing flaws in future products, stress burn-in still serves the screening purpose. In particular, ICs for high consequence of failures, which include applications on communications, medical systems, and others, need to subject themselves to the full screening procedure before they are assembled into a dedicated system.

Perhaps an optimal design methodology will in the future provide us with a fault-free approach coupled with a flaw-free manufacturing process. At least, we hope that bursting the fault-free design approach converges at the rate no slower than the rate of high reliability required by the customers. Until then, burn-in is essential for products that experience high failure rate at a long infant mortality period, such as those that use microelectronics or large software.

In conclusion, quality may not be the final solution but it is a necessary pathway for high productivity and competitiveness. Yield is a quality issue which can be improved through cost-effective burn-in. Therefore, when considering burn-in for yield improvement, we need to optimize the benefit and minimize the cost of burn-in. With the help provided in this book, we can avoid product recalls from the market.

REFERENCES

[1] Aarset, M.V., "How to identify a bathtub hazard rate," *IEEE Transactions on Reliability,* **R-36** (1), 1987, pp 106–108.

[2] Abdel-Ghaly, A.A., Chan, P.Y., and Littlewood, B., "Evaluation of competing software reliability predictions," *IEEE Transactions on Software Engineering,* **SE-12**, Sep. 1986, pp 950–967.

[3] Agarwala, A.S., "Shortcomings in MIL-STD-1629A guidelines for criticality analysis," *Annual Reliability and Maintainability Symposium,* 1990, pp 494–496.

[4] A-Hameed, M.S. and Proschan, F., "Nonstationary shock models," *Stochastic Processes and Their Application,* **1**, 1973, pp 383–404.

[5] A-Hameed, M.S. and Proschan, F., "Shock models with underlying birth process, " *Journal of Applied Probability,* **12**, 1975, pp 18–28.

[6] Alvarez, A.R., *BiCMOS Technology and Applications*, Second Edition, Kluer Academic Publishers, Norwell, 1993.

[7] Amerasekera, A. and Campbell, D.S., *Failure Mechanisms in Semiconduct Devices*, John Wiley & Sons, Chichester, 1987.

[8] Amerasekera, A. and Duvvury, C., *ESD in Silicon Integrated Circuits*, John Wiley & Sons, West Sussex, England, 1995.

[9] Andersen, P.K., Vaeth, and Michael, "Simple parametric and nonparametric models for excess and relative mortality," *Biometrics,* **45**, Jun. 1989, pp 523–535.

[10] Arsenault, J.E. and Roberts, J.A., *Reliability and Maintainability of Electronic Systems*, Computer Science Press, Potomac, Maryland, 1980, pp 131–187.

[11] Asher, H., "MIL-STD-781C: A vicious circle," *IEEE Transactions on Reliability,* **R-36** (4), Oct. 1987, pp 397–402.

[12] Ashour, S.K., "Asymptotic distribution of Bayes estimators in censored type-I samples from a mixed exponential population," *IEEE Transactions on Reliability,* **R-34** (5), Dec. 1985, pp 489–494.

[13] AT&T, *AT&T Reliability Manual,* Edited by Klinger,D.J., Nakada,Y., and Meenedez,M.A., Van Nostrand Rheinhold, New York, 1990.

[14] Bagchi, S.B. and Sarkar, P. , "Bayes interval estimation for the shape parameter of the power distribution," *IEEE Transactions on Reliability,* **R-35** (4), Oct. 1986, pp 396–398.

[15] Bai, D.S. and Chun, Y.R. , "Nonparametric inferences for ramp stress tests under random censoring," *Reliability Engineering & System Safety,* **41** (3), 1993, pp 217–223.

[16] Bailey, C.M. , "Effect of burn-in and temperature cycling on the corrosion resistance of plastic encapsulated integrated circuits," *Proceedings of International Reliability Physics Symposium,* 1977, pp 120–124.

[17] Bain, L.J., "Inferences based on censored sampling from the Weibull or extreme-value distribution," *Technometrics,* **14** (3), 1972, pp 693–702.

[18] Bain, L.J. and Engelhardt, M., *Introduction to Probability and Mathematical Statistics*, PWS-Kent Publishing Company, Boston, MA, 1989.

[19] Balmer, D.W. , "On quickest detection problem with variable monitoring," *Journal of Applied Probability,* **18**, 1981, pp 760–767.

[20] Balanda, K.P. and MacGillivray, H.L. , "Kurtosis: a critical review," *The American Statistician,* **42** (2), May 1988, pp 111–119.

[21] Barlow, R.E. , "Bounds on integrals with applications to reliability problems," *Annals of Mathematical Statistics,* **36**, 1965, pp 565–574.

[22] Barlow, R.E. , "Likelihood ratio tests for restricted families of probability distributions," *Annals of Mathematical Statistics,* **39**, 1968, pp 547–560.

[23] Barlow, R.E. , "Geometry of the total time on test transfer," *Naval Research Logistics Quarterly*, **26**, Feb. 1979, pp 393–402.

[24] Barlow, R.E. , "A Bayes explanation of an apparent failure rate paradox," *IEEE Transactions on Reliability*, **R-34** (2), Jun. 1985, pp 107–108.

[25] Barlow, R.E., Bartholomew, D.J., Bremner, J.M., and Brunk, H.D., *Statistical Inference under Order Restrictions; the Theory and Application of Isotonic Regression*, John Wiley & Sons, Inc., New York, NY, 1972.

[26] Barlow, R.E., Bazovsky, Sr., I., and Wechsler, S., "Classical & Bayes approaches to environmental stress screening (ESS): a comparison," *Annual Reliability and Maintainability Symposium*, 1990, pp 81–84.

[27] Barlow, R.E. and Campo, R., "Total time on test processes and applications to failure data analysis," in *Reliability and Fault Tree Analysis*, Barlow, Fussell, Singpurwalla, Editors, SIAM, Philladelphia, 1975, pp 451–481.

[28] Barlow, R.E., Madansky, A., Proschan, F., and Scheur, E.M., "Statistical estimation procedures for the burn-in process," *Technometrics*, **10** (1), Feb. 1968, pp 51–62.

[29] Barlow, R.E. and Marshall, A.W. , "Tables of bounds for distributions with monotone hazard rate," *Journal of the American Statistical Association*, **60**, 1965, pp 872–890.

[30] Barlow, R.E. and Marshall, A.W., "Bounds on interval probabilities for restricted families of distributions," *Proceedings of the Fifth Berkeley Symposium on Mathematical Statistics and Probability*, **III**, 1967, pp 229–257.

[31] Barlow, R.E. , Marshall, A.W., and Proschan, F., "Properties of probability distributions with monotone hazard rate," *Annals of Mathematical Statistics*, **34**, 1963, pp 375–389.

[32] Barlow, R.E. and Proschan, F. , *Mathematical Theory of Reliability*, SIAM series in applied mathematics, John Wiley & Sons, 1965.

[33] Barlow, R.E. and Proschan, F., *Statistical Theory of Reliability and Life Testing*, Holt, Rinehart and Winston, Inc., New York, NY, 1981.

[34] Bartholomew, D.J., "The sampling distribution of an estimate arising in life testing," *Technometrics*, **5** (3), 1963, pp 361–374.

[35] Basu, A.P., "Estimates of reliability for some distribution useful in life test," *Technometrics*, **6**, 1964, pp 215–219.

[36] Basu, A.P., and Ebrahimi, N., "Nonparametric accelerated life testing," *IEEE Transactions on Reliability*, **R-31** (5), Dec. 1982, pp 432–435.

[37] Bellcore, *Reliability Prediction Procedure for Electronic Equipment*, Technical Reference TR-NWT-000332, Issue 3; Information Exchange Management, Bellcore, Morristown, NJ, 1990.

[38] Bennett, G.K. and Martz, H.F., "An empirical Bayes estimation for the scale parameter of the two-parameter Weibull distribution," *Naval Research Logistics Quarterly*, **20**, 1973, pp 387–393.

[39] Bergman, B., "On age replacement and the total time on test concept," *Scand J. Statist* **6**, 1979, pp 161–168.

[40] Bergman, B., "On reliability theory and its applications," *Scand. J. Statis.* **12**, 1985, pp 1–41.

[41] Bergman, B. and Klefsjö, B., "A graphical method applicable to age-replacement problems," *IEEE Transactions on Reliability*, **R-31** (5), Dec. 1982, pp 478–481.

[42] Bergman, B. and Klefsjö, B., "The total time on test concepts and its use in reliability theory," *Operations Research* **32** (3), May-June 1984, pp 596–606.

[43] Berglundm C.N., "A unified yield model incorporating both defect and parametric effects," *IEEE Transactions on Semiconductor Manufacturing*, **9** (3), Aug. 1996, pp 447–454.

[44] Bezat, A.G. and Montague, L.L., "The effect of endless burn-in on reliability growth projections," *Annual Reliability and Maintainability Symposium*, 1979, pp 392–397.

[45] Bhattacharya, S.K., "Bayesian approach to life testing and reliability estimation," *Journal of the American Statistical Association*, **62**, 1967, pp 48–62.

[46] Bickel, P.J. and Doksum, K.A., "Tests for monotone failure rate based on normalized spacings," *Annals of Mathematical Statistics*, **40**, 1969, pp 1216–1235.

[47] Birnbaum, Z.W., Esary, J.D. and Marshall, A.W., "A stochastic characterization of wear out for components and systems," *Annals of Mathematical Statistics*, **37** (4), 1966, pp 816–826.

[48] Blanks, H.S., "The temperature dependence of component failure rate," *Microelectronics and Reliability*, **20**, 1980, pp 297–307.

[49] Block, H.W. and Savits, T.H., *Burn-in : Series in Reliability and Statistics*, University of Pittsburgh Technical Report No.94-02, Pittsburgh, 1996.

[50] Bologna, S. and Clarotti, C.A., "Expert opinions in software reliability analysis" *Proceedings of the International School of Physics: Accelerated Life Testing and Experts' Opinions in Reliability*, Clarotti, C.A., Lindley, D.V., editors, pp 132–144; North-Holland, 1986.

[51] Bratley, P., Fox, B.L., and Schrage, L.E., *A Guide to Simulation*, Springer-Verlag, New York, 1987.

[52] British Telecom, *Handbook of Reliability Data for Electronic Components used in Telecommunications Systems* (Issue 3), British Telecom plc, Jan. 1984.

[53] Buchanan, J., *CMOS/TTL Digital Systems Design*, McGraw-Hill, New York, NY, 1990.

[54] Bukowski, J.V., Johnson, D.A., and Goble, W.M., "Software-reliability feedback: a physics-of-failure approach," *Annual Reliability and Maintainability Symposium*, 1992, pp 285–289.

[55] Burridge, J. , "Empirical Bayes analysis of survival time data," *Journal of the Royal Statistical Society*, **B**, **43** (1), 1981, pp 65–75.

[56] Burte, H.M. and Copolla, A., "Unified life cycle engineering," *Annual Reliability and Maintainability Symposium*, 1987, pp 8–14.

[57] Bury, K.V., "On the reliability analysis of a two-parameter Weibull process," *INFOR*, **110**, 1972, pp 129–139.

[58] Byers, J.K., Skeith, R.W. and Springer, M.D., "Bayesian confidence limits for the reliability of mixed cascade and parallel independent exponential subsystems," *IEEE Transactions on Reliability*, **R-23** (2), Jun. 1974, pp 104–108.

[59] Canavos, G.C., "An empirical Bayes approach for the Poisson life distribution," *IEEE Transactions on Reliability*, **R-22** (2), Jun. 1973, pp 91–96.

[60] Canavos, G.C. and Tsokos, C.P., "Ordinary and empirical Bayes approach to estimation of reliability in the Weibull life testing model," *Proceedings of the 6th Conference on the Design of Experiments in Army Research Development and Testing*, Virginia, 1970, pp 379–393.

[61] Canavos, G.C. and Tsokos, C.P., "A study of an ordinary and empirical Bayes approach to reliability estimation in the gamma life testing model," *Proceedings of Annual Symposium on Reliability*, 1971, pp 343–349.

[62] Canavos, G.C. and Tsokos, C.P., "Bayesian estimation of life parameters in the Weibull distribution," *Operations Research*, **21**, 1973, pp 755–763.

[63] Canfield, R.V. , "A Bayesian approach to reliability estimation using a loss function," *IEEE Transactions on Reliability*, **R-19** (1), Feb. 1970, pp 13–16.

[64] Canfield, R.V. , "Cost effective burn-in and replacement times," *IEEE Transactions on Reliability*, **R-24** (2), Jun. 1975, pp 154–156.

[65] Canfield, R.V. and Borgman, L.E. , "Some distributions of time to failure for reliability applications," *Technometrics*, **17** (2), 1975, pp 263–268.

[66] Carrubba, E.R. , "Integrating life-cycle cost and cost-of-ownership in the commercial sector," *Annual Reliability and Maintainability Symposium*, 1992, pp 101-108.

[67] Casella, G. , "An introduction to empirical Bayes data analysis," *The American Statistician*, **39**, 1985, pp 83–87.

[68] Chandrasekaran, R. , "Optimal policies for burn-in procedures," *OPSEARCH*, **14** (3), 1977, pp 148–160.

[69] Chang, T.C. and Wysk, R.A., *An Introduction to Computer Aided Process Planning Systems*, Prentice Hall, 1995.

[70] Chao, A. and Hwang, W., "Bayes estimation of reliability for special k-out-of-m:g systems," *IEEE Transactions on Reliability*, **R-32** (4), Oct. 1985, pp 370–373.

[71] Chen, K., Papadopoulos, and Tamer, P., "On Bayes estimation for mixtures of two Weibull distributions under type-I censoring," *Microelectronics and Reliability*, **29** (4), 1989, pp 609–617.

[72] Cheng, S.S., "Optimal replacement rate of devices with lognormal failure distributions," *IEEE Transactions on Reliability*, **R-26** (3), Aug. 1977, pp 174–178.

[73] Cheng, S.S., "Optimal burn-in time of lognormal devices in an optical communication system," *Proceedings Optical Communication Conference*, Amsterdam, The Netherlands, 18.3-1–18.5-5, 1979.

[74] Chi, D., *Software Reliability Optimization by Redundancy and Software Quality Management*, Ph.D dissertation, Department of Industrial and Manufacturing Systems Engineering, Iowa Stat University, Ames, Iowa, 1989.

[75] Chi, D. and Kuo, W., "Burn-in optimization under reliability & capacity restrictions," *IEEE Transactions on Reliability,* **38** (2), Jun. 1989, pp 193–198.

[76] Chien, W.T.K. and Kuo, W., "Optimal burn-in simulation on highly integrated circuit systems," *IIE Transactions,* **24** (5), Nov. 1992, pp 33–43.

[77] Chien, W.T.K. and Kuo, W., "Determine optimal burn-in time for highly integrated circuit systems," *Proceedings 1992 World Conference of Nonlinear Analysis*, 1992.

[78] Chien, W.T.K. and Kuo, W., "A nonparametric approach to estimate system burn-in time," *IEEE Transactions on Semiconductor Manufacturing,* **9** (3), Aug. 1996, pp 461–466.

[79] Chien, W.T.K. and Kuo, W., "Optimization of the burn-in times through redundancy allocation," *Proceedings of the 2nd Annual IIE Research Conference*, Los Angeles, California, 1993, pp 579–583.

[80] Chien, W.T.K. and Kuo, W., "The variance reduction of the simulation for the Dirichlet distributions," *The Second International Conference on Reliability, Maintainability, and Safety*, Beijing, China, 1994 Jun. 7–10, pp 252–257.

[81] Chien, W.T.K. and Kuo, W., "Modeling and maximizing burn-in effectiveness," *IEEE Transactions on Reliability,* **44** (1), Mar. 1995, pp 19–25.

[82] Chien, W.T.K. and Kuo, W. , "A nonparametric Bayes approach to decide system burn-in time," *Naval Research Logistics Quarterly,* 1997, to appear.

[83] Chou, K. and Tang, K., "Burn-in time and estimation of change-point with Weibull-exponential mixture distribution," *Decision Sciences,* **23** (4), Jul-Aug 1992, pp 973–990.

[84] Christou, A., *Integrating Reliability into Microelectronics Manufacturing*, John Wiley & Sons, Chichester, 1994.

[85] Cohen, A.C., "Maximum likelihood estimation in the Weibull distribution based on complete and censored samples," *Technometrics,* **7** (4), 1965, pp 579–588.

[86] Collas, G., "Prediction for system reliability and availability," *Annual Reliability and Maintainability Symposium,* 1989, pp 337–341.

[87] Colombo, A.G., Costantini, D., and Jaarsma, R.J., "Bayes nonparametric estimation of time-dependent failure rate," *IEEE Transactions on Reliability,* **R-34** (2), Jun. 1985, pp 109–112.

[88] Colton, J.S. and Dascanio, J.L., II , "An integrated, intelligent design environment," *Engineering with Computers,* **7**, 1991, pp 11-22.

[89] Confer, R., Canner, J., and Trostle, T. , "Use of highly accelerated life test (HALT) to determine reliability of multilayer ceramic capacitors," *Proceedings of the Electronic Components & Technology Conference*, 1991, pp 320–322.

[90] Connor, R.J. and Mosimann, J.E., "Concepts of independence for proportions with a generalization of the Dirichlet distribution," *Journal of the American Statistical Association,* **64**, 1969, pp 194–206.

[91] Corsi, F. and Martino, S., "Defect level as a function of fault coverage and yield," *Proceedings of European Test Conference*, 1993, pp 507–508.

[92] Couture, D.J. and Martz, H.F. Jr. , "Empirical Bayes estimation in the Weibull distribution," *IEEE Transactions on Reliability,* **R-21** (2), May 1972, pp 75–83.

[93] Cozzolino, J.M. , "The optimal burn-in testing of repairable equipment," *Naval Research Logistics Quarterly,* **17** (2), 1970, pp 167–181.

[94] Crook, D.L., "Evolution of VLSI reliability engineering," *Proceedings of International Reliability Physics Symposium,* 1990, pp 2–11.

[95] Crow, L.H. and Shimi, I.N. , "Maximum likelihood estimation of life-distribution from renewal testing," *Annals of Mathematical Statistics,* **43**, 1972, pp 1827–1838.

[96] Cunningham, J.A., "The use and evaluation of yield models in integrated circuit manufacturing," *IEEE Transactions on Semiconductor Manufacturing,* **3** (2), 1990, pp 60–71.

[97] Cunningham, S.P., Spanos, C.J., and and Voros, K., "Semiconductor yield improvement: results, and best practices," *IEEE Transactions on Semiconductor Manufacturing,* **8** (2), May 1995, pp 103–109.

[98] D'Agostino, R.B. , "Linear estimation of the Weibull parameters," *Technometrics,* **13** (1), 1971, pp 171–182.

[99] D'Agostino, R.B. and Stephens, M.A. (Editors), "Tests based on EDF statistics," by Michael A. Stephens in *Goodness-of-fit techniques*, Marcel Dekker, New York, NY, 1986.

[100] Dance, D. and Jarvis, R., "Using yield models to accelerate learning curve progress," *IEEE Transactions on Semiconductor Manufacturing,* **5** (1), Feb. 1992, pp 41–45.

[101] Dasgupta, A., Barker, D., and Pecht, M. , "Reliability prediction of electronic packages," *Annual Reliability and Maintainability Symposium*, 1990, pp 323–330.

[102] Degraeve, R. *et al.*, "On the field dependence of intrinsic and extrinsic time-dependent dielectric breakdown," *Proceedings of International Reliability Physics Symposium*, 1996, pp 44–54.

[103] Dellin, T.A., *et al.*, "Wafer level reliability," *SPIE Microelectronics Manufacturing and Reliability*, 1992, pp 144–154.

[104] Dhillon, B.S. , "Life distributions," *IEEE Transactions on Reliability*, **R-30** (5), Dec. 1981, pp 457–460.

[105] Dhillon, B.S. and Singh, C., *Engineering Reliability*, John Wiley & Sons, Inc., New York, 1981.

[106] Dinse, G.E., "An alternative to Efron's redistribution-of-mass conn of the Kaplan-Meier estimator," *The American Statistician*, **39** (1), Nov. 1985, pp 290.

[107] Disch, D., "Bayesian nonparametric inference for effective doses in a quantal-response experiment," *Biometrics* **37**, 1981, pp 713–722.

[108] Donovan, T.J. , "Accelerated environmental stress screening & reliability growth testing of the B-52 infrared camera," *Annual Reliability and Maintainability Symposium*, 1993, pp 510–514.

[109] Downs, T., "An approach to the modeling of software testing with some applications," *IEEE Transactions on Software Engineering*, **SE-11** (4), 1985, pp 375–386.

[110] Duran, B.S. and Booker, J.M., "A Bayes sensitivity analysis when using the beta distribution as a prior," *IEEE Transactions on Reliability*, **R-37** (2), Jun. 1988, pp 239–247.

[111] Durbin, J., *Distribution Theory for Tests Based on the Sample Distribution Function*, Regional conference series in applied mathematics, **9**, Society for Industrial and Applied Mathematics, Philadelphia, Pa, 1973.

[112] Efron, B., "The two sample problem with censored data," *Proceedings of the 5th Berkeley Symposium*, **4**, Berkeley: University of California Press, 1967, pp 831–853.

[113] Efron, B. and Morris, C. , "Limiting the risk of Bayes and empirical Bayes estimators – Part I: The Bayes case," *Journal of the American Statistical Association*, **66** (336), Dec. 1971, pp 807–815.

[114] Efron, B. and Morris, C., "Limiting the risk of Bayes and empirical Bayes estimators – Part II: The Empirical Bayes case," *Journal of the American Statistical Association*, **67** (337), Mar. 1972, pp 130–139.

[115] Efron, B. and Morris, C., "Stein's paradox in statistics," *Scientific American*, **236** (5), 1977, pp 119–127.

[116] Efron, B. and Tibshirani, R.J., *An Introduction to the Bootstrap*, Monographs on Statistics and Applied Probability 57, Chapman & Hall, New York, NY, 1993.

[117] Einspuruch, N.G. and Hilbert J.L., *Application Specific Integrated Circuit(ASIC) Technology*, Academic Press, Inc., San Diego, CA, 1991.

[118] English, J.R., Yan, L., and Landers, T.L., "A modified bathtub curve with latent failures," *Annual Reliability and Maintainability Symposium*, 1995, pp 217–222.

[119] Elkings, J.D. and Sweetland, R.L. , "Burn-in forever? There must be a better way!," *Annual Reliability and Maintainability Symposium*, 1978, pp 286–293.

[120] Elperin, T. and Gertsbakh, I. , "Bayes credibility estimation of an exponential parameter for random censoring & incomplete information," *IEEE Transactions on Reliability*, **39** (2), Jun. 1990, pp 204–208.

[121] Elsayed, E.A., *Reliability Engineering*, Addison Wesley Longman, Inc., Reading, MA, 1996.

[122] El-Kareh, B., Ghatalia, A., and Satya, A.V.S., "Yield management in microelectronic manufacturing," *Proceedings of the 45th Electronic Components Conference*, 1995, pp 58–63.

[123] El-Sayyad, G.M. , "Estimation of the parameter of an exponential distribution," *Journal of the Royal Statistical Society*, **29**, 1967, pp 525–532.

[124] Engleman, J.H. , "A Bayesian time-to-failure distribution," *Proceedings of Annual Symposium on Reliability*, 1971, pp 350–355.

[125] Epstein, B. , "Some applications of the Mellin transformation in statistics," *Annals of Mathematical Statistics*, **12**, 1948, pp 370–379.

[126] Epstein, B. , "Truncated life tests in the exponential case," *Annals of Mathematical Statistics*, **25**, 1954, pp 555–564.

[127] Epstein, B. , "The exponential distribution and its role in life-testing," *Industrial Quality Control*, **15**, 1958, pp 2–7.

[128] Epstein, B. , "Statistical life test acceptance procedures," *Technometrics*, **2**, 1960, pp 435–446.

[129] Epstein, B. , "Testing for the validity of the assumption that the underlying distribution of life is exponential," *Technometrics*, **2** (2), 1960, pp 83–101, pp 167–183.

[130] Epstein, B., "Estimation from life test data," *Technometrics*, **2** (4), 1960, pp 447–454.

[131] Epstein, B. and Sobel, M., "Sequential life tests in the exponential case," *Annals of Mathematical Statistics*, **26**, 1955, pp 82–93.

[132] Erto, P., "New practical Bayes estimators for the 2-parameter Weibull distribution," *IEEE Transactions on Reliability*, **R-31** (2), Jun. 1982, pp 194–197.

[133] Esary, J.D. and Proschan, F., "Relationship between system failure rate and component failure rate," *Technometrics*, **5**, 1963, pp 183–189.

[134] Esary, J.D., Marshall, A.W., and Proschan F., "Some reliability applications of the hazard transform," *SIAM Journal on Applied Mathematics*, **18** (4), 1970, pp 849–860.

[135] Esary, J.D., Marshall, A.W., and Proschan F., "Shock models and wear processes," *Annals of Probability*, **1** (4), 1973, pp 627–649

[136] Eubank, R.L., "Approximate regression models and splines," *Communications in Statistics: Theory and Methods*, **13**, 1984, pp 433-484.

[137] Eubank, R.L., *Spline Smoothing and Nonparametric Regression*, Marcel Dekker, New York, NY, 1988.

[138] Fard, N.S. and Dietrich, D., "A Bayes reliability growth model for a development testing program," *IEEE Transactions on Reliability*, **36** (5), Dec. 1987, pp 568–572.

[139] Fayette, D.F, "MIMIC QML status," *Advanced Microelectronics Technology Qualification, Reliability and Logistics Workshop*, Aug. 13-15, 1991, Seattle, WA, pp 245–261.

[140] Ferguson, T.S., "A Bayesian analysis of some nonparametric problems," *Annals of Statistics*, **1** (2), 1973, pp 209–230.

[141] Ferguson, T.S. , "Prior distributions on the space of probability measures," *Annals of Statistics* **2** (4), 1974, pp 615–629.

[142] Ferguson, T.S., "Sequential estimation with Dirichlet process priors," in *Statistical Decision Theory and Related Topics III* **1**, Shanti S. Gupta and James O. Berger, editors, Academic Press, New York, 1982, pp 385–401.

[143] Ferguson, T.S. , "Bayesian density estimation by mixtures of normal distributions," in *Recent Advances in Statistics*, Rizvi, M.H., Rustagi, J., and Siegmund, D., Editors, Academic Press, 1983, pp 287–302.

[144] Ferguson, T.S., Phadia, E.G. and Tiwari, R.C., "Bayesian nonparametric inference," in *Current Issues in Statistical Inference: Essays/ D. Basu*, Malay Ghosh, Pramod K. Pathak, IMS, 1992, pp 127–150.

[145] Ferrara, K.C., Keene, S.J., and Lane, C., "Software reliability from a system perspective," *Annual Reliability and Maintainability Symposium*, 1989, pp 332–336.

[146] Ferris-Prabhu, A.V., "Defect size variations and their effect on the critical area of VLSI devices," *IEEE Journal of Solid State Circuits*, **SC-20** (4), Aug. 1985, pp 878–880.

[147] Ferris-Prabhu, A.V., "Models for Defects and Yield," in *Defect and Fault Tolerance in VLSI Systems*, edited by Koren, I., 1989, pp 33–46.

[148] Ferris-Prabhu, A.V., *Introduction to Semiconductor Device Yield Modeling*, Artech House, Boston, MA, 1992.

[149] Fiacco, A.V., *Introduction to Sensitivity and Stability Analysis in Nonlinear Programming*, Academic Press, New York, 1983, pp 155–193.

[150] Finkelstein, C., *An Introduction to Information Engineering: From Strategic Planning to Information Systems*, Addison-Wesley Publishing Company, Reading, MA, 1989.

[151] Flood, J.L., "Reliability of plastic encapsulated integrated circuits," *Proceedings of International Reliability Physics Symposium*, 1972, pp 95–99.

[152] Foulk, P.W., *CAD of Concurrent Computers*, John Wiley and Sons, Inc., New York, 1985.

[153] Foster Jr., W.M., "Thermal verification testing of commercial printed-circuit boards," *Annual Reliability and Maintainability Symposium*, 1992, pp 189–195.

[154] Fox, B., "Total annual cost, a reliability criterion". *Proceedings Symposium Reliability and Quality Control*, 1964, pp 266–273.

[155] Friedman, M.A. and Tran, P., "Reliability techniques for combined hardware/software systems," *Annual Reliability and Maintainability Symposium,* 1992, pp 290–293.

[156] Fujii, S. and Sandoh, H., "Bayes reliability assessment of a 2-unit hot-standby redundant system," *IEEE Transactions on Reliability,* **R-33** (4), Oct. 1984, pp 297–300.

[157] Fuqua, N.B., *Reliability Engineering for Electronic Design,* Marcel Dekker, New York, 1987.

[158] *GAMS Users Manual,* GAMS Development Corporation, Redwood city, CA, 1988.

[159] Garrison, R.H., *Managerial Accounting,* Business Publications, Inc., Plano, Texas, 1985.

[160] Gaver, D.P. and O'muricheartaigh, I.G., "Robust empirical Bayes analysis of event rates," *Technometrics,* **29** (1), Feb. 1987, pp 1-15.

[161] Geurts, J.H.J., "On the small-sample performance of Efron's and of Gill's version of the product limit estimator under non-proportional hazards," *Biometrics,* **43**, Sep. 1987, pp 683–692.

[162] Gibbons, J.D., *Nonparametric Statistics : An Introduction,* Sage university papers series, Quantitative applications in the social sciences; no. 90; Sage Publications, Newbury Park, CA, 1993.

[163] Gimlin, D.R. and Breipohl, A.M. , "Bayesian acceptance sampling," *IEEE Transactions on Reliability,* **R-21** (3), Aug. 1972, pp 176–180.

[164] Giuntini R.E. and Giuntini, M.E., "Simulating a Weibull posterior using Bayes inference," *Annual Reliability and Maintainability Symposium,* 1993, pp 49–55.

[165] Ghatalia, A. and El-Kareh, B., *Yield Management in Microelectronic Manufacturing,* Short Course Notes of The National Alliance for Photonics Education in Manufacturing, Austin, TX, Oct. 1996.

[166] Goel, A.L. and Okumoto, K., "time-dependent error-detection rate model for software reliability and other performance measures," *IEEE Transactions on Reliability,* **R-28** (3), Aug. 1979, pp 206–211.

[167] Green, P.J. and Silverman, B.W., *Nonparametric Regression and Generalized Linear Models: A Roughness Penalty Approach,* Monographs on Statistics and Applied Probability 58, Chapman & Hall, New York, NY, 1994.

[168] Grosh, D.L., *A Primer of Reliability Theory,* John Wiley & Sons, Inc., New York, 1989.

[169] Guida, M., Calabria, R. and Pulcini, G. , "Bayes inference for a non-homogeneous Poisson process with power intensity law," *IEEE Transactions on Reliability,* **38** (5), Dec. 1989, pp 603–609.

[170] Gupta, R.C. and Michalek, J.E., "Determination of reliability function by the TTT transform," *IEEE Transactions on Reliability,* **R-34** (2), Jun. 1985, pp 175–176.

[171] Haim M. and Porat, Z., "Bayes reliability modeling of a multistate consecutive K-out-of-n: f system," *Annual Reliability and Maintainability Symposium,* 1991, pp 582–586.

[172] Hall, P.A. and Resnick, M., "Standards," in *Software Engineer's Reference Book* (**50**), McDermid, J.A. Editor, Butterworth-Heinemann Ltd., Jordan Hill, Oxford, 1991.

[173] Hallberg, O. , "Failure-rate as a function of time due to log-normal life distributions of weak parts," *Microelectronics and Reliability,* **16**, 1977, pp 155–158.

[174] Hansen, C.K., "Effectiveness of yield-estimation and reliability-prediction based on wafer test-chip measurements," *Annual Reliability and Maintainability Symposium,* 1997, pp 142–148.

[175] Hansen, C.K. and Thyregod, P., "Modeling and estimation of wafer yields and defect densities from microelectronics test structure data," *Quality and Reliability Engineering International,* **12**, 1996, pp 9–17.

[176] Hansen, M.D., "Software system safety and reliability," *Annual Reliability and Maintainability Symposium,* 1988, pp 214–217.

[177] Harris, C. and Singpurwalla, N. , "Life distributions derived from stochastic hazard functions," *IEEE Transactions on Reliability,* **R-17** (2), Jun. 1968, pp 70–79.

[178] Hart, L. , "A Bayes approach to simultaneous evaluation of similar assemblies," *IEEE Transactions on Reliability,* **38** (4), Oct. 1989, pp 483–484.

[179] Hart, L. , "Reliability on modified designs: a Bayes analysis of an accelerated test of electronic assemblies," *IEEE Transactions on Reliability,* **39** (2), Jun. 1990, pp 140–144.

[180] Harter, H.L., and Moore, A.H. , "Maximum-likelihood estimation of the parameters of gamma and Weibull populations from complete censored samples," *Technometrics,* **7** (4), 1965, pp 639–643.

[181] Harter, H.L., and Moore, A.H. , "Iterative maximum-likelihood estimation of the parameters of normal populations from singly and double censored samples," *Biometrika*, **53**, 1966, pp 205–213.

[182] Harter, H.L., and Moore, A.H. , "Maximum-likelihood estimation, from censored samples, of the parameters of a logistic distribution," *Journal of the American Statistical Association*, **62**, 1967, pp 675–684.

[183] Harter, H.L., and Moore, A.H. , "Maximum-likelihood estimation, from doubly censored samples, of the parameters of the first asymptotic distribution of extreme values," *Journal of the American Statistical Association*, **63**, 1968, pp 889–901.

[184] Higgins, J.J. and Tsokos, C.P. , "On the behavior of some quantities used in Bayesian reliability demonstration tests," *IEEE Transactions on Reliability*, **R-25** (4), Oct. 1976, pp 261–264.

[185] Higgins, J.J. and Tsokos, C.P. , "Comparison of Bayesian estimates of failure intensity for fitted priors of life data," in *The Theory and Applications of Reliability*, **II**, Tsokos, C.P. and Shimi, I.N., Editors., Academic Press, New York, NY, 1977, pp 75–92.

[186] Higgins, J.J. and Tsokos, C.P. , "Modified method-of-moment in empirical Bayes estimation," *IEEE Transactions on Reliability*, **R-28** (1), Apr. 1979, pp 27–31.

[187] *Hindsight/C User Manual Installation and Tutorial*, Advanced Software Automation, Inc., Santa Clara, CA, 1993.

[188] *Hindsight/C User Manual*, Advanced Software Automation, Inc., Santa Clara, CA, 1992.

[189] Hnatek, E.R., "A realistic view of VLSI burn-in," *Evaluation Engineering*, **28** (2), 1989, pp 80+.

[190] Hnatek, E.R., *Integrated Circuit Quality and Reliability*, second edition, Marcel dekker, New York, 1995.

[191] Hoel, P.G. , *Introduction to Mathematical Statistics*, fifth edition, John Wiley & Sons, Inc, 1984.

[192] Hogg, R.V. and Craig, A.T. , *Introduction to Mathematical Statistics*, fourth edition, Macmillan Publishing Co., Inc, 1986.

[193] Hollander, M. and Proschan, F. , "Testing for the mean residual life," *Biometrika*, **62** (3), 1975, pp 585–593.

[194] Hu, C., "Future CMOS scaling and reliability," *Proceedings of the IEEE*, **81** (5), May 1993, pp 682–689.

[195] Hughes, J., "A Practical assessment of current plastic encapsulated microelectronic devices," *Quality and Reliability Engineering International*, **5**, 1989, pp 125–129.

[196] Hunter, J.S., "Statistical Design Applied to Product Design," *Journal of Quality Technology* **17** (4), Oct. 1985, pp 210–221.

[197] Huston, H.H. and Clarke C.P., "Reliability defect detection and screening during processing - theory and implementation," *Proceedings of International Reliability Physics Symposium*, 1992, pp 268–275.

[198] Ignizio, J.P., *Goal Programming and Extensions*, Lexington Book, Massachuetts, 1976.

[199] Ireson, W.G., Coombs, Jr., C.F., and Moss R.Y., *Handbook of Reliability Engineering and Management*, Second Edition, McGraw-Hill, New York, 1996.

[200] Jarvis, W.H. *Bayesian Analysis for Cost-Optimal Burn-In testing*, Ph.D. Dissertation, George Washington University, 1996.

[201] Jeffreys, H. , *Theory of Probability*, Oxford Clarendon Press, third Edition, 1983.

[202] Jelinski, Z. and Moranda, P.B., "Software reliability research," McDonnell Douglas Astronautics paper ED 1808, presented at the Conf. on Statistical Methods for the Evaluation of Computer System Performance, Brown University, Providence, RI, Nov. 1971, pp 465–484; also *Statistical Computer Performance Evaluation*, W. Freiberger, Ed., 1972; Academic.

[203] Jensen, F. and Petersen, N.E. , *Burn-in, An Engineering Approach to Design and Analysis of Burn-in Procedures*, John Wiley & Sons, Inc., New York, 1982.

[204] Jones, T.H. , *Electronic Components Handbook*, Reston Publishing Company, Inc., Reston, Virginia, 1978.

[205] Johns, M.V., Jr. and Lieberman, G.J. , "An exact asymptotically efficient confidence bound for reliability in the case of the Weibull distribution," *Technometrics*, **8**, 1966, pp 135–175.

[206] Johnson, N.L. and Kotz, S. , *Distributions in Statistics: Continuous Univariate Distributions-2*, John Wiley & Sons, New York, NY, 1970.

[207] Johnson, E.G. and Routledge, R.D., "The line transect method: a nonparametric estimator based on shape restrictions," *Biometrics*, **41**, Sep. 1985, pp 669–679.

[208] Kackar, R.N. , "Off-line quality control, parameter design, and the Taguchi method," *Journal of Quality Technology* **17** (4), Oct. 1985, pp 176–188.

[209] Kalbfleisch, J.D. and Ross L.P., *The Statistical Analysis of Failure Time Data*, John Wiley & Sons, New York, NY, 1980.

[210] Kaplan, S., Cunha, G.D.M., Dykes, A.A., and Shaver, D., "A Bayesian methodology for assessing reliability during product development," *Annual Reliability and Maintainability Symposium*, 1990, pp 205–209.

[211] Kaplan, E.L. and Meier, P. , "Nonparametric estimation from incomplete observations," *Journal of the American Statistical Association*, **53**, 1958, pp 457-481.

[212] Kapur, K.C., and Lamberson, L.R., *Reliability in Engineering Design*, John Wiley & Sons, New York, NY, 1977.

[213] Karlin, S. , *Total Positivity*, I, Stanford University Press, Stanford, California, 1968.

[214] Keene, S.J., "Assuring software safety," *Annual Reliability and Maintainability Symposium*, 1992, pp 274–279.

[215] Kim, T., Kuo, W., and Chien, W.T.K., "A relation model of yield and reliability for the gate oxide failures," *Annual Reliability and Maintainability Symposium*, 1998 to appear.

[216] Klefsjö, B. , "On aging properties and the total time on test transforms," *Scand J. Statist* **9**, 1982, pp 37–41.

[217] Klefsjö, B. , "On aging properties and the total time on test transforms," *Reliability Engineering* **5**, 1986, pp 231–241.

[218] Kolarik, W., Davenport, J., Fant, E., and McCoun, K. , "Early design phase life cycle reliability modeling," *Annual Reliability and Maintainability Symposium*, 1987, pp 335–340.

[219] Koss, W.E., "Software-reliability metrics for military systems," *Annual Reliability and Maintainability Symposium*, 1988, pp 190–194.

[220] Kubat, P. and Koch, H.S., "Pragmatic testing protocols to measure software reliability," *IEEE Transactions on Reliability*, **R-32** (4), Oct. 1983, pp 338–341.

[221] Kumar, S. and Tiwari, R.C. , "Bayes estimation of reliability under a random environment governed by a Dirichlet prior," *IEEE Transactions on Reliability*, **38** (2), Jun. 1989, pp 218–223.

[222] Kunitz, H. , "A new class of bathtub-shaped hazard rates and its application in a comparison of two test-statistics," *IEEE Transactions on Reliability*, **38** (3), Aug. 1989, pp 351–354.

[223] Kuo, L., "Computations of mixtures of Dirichlet processes," *SIAM Journal on Scientific and Statistical Computing*, **7**, 1986, pp 60–71.

[224] Kuo, W., "Software reliability estimation: a realization of competing risk," *Microelectronics and Reliability*, **23** (2), 1983, pp 249–260.

[225] Kuo, W., "Reliability enhancement through optimal burn-in," *IEEE Transactions on Reliability*, **R-33** (2), Jun. 1984, pp 145–156.

[226] Kuo, W., "Modeling and management of software reliability: a tutorial," in *Handbook of Industrial Engineering*, McGraW Hill, 1990.

[227] Kuo, W., (Editor) , *Quality through Engineering Design*, Elsevier Scientific Pub., Amsterdam, 1993.

[228] Kuo, W. , "Incompatibility in evaluating large-scale systems reliability," *IEEE Transactions on Reliability*, **43** (4), Dec. 1994, pp 659–660.

[229] Kuo, W. and Oh, L.(Guest Editors), "Design for Reliability," *IEEE Transactions on Reliability*, **44** (2), 1995, pp 170–171.

[230] Kuo, W. and Kuo, Y. , "Facing the headaches of early failures: a state-of-the-art review of burn-in decisions," *Proceedings of the IEEE*, **71** (11), Nov. 1983, pp 1257–1266.

[231] Kuo, W. and Lingraji, B.P., "Application of an availability model for evaluation of alternatives," presented at 1981 Fall ORSA/TIMS Joint National Meeting in Houston, USA.

[232] Kuper, F., *et al.*, "Relation between yield and reliability of integrated circuits: experimental results and application to continuous early failure rate reduction programs," *Proceedings of International Reliability Physics Symposium*, 1996, pp 17–21.

[233] Langberg, N., Leon, R. V., and Proschan, F. , "Characterization of nonparametric classes of life distributions," *Annals of Probability*, **8** (6), 1980, pp 1163–1170.

[234] LaPadula, L.J., "Engineering of Quality Software Systems," Report RADC-TR-74-325, Vol. VIII: Software Reliability Modeling and Measurement Techniques, 1973.

[235] Laska, E.M. and Meisner, M.J., "Nonparametric estimation and testing in a cure model," *Biometrics*, **48**, Dec. 1992, pp 1223–1234.

[236] Launer, R.L., "Graphical techniques for analyzing failure data with percentile residual-life functions," *IEEE Transactions on Reliability*, **42** (1), Mar. 1993, pp 71–75.

[237] Lawless, J.F. , *Statistical Models and Methods for Lifetime Data*, John Wiley & Sons, New York, NY, 1982.

[238] Lawrence, M.J. , "An investigation of the burn-in policies," *Technometrics*, **8** (1), 1966, pp 61–71.

[239] Lawrence, J.D. , "Memory burn-in with test," *Proceedings Test Conference*, Philadelphia, PA, 1980, pp 489–493.

[240] Lee, J.C., Chen, I.C., and Hu, C., "Modeling and characterizing of gate oxide reliability," *IEEE transaction on Electron Devices*, **35** (12), 1988, pp 2268–2277.

[241] Lee, M.H., *Strong Consistent Modified Maximum Likelihood Estimation of U-Shaped Hazard Function*, Ph.D. dissertation, Iowa State University, 1987.

[242] Lee, R.A , "Reliability and TQM," *Reliability Society Newsletter* **39**, (1), Jan. 1993, p 7.

[243] Lee, S.C.S. and Locke, C. , "On a class of tests of exponentiality," *Technometrics*, **22** (4), Nov. 1980, pp 547–554.

[244] Leemis, L.M., "Nonparametric estimation of the cumulative intensity function for a nonhomogeneous Poisson process," *Management Science*, **37**, July 1991, pp 886–900.

[245] Leemis, L.M. and Beneke, M. , "Burn-in models and methods: a review," *IIE Transactions*, **22** (2), Jun. 1990, pp 172–180.

[246] Lemon, G.H. , "An empirical Bayes approach to reliability," *IEEE Transactions on Reliability*, **R-21** (3), Aug. 1972, pp 155–158.

[247] Leonard, C.T., "Passive cooling for avionics can improve airplane efficiency and reliability," NAECON, 1987.

[248] Li, T.F. , "Empirical Bayes approach to reliability estimation for the exponential distribution," *IEEE Transactions on Reliability*, **R-33** (3), Aug. 1984, pp 233–236.

[249] Lieberman, G.J., and Ross, S.M. , "Confidence intervals for independent exponential series systems," *Journal of the American Statistical Association*, **66**, 1971, pp 837–840.

[250] Lin, H. and Kuo, W., "Reliability cost in software life-cycle models," *Annual Reliability and Maintainability Symposium*, 1987, pp 364–368.

[251] Lindgren, B.W. , *Statistical Theory*, Third edition, Macmillan, New York, 1968.

[252] Lindley, D.V. , "Sequential sampling: two decision problems with linear losses for binomial and normal random variables," *Biometrika*, **52**, 1961, pp 507–532.

[253] Lingappaiah, G.S. , "Shifting shape-parameter in life tests: a Bayes analysis," *IEEE Transactions on Reliability*, **R-32** (3), Aug. 1983, pp 317–322.

[254] Littlewood, B., "Stochastic reliability- growth: a model for fault-removal in computer- programs and hardware- designs," *IEEE Transactions on Reliability*, **R-30** (4), Oct. 1981, pp 313–320.

[255] Littlewood, B. , "Rationale for a modified Duane model," *IEEE Transactions on Reliability*, **R-33** (2), Jun. 1984, pp 157–159.

[256] Littlewood, B. and Verrall, J.L., "A Bayesian reliability growth model for computer software," *Applied Statistics*, **22**, 1973, pp 332–346.

[257] Lo, A.Y., "On a class of Bayesian nonparametric estimates: I. Density estimates," *Annals of Statistics*, **12** (1), 1984, pp 351–357.

[258] Lochner, R.H., "A generalized Dirichlet distribution in Bayesian life testing," *Journal of the Royal Statistical Society*, B, **37**, 1975, pp 103–113.

[259] Luthra, P. , "MIL-HDBK-217: What is wrong with it," *IEEE Transactions on Reliability*, **39** (5), Dec. 1990, p 518.

[260] Lynch, T., Pham, H., and Kuo, W., "Modeling software-reliability with multiple failure-types and imperfect debugging," *Annual Reliability and Maintainability Symposium*, 1994, pp 235–241.

[261] MacFarland, W.J. , "Bayes' equation, reliability, and multiple hypothetic testing," *IEEE Transactions on Reliability*, **R-21** (3), Aug. 1972, pp 136–147.

[262] Mann, N.R., and Grubbs, F.E. , "Approximately optimum confidence bounds on series system reliability for exponential time to failure data," *Biometrika*, **59** (1), 1972, pp 191–204.

[263] Marcus, R. and Blumenthal, S. , "A sequential screening procedure," *Technometrics*, **16** (2), 1974, pp 229–234.

[264] Marko, D.M. and Schoonmaker, T.D. , "Optimizing spare module burn-in," *Annual Reliability and Maintainability Symposium*, 1982, pp 83–86.

[265] Marshall, A.W. and Proschan, F. , "Classes of distributions applicable in replacement, with renewal theory implications," *Proceedings of the 6th Berkeley Symposium on Mathematical Statistics and Probability*, I, University of California Press, Berkeley, CA, 1972, pp 395–415.

[266] Martz, H.F., and Lian, M.G. , "A survey and comparison of several empirical Bayes estimators for the binomial parameter," *Journal of Statistical Computation and Simulation*, **3**, 1974, pp 165–178.

[267] Martz, H.F. and Lian, M.G. "Bayes and empirical Bayes point and interval estimation of reliability for the Weibull model," in *The Theory and Applications of Reliability*, II, Tsokos, C.P. and Shimi, I.N., Editors, Academic Press, New York, NY, 1977, pp 203–233.

[268] Martz, H.F., and Waller, R.A. , *Bayesian Reliability Analysis*, John Wiley & Sons, 1982.

[269] Martz, H.F., and Zimmer, W.J. , "A nonparametric Bayes empirical Bayes procedure for estimating the percent nonconforming in accepted Lots," *Journal of Quality Technology*, **22** (2), Apr. 1990, pp 92–104.

[270] Mauldin, R.D., Sudderth, W.D., and Williams, S.C., "Polya trees and random distributions," *Annals of Statistics* **20** (3), 1992, pp 1203–1221.

[271] Mazzuchi, T.A. and Singpurwalla N.D., "A Bayesian approach for inference for monotone failure rates," *Statistics and Probability Letters*, **37**, 1985, pp 135–141.

[272] Mazzuchi, T.A. and Soyer, R. , "Dynamic models for statistical inference from accelerated life tests". *Annual Reliability and Maintainability Symposium*, 1990, pp 67–70.

[273] Mazzuchi, T.A., Soyer, R., and Spring, R.V. , "The proportional hazards model in reliability," *Annual Reliability and Maintainability Symposium*, 1989, pp 252–255.

[274] Mazzuchi, T.A. and Soyer R., "A Bayesian attribute reliability growth model," *Annual Reliability and Maintainability Symposium*, 1991, pp 322–324.

[275] McCahon, C.S., Hwang, C.L., and Tillman, F.A. , "A multiple attribute of Bayesian availability estimators," *IEEE Transactions on Reliability*, **R-32** (5), Dec. 1983, pp 496–503.

[276] McPherson, J.W. , "Stress dependent activation energy," *Proceedings of International Reliability Physics Symposium*, 1986, pp 12–17.

[277] Meeker, W.Q. and Escobar, L.A. , "Recent and future research on practical methods for accelerated testing," in *Quality through Engineering Design*, Kuo, W., Editor, Elsevier Scientific Publishers, Amsterdam, 1993.

[278] Meeker, W.Q. and Escobar, L.A. , *Statistical Methods for Reliability Data*, John Wiley, N.Y., 1998.

[279] Mi, J., "Maximization of a survival probability and its applications," *Journal of Applied Probability*, **3** (4), 1994, pp 1026–1033.

[280] Michalka, T.L., Varshney, R.C., and Meindl, J.D., "A discussion of yield modeling with defect clustering, circuit repair, and circuit redundancy," *IEEE Transactions on Semiconductor Manufacturing*, **3** (3), Aug. 1990, pp 116–127.

[281] Miller, D.R., "Exponential order statistic models of software reliability growth," *IEEE Transactions on Software Engineering*, **SE-12** (1), Jan. 1986, pp 12–24.

[282] Miller, D.R. and Sofer, A., "Completely monotone regression estimates of software failure rates," *Proceedings: 8th International Conference on Software Engineering*, 1985, pp 343–348.

[283] Miller, D.R. and Sofer, A., "Least-square regression under convexity and higher-order difference constraints with application to software reliability," *Advances in Order Restricted Statistical Inference*, Edited by Brillinger, *et al.*, Lecture Notes in Statistics 37, Springer-Verlag, New York; 1985, pp 91–124.

[284] Miller, R.G., *Survival Analysis*, John Wiley & Sons, New York, NY, 1981.

[285] Miller, R.W. , "Parametric empirical Bayes tolerance intervals," *Technometrics*, **31**, 1989, pp 449–459.

[286] Miller, S.R. and Greyserman, A.G., "A procedure for optimal determination of product burn-in test duration and warranty period," SRI, PRRL-89-TR-030, 1989.

[287] Mills, H.D., "On the Statistical Validation of Computer Programs," IBM Federal Systems Division, Report FSC-72-6015, Gaitherburg, MD, 1972.

[288] Montgomery, D.C., *Design and Analysis of Experiments*, Second Edition, John Wiley & Sons, Inc., New York, 1984.

[289] Montgomery, D.C., *Introduction to Statistical Quality Control*, Second Edition, John Wiley & Sons, Inc., New York, 1991.

[290] Morgan, T.M., "Nonparametric estimation of duration of accrual and total study length for clinical trials," *Biometrics*, **43**, Dec. 1987, pp 903–912.

[291] Morris, C.N. , "Parametric empirical Bayes inference: theory and applications," *Journal of the American Statistical Association*, **78** (381), Mar. 1983, pp 47–65.

[292] Morris, S.F. and Reily, J.F., "MIL-HDBK-217 – a favorite target," *Annual Reliability and Maintainability Symposium*, 1993, pp 503–509.

[293] Mudholkar, G.S., Srivastava, D.K., "Exponentiated Weibull family for analyzing bathtub failure-rate data," *IEEE Transactions on Reliability*, **42** (2), Jun. 1993, pp 299-302.

[294] Murphy, B.T., "Cost-size optima of monolithic integrated circuit," *Proceedings of the IEEE*, **52** (12), Dec. 1964, pp 1537–1545.

[295] Musa, J.D., Iannino, A. and Okumoto, K., *Software Reliability*, McGraw-Hill, New York, NY, 1990.

[296] Nelson, W. and Kielpinski, T. , "Theory for optimum censored accelerated life tests for normal and lognormal distributions," *Technometrics*, **18**, 1976, pp 105–114.

[297] Nelson, W., Morgan, C.B., and Caporal, P. , *STATPAC Simplified-A introduction to STATPAC, a general statistical package for data analysis and fitting models to data*, TIS Report 83CRD146, General Electric Company, Corporate Research and Development, 1983.

[298] Nelson, W. , *Accelerated Testing: Statistical Models, Test Plans, and Data Analyses*, John Wiley & Sons, Inc., New York, 1990.

[299] Nevins, J.L. and Whitney, D.E., *Concurrent Design of Products and Processes: A Strategy for the Next Generation in Manufacturing*, McGraw-Hill, Inc., New York, 1989.

[300] Nguyen, D.G. and Murthy, D.N.P. , "Optimal burn-in time to minimize cost for products sold under warranty," *IIE Transactions*, **14** (3), 1982, pp 167–174.

[301] Ninomiya, T. and Harada, K., "Multilayer debugging process (a new method of screening)," *IEEE Transactions on Reliability*, **R-21**, 1972, pp 224–229.

[302] Oakes, D., "A note on the Kaplan-Meier estimator," *The American Statistician*, **47**, Feb. 1993, pp 39–40.

[303] O'Connor, P.D.T., *Practical Reliability Engineering*, Heyden & Son Ltd, London, Great Britain, 1981.

[304] O'Connor, P.D.T. , "Microelectronic system reliability prediction," *IEEE Transactions on Reliability*, **R-32** (1), Apr. 1983, pp 9–13.

[305] Ohr, S.A. , *CAE: A Survey of Standards Trends and Tools*, John Wiley & Sons, Inc., New York, 1990.

[306] Ohtera, H. and Yamada, S., "Optimal allocation & control problems for software- testing resources," *IEEE Transactions on Reliability*, **39** (2), Jun. 1990, pp 171–176.

[307] Okabe, T., Nagata, M., and Shimada, S., "Analysis on yield of integrated circuits and a new expression for the yield," *Electrical Engineering in Japan*, **92** (6), Dec. 1972, pp 135–141.

[308] Olsson, C., "Reliability of plastic encapsulated integrated circuits," *Proceedings of International Reliability Physics Symposium*, 1972, pp 53–72.

[309] Padgett, W.J. , "Bayes estimation of reliability for the inverse gaussian model," *IEEE Transactions on Reliability*, **R-30** (4), Oct. 1981, pp 384–385.

[310] Padgett, W.J. , "On Bayes estimation of reliability for the Birnbaum-Saunders fatigue life model," *IEEE Transactions on Reliability*, **R-31** (5), Dec. 1982, pp 436–438.

[311] Padgett, W.J. and Robinson, J.A. , "Empirical Bayes estimators of reliability for lognormal failure model," *IEEE Transactions on Reliability*, **R-27** (5), Dec. 1978, pp 332–336.

[312] Padgett, W.J. and Tsokos, C.P. , "Bayes estimation of reliability for the lognormal failure model," in *The Theory and Applications of Reliability*, **II**, Tsokos, C.P. and Shimi, I.N., Editors., Academic Press, New York, NY, 1977, pp 133–161.

[313] Padgett, W.J. and Tsokos, C.P., "On Bayes estimation of reliability for mixtures of life distributions," *SIAM Journal on Applied Mathematics*, **34**, Jun. 1978, pp 692–703.

[314] Pantic, D. , "Benefits of integrated-circuit burn-in to obtain high reliability parts," *IEEE Transactions on Reliability*, **R-35** (1), Feb. 1986, pp 3–6.

[315] Papadopoulos, A.S. , "The Burr distribution as a failure model from a Bayesian approach," *IEEE Transactions on Reliability*, **R-27** (5), Dec. 1978, pp 369–371.

[316] Papadopoulos, A.S. and Padgett, W.J. , "On Bayes estimation for mixtures of two exponential- life- distributions from right-censored samples," *IEEE Transactions on Reliability,* **R-35** (1), Apr. 1986, pp 102–105.

[317] Papadopoulos, A.S. and Tiwari, R.C., "Comparison of Bayesian nonparametric estimates of the reliability with rival estimates," *Microelectronics and Reliability,* **32** (1/2), 1992, pp 233–240.

[318] Papadopoulos, A.S. and Tsokos, C.P. , "Bayesian reliability estimates of the binomial failure model," *JUSE,* **21**, 1974, pp 9–26.

[319] Papadopoulos, A.S. and Tsokos, C.P. , "Bayesian confidence bounds for the Weibull failure model," *IEEE Transactions on Reliability,* **R-24** (1), Apr. 1975, pp 21–26.

[320] Papadopoulos, A.S. and Tsokos, C.P. , "Bayesian analysis of the Weibull failure model with unknown scale and shape parameters," *Statistica,* 1976.

[321] Park, D.H., "Testing whether failure rate changes its trend," *IEEE Transactions on Reliability,* **37** (4), Oct. 1988, pp 375–378.

[322] Park, K.S., "Effect of burn-in on mean residual life," *IEEE Transactions on Reliability,* **R-34** (5), Dec. 1985, pp 522–523.

[323] Parzen, E., "Density Quantile Estimation Approach to Statistical data Modeling," in *Smoothing Techniques for Curve Estimation: Proceedings of Workshop Held in Heidelberg,* April 2-4, 1979, pp 155–180, Gasser, T.A. and Rosenblatt, M., Editors, Springer-Verlag, New York, NY, 1979.

[324] Pathak, P.K., Singh, A.K., and Zimmer, W.J. , "Bayes estimation of hazard & acceleration in accelerated testing," *IEEE Transactions on Reliability,* **40** (5), Dec. 1991, pp 615–621.

[325] Perlstein, H.J., Littlefield, J.W., and Bazovsky, I., "The quantification of environmental stress screening," *Proceedings of The Institute of Environmental Sciences,* 1987, pp 202–208.

[326] Perlstein, H.J. and Littlefield, J.W., "ESS quantification for complex systems," *The Journal of Environmental Sciences,* 1989, pp 49–57.

[327] Pham, T.G. and Turkkan, N. , "Bayes binomial sampling by attributes with a general-beta prior distribution," *IEEE Transactions on Reliability,* **41** (2), Jun. 1992, pp 310–316.

[328] Pillo, G.D. and Grippo, L., "Exact penalty functions in constrained optimization," *SIAM Journal on Control and Optimization* **27** (6), Nov. 1989, pp 1333–1360.

[329] Plesser, K.T. and Field, T.O., "Cost-optimized burn-in duration for repairable electronic system," *IEEE Transactions on Reliability,* **R-26** (3), Aug. 1977, pp 195–197.

[330] Pohl, E.A. and Dietrich, D.L., "Environmental stress screening strategies for complex system: a 3-level mixed distribution model," *Microelectronics and Reliability,* **35** (4), 1995, pp 637–656.

[331] Pohl, E.A. and Dietrich, D.L., "Environmental stress screening strategies for multicomponent systems with Weibull failure-times and imperfect failure detection," *Annual Reliability and Maintainability Symposium,* 1995, pp 223–232.

[332] Prairie, R.R. and Zimmer, W.J., "An iterative Bayes procedure for reliability assessment," *Annual Reliability and Maintainability Symposium,* 1990, pp 10–14.

[333] Prendergast, J.G., "Reliability and quality correlation for a particular failure mechanism," *Proceedings of International Reliability Physics Symposium,* 1993, pp 87–93.

[334] Price, J.E., "A new look at yield of integrated circuits," *Proceedings of the IEEE,* **58** (8), Aug. 1970, pp 1290–1291.

[335] Prince, B., *Semiconductor Memories: A Handbook of Design, Manufacturing, and Application,* 2nd Ed., John Wiley & Sons, New York,

[336] Proschan, F. and Pyke, R., "Tests for monotone failure rate," *Proceedings of the 5th Berkeley Symposium on Mathematical Statistics and Probability,* **III**, University of California Press, Berkeley, CA, 1967, pp 293–312.

[337] Pugh, E.L., "The best estimate of reliability in the exponential case," *Operations Research,* **11**, 1963, pp 57–61.

[338] Raheja, D.G., "Software reliability growth process – a life cycle approach," *Annual Reliability and Maintainability Symposium,* 1989, pp 52–55.

[339] Raiffa, H. and Schlaiffer, R., *Applied Statistical Decision Theory,* Graduate School of Business Administration, Harvard University, Massachusetts, 1961.

[340] Ramsey, F., "A Bayesian approach to bioassay," *Biometrics,* **28**, 1972, pp 841–858.

[341] Rawicz, A.H. , "Burn-in of incandescent sign lamps," *IEEE Transactions on Reliability,* **R-35** (4), Oct. 1986, pp 375–376.

[342] Rawicz, A.H. and Jiang, H.X., "Diagnostic expert-system for mechanical reliability in heavy trucks," *Annual Reliability and Maintainability Symposium,* 1992, pp 426–431.

[343] Reich, P.G., "Problems in the investigation of reliability-associated life-cycle costs of military airborne systems," AGARD Lecture Series No. 100: *Methodology for Control of Life Cycle Costs for Avionics Systems,* 1979, pp 5-1–5-5.

[344] Reddy, R. and Dietrich, D., "A two-level ESS model: a mixed distribution approach," *IEEE Transactions on Reliability,* **43**, 1994, pp 85–90.

[345] Rickers, H.C. and Manno, P.F., "Micro processor and LSI microcircuit reliability model," *IEEE Transactions on Reliability,* **R-29** (3), Aug. 1980, pp 196–202.

[346] Robbins, H., "An empirical Bayes approach to statistics," Proceedings Third Berkeley Symposium on Math. Statistics and Probability, **I**, 1955, pp 157–163.

[347] Robbins, H. , "The empirical Bayes approach to statistical decision problems," *Annals of Mathematical Statistics,* **35**, 1964, pp 1–20.

[348] Robinson, D.G. and Dietrich, D., "A new nonparametric growth model," *IEEE Transactions on Reliability,* **R-34** (4), Oct. 1987, pp 411–418.

[349] Robinson, D.G. and Dietrich, D. , "A system-level reliability-growth model," *Annual Reliability and Maintainability Symposium,* 1988, pp 243–247.

[350] Robinson, D.G. and Dietrich, D. , "A nonparametric-Bayes reliability growth model," *IEEE Transactions on Reliability,* **38** (5), Dec. 1989, pp 591–598.

[351] Rodriguez, F.W. and Welchsler, S. , "A discrete explanation of a failure-rate paradox," *IEEE Transactions on Reliability,* **42** (1), Mar. 1993, pp 132–133.

[352] Romanchik, D., "Why Burn-in ICs ?," *Test & Measurement World,* Oct. 1992, pp 85–86, 88.

[353] Root B.J. and Turner, T., "Wafer level electromigration tests for production monitoring," *Proceedings of International Reliability Physics Symposium,* 1985, pp 100–107.

[354] Rosenbaum, E., "BERT:IC reliability simulator," *International Wafer Level Reliability Workshop Final Report,* 1991, pp 87–100.

[355] Ross, P.J. , *Taguchi Techniques for Quality Engineering: Loss Function, Orthogonal Experiments, Parameter and Tolerance Design,* McGraw-Hill, Inc., New York, NY, 1988.

[356] Ross, S.M., *Introduction to Probability Models,* Academic Press Inc., London, 1989.

[357] Ryerson, C.M. and Ryerson, C., "Thought provoking gems from my reliability experience," *Annual Reliability and Maintainability Symposium,* 1989, pp 234–238.

[358] Sabnis, A.G., *VLSI Electronics Microstructure Science V.22 VLSI Reliability,* Academic Press, Inc., San Diego, CA, 1990.

[359] Samuel, E. , "An empirical Bayes approach to the testing of certain parametric hypotheses," *Annals of Mathematical Statistics,* **34**, 1963, pp 1370–1385.

[360] Sarkar, T.K. , "An exact lower confidence bound for the reliability of a series system where each component has an exponential time to failure distribution," *Technometrics,* **13** (3), 1971, pp 535–546.

[361] Schäbe, H. , "Bayes estimates under asymmetric loss," *IEEE Transactions on Reliability,* **40** (1), Apr. 1991, pp 63–67.

[362] Schafer, R.F. and Feduccia, A.J. , "Prior distributions fitted to observed reliability data," *IEEE Transactions on Reliability,* **R-21** (3), Aug. 1972, pp 148-154.

[363] Schick, G.J. and Wolverton, R.W., "An analysis of competing software reliability models," *IEEE Transactions on Software Engineering,* **SE-4**, Mar. 1978, pp 104–120.

[364] Schmoyer, R.L., "Nonparametric analyses for two-level single-stress accelerated life tests," *Technometrics,* **33**, May 1991, pp 175–186.

[365] Schneider, V., "Some experimental estimators for developmental and delivered errors in software development projects," *Proc. COMPSAC,* 1980, pp 495–498.

[366] Schneidewind, N.F., "Software reliability model with optimal selection of failure data," *IEEE Transactions on Software Engineering,* **19**, Nov. 1993, pp 1095–1104.

[367] Schoenfelder, J.R., Shachtman, R.H., and Johnston, G.J., "The comparison of life table and Markov chain techniques follow-up studies," *Operations Research,* **33**, Jan.-Feb. 1985, pp 126–133.

[368] Schoonmaker, T.D. and Marko, D.M. , "Predicting burn-in performance for a new design," *Annual Reliability and Maintainability Symposium,* 1983, pp 423–428.

[369] Schroen, W.H., "Process testing for reliability control," *Proceedings of International Reliability Physics Symposium,* 1978, pp 81–87.

[370] Selby, M.A. and Shoukri, M.M. , "Bayes comparison of 2 lognormal reliability functions," *IEEE Transactions on Reliability*, **39** (3), Aug. 1990, pp 336–341.

[371] Seth S.C. and Agrawal, V.D., "On the Probability of fault occurrence," in *Defect and Fault Tolerance in VLSI Systems*, Koren I. (Editor), Plenum Press, New York, 1989, pp 47–52.

[372] Sethuraman, J. and Singpurwalla, N.D., "Testing of hypotheses for distributions in accelerated life tests," *Journal of the American Statistical Association*, **77**, 1982, pp 204–208.

[373] Sethuraman, J. and Tiwari, R.C., "Convergence of Dirichlet measures and the interpretation of their parameters," in *Statistical Decision Theory and Related Topics III 2*, Gupta, S.S. and Berger, J.O., Editors, Academic press, 1982, pp 305–315.

[374] Shaked, M. and Singpurwalla, N.D. "Nonparametric estimation and goodness of fit testing of hypothesis for distribution in accelerated life testing," *IEEE Transactions on Reliability*, **R-31** (1), Apr. 1982, pp 69–74.

[375] Shaked, M., Zimmer, W.J., and Ball, C.A. , "A nonparametric approach to accelerated life testing," *Journal of the American Statistical Association*, **74**, 1979, pp 694–699.

[376] Shalvoy, C.E. , "Finding the most effective stress-screening strategy," *Electronics Test*, **12** (3), Mar. 1989, pp 42–46.

[377] Sharma, K.K. and Bhutani, R.K. , "A comparison of classical and Bayes risks when the quality varies randomly," *Microelectronics and Reliability*, **32** (4), 1992, pp 493–495.

[378] Sheng, Z. and Fan, D. , "Bayes attribute acceptance-sampling plan," *IEEE Transactions on Reliability*, **41** (2), Jun. 1992, pp 307–309.

[379] Shideler, J.A., "A model of a wafer level reliability program for a large corporation," *International Wafer Level Reliability Workshop 1991*, Trapp, O.D., Editor, Stanford University, Stanford, CA, 1991. pp 75–85.

[380] Shideler, J.A., *et al.*, "A systematic approach to wafer level reliability," *Solid State Technology*, March 1995, p 47+.

[381] Shimi, I and Tsokos, C.P. , "The Bayesian and nonparametric approach to reliability studies: a survey of recent work," in *The Theory and Applications of Reliability*, I, Tsokos, C.P. and Shimi, I.N., Editors, Academic Press, New York, NY, 1977, pp 5–47.

[382] Shingo, S., *Study of the Toyota Production System from Industrial Engineering Viewpoint*, Productivity Press, Cambridge, MA, 1989.

[383] Shooman, M.L., "Software reliability: a historical perspective," *IEEE Transactions on Reliability*, **R-33** (1), Apr. 1984, pp 48–55.

[384] Shoukri, M.M. , "Simple Bayes test of equality of exponential means," *IEEE Transactions on Reliability*, **R-36** (5), Dec. 1987, pp 613–616.

[385] Simmons, D.B., Ellis, N.C., Fujihara, H., and Kuo, W., *Software Measurement, A Visualization Toolkit for Project Control & Process Improvement*, Prentice-Hall, 1998.

[386] Singh, A. and Singh, A. , "Empirical Bayes estimation of mean life for a censored sample, constant hazard rate model," *IEEE Transactions on Reliability*, **R-35** (4), Oct. 1986, pp 399–402.

[387] Singpurwalla, N.D. and Wong, W.H., "Estimation of the failure rate – a survey of nonparametric methods, Part I: non-Bayesian methods," *Communications in Statistics:Theory and Methods*, **12** (5), 1983, pp 559–588.

[388] Sinha, S.K. , "Bayes estimation of the reliability function of normal distribution," *IEEE Transactions on Reliability*, **R-34** (4), Oct. 1985, pp 360–362.

[389] Sinnadurai, N., "Plastic Packaging is Highly Reliable," *IEEE Transactions on Reliability*, **45** (2), 1996, pp 184–193.

[390] Smith, A.F.M., "A Bayesian note on reliability growth during a development testing Program," *IEEE Transactions on Reliability*, **R-26** (5), Dec. 1977, pp 346–347.

[391] Soden, J.M. and Anderson R.E., "IC failure analysis: techniques and tools for quality and reliability improvement," *Proceedings of the IEEE*, **81** (5), May 1993, pp 703–715.

[392] Sofer, A. and Miller, D.R., "A nonparametric software-reliability growth model," *IEEE Transactions on Reliability*, **40** (3), Aug. 1991, pp 329–337.

[393] Soland, R.M. , "Bayesian analysis of the Weibull process with unknown scale parameter and its application to acceptance sampling," *IEEE Transactions on Reliability*, **R-17**, (2), Jun. 1968, pp 84–90.

[394] Soland, R.M. , "Bayesian analysis of the Weibull process with unknown scale and shape parameters," *IEEE Transactions on Reliability*, **R-18** (4), Nov. 1969, pp 181–184.

[395] Springer, M.D. and Thompson, W.E. , "Bayesian confidence limits for the product of N binomial parameters," *Biometrika*, **53**, 1966, pp 611–613.

[396] Srinivasan, C. and Zhou, M., "A note on pooling Kaplan-Meier estimators," *Biometrics*, **49**, Sep. 1993, pp 861–864.

[397] Stacy, E.W. , "A generalization of the gamma distribution," *Annals of Mathematical Statistics*, **33**, 1962, pp 1187–1192.

[398] Stapper, C.H., "Modeling of integrated circuit defects sensitivities," *IBM Journal of Research and Development*, **27**, 1983, pp 549–557.

[399] Stapper, C.H., "Modeling of defects in integrated circuit photolithographic patterns," *IBM Journal of Research and Development*, **28**, 1984, pp 461–475.

[400] Stapper, C.H., "The effects of wafer to wafer defect density variations on integrated circuit defect and fault distributions," *IBM Journal of Research and Development*, **29**, 1985, pp 87–97.

[401] Stapper, C.H., "On yield, fault distributions and clustering of particles," *IBM Journal of Research and Development*, **30**, May 1986, pp 326–338.

[402] Stapper, C.H., "Fact and fiction in yield modeling," *Microelectronics Journal*, **20**, (1/2), 1989, pp 129–151.

[403] Stapper, C.H., "Large-area fault clusters and fault tolerance in VLSI circuits:a review," *IBM Journal of Research and Development*, **33**, 1989, pp 162–173.

[404] Stapper, C.H., Armstrong, F.M., and Saji, K., "Integrated circuit yield statistics," *Proceedings of the IEEE*, **71** (4), Apr. 1983, pp 453–470.

[405] Stapper, C.H. and Rosner, R.J., "Integrated circuit yield management and yield analysis: development and implementation," *IEEE Transactions on Semiconductor Manufacturing*, **8** (2), May 1995, pp 95–102.

[406] Stephenson, W.R., Hulting, F.L. and Moore, K. , "Posterior probabilities for identifying active effects in unreplicated experiments," *Journal of Quality Technology*, **21** (3), Jul. 1989, pp 202–206.

[407] Stewart, L.T. and Johnson, J.D. , "Determining optimum burn-in and replacement times using Bayesian decision theory," *IEEE Transactions on Reliability*, **R-21** (3), Aug. 1972, pp 170–175.

[408] Sturm, G.W., Feltz, C.J., and Yousry, M.A. , "An empirical Bayes strategy for analyzing manufacturing data in real time" , *Quality and Reliability Engineering International*, **7**, 1991, pp 159–167.

[409] Sultan, T.I., "The effect of stress screening process on yield and cost," *Microelectronics and Reliability*, **29** (5), 1989, pp 695–699

[410] Susarla, V. and Van Ryzin, J., "Nonparametric Bayesian estimation of survival curves from incomplete observations," *Journal of the American Statistical Association*, **71** (356), Dec. 1976, pp 897–902.

[411] Swanson, D.E., "Forty years and looking forward," *Semicond. Int.*, **11** (1), Jan. 1988, p13.

[412] Sze, S.M. (editor), *VLSI Technology*, McGraw-Hill, New York, NY, 1983.

[413] Taguchi, G., *Introduction to Quality Engineering: Designing Quality into Products and Processes*, Asian Productivity Organization, Tokyo, Japan, 1986.

[414] Takeda, E., *et al.*, "VLSI reliability challenges: from device physics to wafer scale systems," *Proceedings of the IEEE*, **81** (5), May 1993, pp 653–674.

[415] Tam, S.M., "Demonstrated Reliability of Plastic-Encapsulated Microcircuits for Missile Applications," *IEEE Transactions on Reliability*, **44** (1), 1995, pp 8–13.

[416] Tanner, M.A., "A note on the variable kernel estimator of the hazard function from randomly censored data," *Annals of Statistics*, **11** (3), Mar. 1983, pp 994–998.

[417] Tanner, M.A. and Wong, W.H., "The estimation of the hazard function from randomly censored data by the kernel method," *Annals of Statistics*, **11** (3), 1983, pp 989–993

[418] Tanner, M.A. and Wong, W.H., "Data-based nonparametric estimation of the hazard function with applications to model diagnostics and exploratory analysis," *Journal of the American Statistical Association*, **79** (385), Mar. 1984, pp 174–182.

[419] Tapia, R.A. and Thompson, J.R., *Nonparametric Probability Density Estimation*, Johns Hopkins University Press, Baltimore, 1978.

[420] Tillman, F.A., Hwang, C.L. and Kuo, W., *Optimization of Systems Reliability*, Marcel Dekker, New York, NY, 1988.

[421] Tobias, P.A. and Trindade, D.C., *Applied Reliability*, Van Nostrand Company, New York, NY, 1995.

[422] Trachtenberg, M., "The linear software reliability model and uniform testing," *IEEE Transactions on Reliability*, **R-34** (1), Apr. 1985, pp 8–16.

[423] Tsai, W., Jewell, N.P., and Wang, M., "A note on the product-limit estimator under right censor and left truncation," *Biometrika*, **74**, Dec. 1987, pp 883–886.

[424] Tsokos, C.P. and Canavos, G.C. , "Bayesian concepts for the estimation of reliability in the Weibull life-testing model," *International Statistical Review*, **40** (2), 1972, pp 153–160.

[425] Tsokos, C.P. and Rao, A.N.V. , "Robustness studies in Bayesian developments," *Proceedings of the 20th Conference on the Design of Experiments in Army Research Development and Testing*, ARO, 75-2, 1974, pp 273–302.

[426] Tyagi, R.K., Kumar, S., Tiwari, R.C. and Bhattarya, S.K., "Parametric empirical Bayes approach to reliability analysis for the geometric life-model," *Microelectronics and Reliability*, **32** (9), 1992, pp 1271–1282.

[427] Turner, T.E., "Wafer level reliability: process control for reliability," *Microelectronics and Reliability*, **36** (11/12), 1996, pp 1839–1846.

[428] Upadhyay, S. K. and Pandey, M., "Prediction limits for an exponential distribution: A Bayes predictive distribution approach," *IEEE Transactions on Reliability*, **38** (5), Dec. 1989, pp 599–602.

[429] US Department of Defense, *Military Standardization Handbook : reliability Prediction of Electronic Equipment*, MIL-HDBK-217D, US Department of Defense, Washington, Jan. 1982.

[430] Vaccaro, J. , "Reliability physic – an assessment," *A. Symp. Reliab. IEEE*, 1970, pp 348–363.

[431] Vander Pol, J., Kuper, F., and Ooms, E., "Relation between yield and reliability of integrated circuits and application to failure rate assessment and reduction in the one digit fit and ppm reliability era," *Microelectronics and Reliability*, **36** (11/12), 1996, pp 1603–1610.

[432] Vannoy, E.H., "Improving "MIL-HDBK-217 Type" models for mechanical reliability prediction," *Annual Reliability and Maintainability Symposium*, 1990, pp 341–345.

[433] Vollertsen, R. -P., "Statistical modeling of time dependent oxide breakdown distributions," *Microelectronics and Reliability*, **33** (11/12), 1993, pp 1665–1677.

[434] Waldron, W.B. and Waldron, K.J., "Conceptual CAD tools for mechanical designers," *Proceedings of the 1988 ASME International Computers in Engineering Conference and Exhibition*, Jul. 31– Aug. 4, 1988, pp 203–209.

[435] Wallmark, T.J., "Design considerations for integrated electron devices," *Proceedings of IRE*, **48**, Mar. 1960, pp 293–300.

[436] Wang, Y.H. and Chang, S.A., "A new approach to the non-parametric tests for exponential distribution with unknown parameters," In *The Theory and Applications of Reliability*, **II**, Tsokos, C.P. and Shimi, I.N., Editors, Academic Press, New York, NY, 1977, pp 235–258.

[437] Washburn, L.A. , "Determination of optimum burn-in time: a composite criterion," *IEEE Transactions on Reliability*, **R-19** (4), Nov. 1970, pp 134–140.

[438] Watson, G.F., "Plastic-packaged ICs in military equipment," *IEEE Spectrum*, Feb. 1991, pp 62–63.

[439] Watson, G.S. and Wells, W.T., "On the possibility of improving the mean useful life of items by eliminating those with short lives," *Technometrics*, **3** (2), 1961, pp 281–298.

[440] Wei, X. , "Testing whether one distribution is more IFR than another," *Microelectronics and Reliability*, **32** (1-2), Jan.-Feb. 1992, pp 271–273.

[441] Wei, X., "Test exponentiality against a monotone hazard function alternative based on TTT transformation," *Microelectronics and Reliability*, **32** (5), May 1992, pp 607–610.

[442] Weiss, G.H. and Dishon, M., "Some economic problems related to burn-in programs," *IEEE Transactions on Reliability*, **R-20** (3), Aug. 1971, pp 190–195.

[443] Wilson, M.A., "Experience with Bayesian reliability measurement of large systems," *IEEE Transactions on Reliability*, **R-21** (3), Aug. 1972, pp 181–185.

[444] Winterbottom, A. , "Approximate Bayesian intervals for the reliability of series systems from mixed subsystem test data," *Naval Research Logistics Quarterly*, **30**, 1983, pp 313–317.

[445] Williams, T.W. and Brown, N.C., "Defect level as a function of fault coverage," *IEEE Transactions on Computers*, **C-30**, 1981, pp 508–509.

[446] Woods, M.H. and Euzent, B.L, "Reliability in MOS integrated circuits," *IEDM*, 1984, pp 50–55.

[447] Wurnik, F. and Pelloth, W., "Functional burn-in for integrated circuits," *Microelectronics and Reliability*, **30** (2), 1990, pp 265–274.

[448] Xie, M., "A shock model for software failures," *Microelectronics and Reliability,* **27** (4), 1987, pp 717–724.

[449] Xie, M., *Software Reliability Modelling,* World Scientific, Singapore, 1991.

[450] Xu, Z., Kuo, W., and Lin, H., "Optimization limits in improving system reliability," *IEEE Transactions on Reliability,* **39** (1), Apr. 1990, pp 51–60.

[451] Yamada, S., Ohba, M., and Osaki, S., "S-shaped reliability growth modeling for software Error detection," *IEEE Transactions on Reliability,* **R-32** (5), Dec. 1983, pp 475–478.

[452] Yan, L., *Economic Cost Modeling of ESS and Burn-in,* Ph.D. Dissertation, University of Arkansas, 1995.

[453] Yates III, W.D. and Shaller, D.A., "Reliability engineering as applied to software," *Annual Reliability and Maintainability Symposium,* 1990, pp 425–429.

[454] Yousry, M.A., *et al.*, "Process monitoring in real time: empirical Bayes approach – discrete case," *Quality and Reliability Engineering International,* **7**, 1991, pp 123–132.

[455] Zacks, S. , "Bayes estimation of the reliability of series and parallel systems of independent exponential components," in *The Theory and Applications of Reliability,* **II**, Tsokos, C.P. and Shimi, I.N., Editors., Academic Press, New York, NY, 1977, pp 55–74.

[456] Zacks, S., "Survey of classical and Bayesian approaches to the change-point problems: fixed sample and sequential procedures of testing and estimation," in *Recent Advances in Statistics,* Rizvi, M.H., Rustagi, J., and Siegmund, D., Editors, 1983, pp 245–269.

[457] Zacks, S. and Barzily, Z. , "Detecting a shift in the probability of success in a sequence of bernoulli trials," *Journal of Statistical Planning and Inference,* **5**, 1981, pp 107–119.

[458] Zacks, S. , *Parametric Statistical Inference: Basic Theory and Modern Approaches,* Pergamon Press, Oxford, 1981.

[459] Zacks, S. , "The probability distribution and the expected value of a stopping variable associated with one-sided CUSUM procedures," *Commun. Statist., A,* **10**, 1981, pp 2245–2258.

[460] Zaino, N.A. and Berke, T.M. , "Determining the effectiveness of run-in: a case study in the analysis of repairable-system data," *Annual Reliability and Maintainability Symposium,* 1992, pp 59–70.

[461] Zheng, Z., "A time sequential plan for nonparametric testing of hypotheses with censored data," *Biometrika,* **75**, Sep. 1988, pp 607–610.

[462] Zimmerman, W. , "Screening tests to monitor early life failure," *Annual Reliability and Maintainability Symposium,* 1983, pp 443-447.

[Other Related References]

[463] Alexanian, I.T. and Brodie, D.E. , "A method for estimating the reliability of ICS," *IEEE Transactions on Reliability,* **R-26** (5), Dec. 1977, pp 359–361.

[464] Amster, S.J. and Hooper, J.H. , "Accelerated life tests with measured degradation data and growth curve models," paper presented at the American Statistical Association Annual Meeting, Toronto, Canada, 1983.

[465] Amster, S. and Hooper, J. , "Statistical methods for reliability improvement," *AT&T Technical Journal,* March/April, **65** (2), 1986, pp 69–76.

[466] Andersen, P.K., Borgan, Gill, R.D., and Keiding, N., *Statistical methods based in counting processes,* Springer-Verlag, New York, NY, 1992.

[467] Anderson, J.A. and Senthilselvan, A., "Smooth estimates for the hazard function," *Journal of the Royal Statistical Society,* B, **42**, 1980, pp 322–327.

[468] Anderson, J.A. and Blair, V., "Penalized maximum likelihood estimation in logistic regression and discrimination," *Biometrika,* **69**, 1981, pp 123–136.

[469] Anon, "Burn-in and test systems keep pace with new devices," *Evaluation Engineering,* **27** (9), Sep. 1988, pp 71–73.

[470] Ascher, H. and Feingold, H. , ""Bad-as-old" analysis of system failure data," Annals of Assurance Science: Proceedings of the Eighth Reliability and Maintainability Conference, 1969, pp 49–62.

[471] Askeland, D.R., *The Science and Engineering of Materials,* Second Edition, PWS-Kent Publishing Company, Boston, Massachusetts, 1989.

[472] Atkinson, A.C. and Whittaker, J., "The generation of beta random variables with one parameter greater than and one parameter less than 1," *Applied Statistics, Algorithm* #134, 1979, pp 90–91.

[473] Bai, D.S., Kim, M.S. and Lee, S.H. , "Optimum simple step-stress accelerated life tests with censoring," *IEEE Transactions on Reliability,* **38** (5), Dec. 1989, pp 528–532.

[474] Barlow, R.E., Pereira, C.A.B., and Wechsler, S., "A Bayesian approach to environmental stress screening," *Naval Research Logistics Quarterly,* **41**, 1994, pp 215–228.

[475] Barlow, R.E., *et al.*, "Estimation procedures for the "burn-in process"," *Technometrics,* **10**, 1968, pp 51–62.

[476] Barlow, R.E. and Scheuer, E.M. , "Estimation from accelerated life testings," *Technometrics,* **13**, Feb. 1971, pp 145–159.

[477] Barry, D. , "Nonparametric Bayesian regression," *Annals of Statistics,* **14**, 1986, pp 934–953.

[478] Basin, S.L., *Estimation of Software Error rate via Capture-Recapture Sampling,* Science Applications, Inc., Palo Alto, CA, 1974.

[479] Bayarri, M.J. and DeGroot, M.H. , "Gaining weight: a Bayesian approach," *Bayesian Statistics,* **3**, Bernardo, J.M. *et al.* (Editors), Clarendon, Oxford, 1988, pp 25–44.

[480] Bayarri, M.J. and DeGroot, M.H., "Optimal reporting of predictions," *Journal of the American Statistical Association,* **84**, 1989, pp 214–222.

[481] Becker, R.A., Chambers, J.M., and Wilks, A.R. , *The New S Language: A Programming Environment for Data Analysis and Graphics,* Wadsworth & Brooks/Cole Advanced Books & Software, Pacific Grove, California, 1988.

[482] Bendell, A., Editors, *Software Reliability: A State of the Art Report,* Pergamon Infotech Limited, Berkshire, England, 1986.

[483] Bernardo, J.M. and Giron, F.J., "A Bayesian analysis of simple mixture problems," *Bayesian Statistics* **3**, Bernardo, J.M., *et al.* (Editors), North-Holland, Amsterdam, 1988, pp 67–78.

[484] Bezat, A.G., Norquist, V., and Montague, L.L., "Growth modeling improves reliability predictions," *Annual Reliability and Maintainability Symposium,* 1975, pp 317–322.

[485] Blanks, H.S. , "Accelerated vibration fatigue life testing of leads and soldered joints," *Microelectronics and Reliability,* **15**, 1976, pp 213–219.

[486] Blanks, H.S. , "Electronic reliability: a state-of-the-art survey," *Microelectronics and Reliability,* **20**, 1980, pp 219–245.

[487] Block, H.W., Mi, J., and Savits, T.H., "Burn-in and mixed populations," *Journal of Applied Probability,* **30**, 1993, pp 692–702.

[488] Block, H.W., Mi, J., and Savits, T.H., "Burn-in at the component and system Level," *Lifetime data: Models in Reliability and Survival Analysis,* Kluwer Academic Pub., Dordrecht, 1995, pp 53–58.

[489] Bonato, B., "New burn-in challenges for IC manufacturers," *Evaluation Engineering,* **32**, 1993, pp 108–110.

[490] Boukai, B. , "Bayes sequential procedure for estimation and determination of burn-in time in a hazard rate model with an unknown change-point parameter," *Sequential Analysis,* **6** (1), 1987, pp 37–53.

[491] Brocklehurst, S. and Littlewood, B., "New ways to get accurate reliability measure," *IEEE Transactions on Software Engineering,* Jul. 1992, pp 34–42.

[492] Brooks, W.D. and Motley, R.W., *Analysis of Discrete Software Reliability Models,* Technical Report RADC-TR-80-84, Rome Air Development Center, New York, 1980.

[493] Bruls, E.M.J.G., "Reliability aspects of defect analysis," *IEEE/ETC,* 1993, pp 17–26.

[494] Buchanan, W.B. and Singpurwalla, N.D. , "Some stochastic characterizations of multivariate survival, " in *The Theory and Applications of Reliability,* **I**, Tsokos, C.P. and Shimi, I.N. Editors., Academic Press, New York, NY, 1977, pp 329–348.

[495] Bugaighis, M.M. , "Efficiencies of MLE and BLUE for parameters of an accelerated life-test model," *IEEE Transactions on Reliability,* **37** (2), Jun. 1988, pp 230–233.

[496] Burrows, R.W., "The Role of Temperature in the Environmental Acceptance Testing of Electronic Equipment," pp 42–45.

[497] Campbell, M. , "Monitored burn-in improves VLSI IC reliability," *Computer Design,* Apr. 1985, pp 143–146.

[498] Carey, M.B. and Koenig, R.H. , "Reliability assessment based on accelerated degradation: a case study," *IEEE Transactions on Reliability,* **40** (5), Dec. 1991, pp 499–506.

[499] Chandra, M., Singpurwalla, N.D., and Stephens, M.A., "Kolmogorov statistics for tests of fit for the extreme-value and Weibull distributions," *Journal of the American Statistical Association,* **76** (375), Sep. 1981, pp 729–730.

[500] Charloner, K. and Larntz, K. , *Bayesian Design for Accelerated Life Testing*, Technical Report, University of Minnesota, 1990.

[501] Chatterjee, S. and Chatterjee, S., "On combining expert opinions," *American Journal of Mathematical and Management Science*, **7**, 1987, pp 271–295.

[502] Chenoweth, H.B., "Reliability prediction, in the conceptual phase, of a processor system with its embedded software," *Annual Reliability and Maintainability Symposium,* 1991, pp 416–422.

[503] Chernoff, H. , "Optimal accelerated life designs for estimation," *Technometrics*, **4**, 1962, pp 381–408.

[504] Clarke, J.M. , "No-growth growth curves," *Annual Reliability and Maintainability Symposium,* 1979, pp 407–412.

[505] Clarke, L.A., Podgurski, A., Richardson, D.J., and Zeil, S.J., "A formal evaluation of data flow path selection criteria," *IEEE Transactions on Software Engineering,* **SE-15** (11), Nov. 1989, pp 1318–1332.

[506] Clarotti, C.A. and Spizzichino, F., "Bayes burn-in decision procedures," *Probab. Eng. Inf. Sci.* **4**, 1990, pp 437–455.

[507] Codier, E.O. , "Reliability growth in real life," *Annual Reliability and Maintainability Symposium,* 1968, pp 458–469.

[508] Constantini, C. and Spizzichino, F., "Optimal stopping of life testing: use of stochastic orderings in the case of conditionally exponential life-times," *Stochastic Ordering and Decision under Risk*, Mosler, K. and Scarsini, M. (Editors), Institute of Mathematical Statistics, Lecture Notes, **19**, 1991.

[509] Cooke, R.M., "Statistics in expert resolution: a theory of weights for combining expert opinion," *Stat. in Science: The Foundations of Stat'l Meth. in Biology, Physics and Economics* Roger Cooke and Domenico Costantini, Kluwer, 1990, pp 41–72.

[510] Coulter, N.S., "Software Science and Cognitive Psychology," *IEEE Transactions on Software Engineering,* **SE-9** (2), Mar. 1983, pp 166–171.

[511] Craven, P. and Wahba, G., "Smoothing noisy data with spline functions: estimating the correct degree of smoothing by the method of generalized cross-validation," *Numer. Math.*, 31, 1979, pp 377–403.

[512] Crow, L., "Confidence Interval Procedures for Reliability Growth Analysis," Technical Report 197, US Army Material Systems Analysis Activity, MD, 1977.

[513] Deely, J.J. and Lindley, D.V. , "Bayes empirical Bayes," *Journal of the American Statistical Association,* **76** (376), Dec. 1981, pp 833–841.

[514] DeGroot, M.H., "A Bayesian view of assessing uncertainty and comparing expert opinion," *Journal of Statistical Planning and Inference*, **20**, 1988, pp 295–306.

[515] Delmont, M. and Welby, S. , "Reliability improvement of power inverters through environmental stress simulation," *Annual Reliability and Maintainability Symposium,* 1990, pp 71–80.

[516] Denton, D.L. and Blythe, D.M. , "The impact of burn-in on IC reliability," *Journal of Environmental Sciences*, Jan./Feb. 1986, pp 19–23.

[517] Dinitto, J.R., Lasch, K.B. and Farrell, J.P. , "Prelid burn-in of hybrid circuits," Proceedings 28th Electronic Components Conference, 1978, pp 340–343.

[518] Director, S.W., Maly, W., and Strojwas, A.J., *VLSI Design for Manufacturing: Yield Enhancement*, Kluwer Academic Publishers, Boston, 1990.

[519] Dishon, M. and Weiss, G.H. , "Burn-in programs for repairable systems, " *IEEE Transactions on Reliability,* **R-22** (5), Dec. 1973, pp 265–267.

[520] Dishon, M. and Weiss, G.H. , "A model for burn-in programs for components with eliminatable defects," *IEEE Transactions on Reliability,* **R-25** (4), Oct. 1976, pp 265–267.

[521] Duane, J.T. , "Learning curve approach to reliability monitoring," *IEEE Transactions on Aerospace*, **2** (2), 1964, pp 563–566.

[522] Duran, J.W. and Nitafos, S.C., "An evaluation of random testing," *IEEE Transactions on Software Engineering,* **SE-10** (4), Jul. 1984, pp 438–444.

[523] Embabi, S.H.K., Bellaouar, A., and Elmasry, M.I., *Digital BiCMOS Integrated Circuit Design*, Kluwer Academic Publishers, Norwell, 1993.

[524] Epstein, B. and Sobel, M. , "Life testing," *Journal of the American Statistical Association,* **48**, 1953, pp 486–502.

[525] Epstein, B. and Sobel, M. , "Some theorems relevant to life testing from an exponential distribution," *Annals of Mathematical Statistics*, **25**, 1954, pp 373–381.

[526] Epstein, B. and Tsao, C.K. , "Some tests based on ordered observations from two exponential populations," *Technometrics*, **12**, 1953, pp 399–407.

[527] Escobar, L.A., and Meeker, W.Q., Jr. , "Recent and future research on practical methods for accelerated testing," 1992.

[528] Evans, R.A., "The analysis of accelerated temperature-tests," pp 294–302.

[529] Feduccia, A.J., "Reliability prediction - use it wisely," *IEEE Transactions on Reliability*, **37** (5), Dec. 1988, p 457.

[530] Felician, L. and Zalateu, G., "Validating Halstead's theory for PASCAL programs," *IEEE Transactions on Software Engineering*, **SE-15** (12), Dec. 1989, pp 1630–1632.

[531] Ferris-Prabhu, A.V., "Modeing the critical area in yield forecasts," *IEEE Journal of Solid State Circuits*, **SC-20** (4), Aug. 1985, pp 874–878.

[532] Ferris-Prabhu, A.V., "Role of defect size distribution in yield modeling," *IEEE Transactions on Electron Devices*, **ED-32** (9), sep. 1985, pp 1727–1736.

[533] Ferris-Prabhu, A.V., "Yield implications and scaling laws for submicrometer devices," *IEEE Transactions on Semiconductor Manufacturing*, **1** (2), May 1988, pp 49–61.

[534] Fertig, K.W. and Murthy, V.K. , "models for reliability growth during burn-in: theory and applications," *Annual Reliability and Maintainability Symposium*, 1978, pp 504–509.

[535] Fink, R.W., "Screening for reliability growth," *Proceedings of Annual Symposium on Reliability*, 1971, pp 316–320.

[536] Fitzsimmons, A. and Love, T., "Review and evaluation of software science," *ACM Computing Surveys*, **10**, 1978, pp 3–18.

[537] Foster, J.W. and Craddock, W.T., "Estimating life parameters from burn-in data," *Annual Reliability and Maintainability Symposium*, 1974, pp 206–209.

[538] Foster, R.C., "Why consider screening, burn-in, and 100-percent testing for commercial devices?" , *IEEE Transactions on Manufacturing Technology*, **MFT-5** (3), 1976, pp 58–58.

[539] Foster, R.C., "How to avoid getting burned with burn-in," *Circuits Manufacturing*, **16** (8), 1976, pp 56–61.

[540] Gallace, L., "Practical applications of the Weibull distribution to power hybrid burn-in," *RCA Engineer*, **19** (4), 1973, pp 58–61.

[541] Georgiev, A.A., "A fast algorithm for curve fitting," in *COMPSTAT: Proceedings in Computational Statistics*, 7th symposium held at Rome 1986, Francesco de Antoni, N. Lauro, and Arthur Rizzi, Editors, Physica-Verlag, Heidelberg, 1986.

[542] Gill, G.K. and Kemerer, C.F., "Cyclomatic complexity density and software maintenance productivity," *IEEE Transactions on Software Engineering*, **SE-17**, Dec. 1991, pp 1284–1288.

[543] Gironi, G. and Malberti, P. , "A burn-in program for wear-out unaffected equipments," *Microelectronics and Reliability*, **15**, 1976, pp 227–232.

[544] Glaser, R.E. , "Bathtub and related failure rate characterizations," *Journal of the American Statistical Association*, **75** (371), Sep. 1980, pp 667–672.

[545] Glaser, R.E. , "Estimation for a Weibull accelerated life test model," *Naval Research Logistics Quarterly*, **31**, 1984, pp 559–570.

[546] Glass, R.L., "Persistent software errors," *IEEE Transactions on Software Engineering*, **SE-7** (2), Mar. 1981, pp 162–168.

[547] "Glossary of Electromagnetic Compatibility (EMC)"

[548] Goel, A.L. and Okumoto, K., "A time dependent error detection rate model for a large scale software system," *Proc. 3rd USA-Japan Computer Conf.*, 1978, pp 35–40.

[549] Goel, A.L. and Okumoto, K., "A Markovian model for reliability and other performance measures," *Proc. National Computer Conf.*, 1979, pp 769–774.

[550] Good, I.J. and Gaskins, R., "Nonparametric roughness penalties for probability densities," *Biometrika*, **58**, 1971, pp 255–277.

[551] Good, I.J. and Gaskins, R., "Density estimation and bump hunting by the penalized likelihood method exemplified by scattering and meteorite data (with Comments)," *Journal of the American Statistical Association*, **75**, 1980, pp 42–73.

[552] Gottfried, P. , "Some aspects of reliability growth," *IEEE Transactions on Reliability*, **R-36** (1), Apr. 1987, pp 11–16.

[553] Gralian, D., "Next generation burn-in development," *IEEE Transactions on Components, Packaging, and Manufacturing Technology – Part B : Advanced Packaging*, **17** (2), May 1994, pp 190–196.

[554] Green, A.E. and Bourne, A.J. , *Reliability Technology*, John Wiley & Sons, New York, NY, 1972.

[555] Green, J.E. , "The problems of reliability growth and demonstration with military electronics," *Microelectronics and Reliability*, **12** (6), 1973, pp 513–520.

[556] Gu, C., "Rkpack and Its Applications: Fitting Smoothing Apline Models," Technical Report No. 857, Department of Statistics, University of Wisconsin- Madison, May 1989.

[557] Gu, C., "Adaptive spline smoothing in non Gaussian regression models," *Journal of the American Statistical Association*, **85**, 1990, pp 801–807.

[558] Guess, F.M. and Park, D.H. , "Modeling discrete bathtub and upside-down bathtub mean residual-life functions," *IEEE Transactions on Reliability*, **37** (5), Oct. 1988, pp 545–549.

[559] Guess, F., Walker, Esteban, and Galliant, D. , "Burn-in to improve which measure of reliability," *Microelectronics and Reliability*, **32** (6), 1991, pp 759–762.

[560] Gurland, J. and Sethuraman, J., "How pooling failure data may reverse increasing failure rates," TR#907, University of Wisconsin, 1993.

[561] Hall, F.H., Paul, R.A., and Snow, W.E., "R&M engineering for off-the-shelf critical software," *Annual Reliability and Maintainability Symposium*, 1988, pp 218–226.

[562] Halstead, M.H., *Elements of Software Science*, Elsevier North-Holland, New York, 1977.

[563] Hamada, M. , *Comparing Accelerated Life Test Plans by Their Non-Estimability Probabilities*, Department of Statistics and Actuarial Science, University of Waterloo, 1990.

[564] Hamilton, H.E. , "The ABCs of selecting a memory IC burn-in system," *Test & Measurement World*, Sep. 1991, pp 121–127.

[565] Hamilton, H.E. , "An overview- VLSI burn-in considerations," *Evaluation Engineering*, Feb. 1992, pp 16–20.

[566] Hamer, P., "Types of metrics," in *Software Reliability: A State of the Art Report*, Sec. 8, A. Bendell and P. Mellor (Editors), Pergamon Infotech Limited, Berkshire, England; 1986, pp 95–103.

[567] Hanlon, E. , "A look at decreasing TTL and MOS burn-in costs," *Insulation/Circuits*, Jun. 1977, pp 39–41.

[568] Hanlon, E. , "Burn-in," *Circuits Manufacturing*, **20** (7), 1980, pp 56–68.

[569] Hannaman, D.J., Zamani, N., Dhiman,J., and Buehler, M.G. , "Error analysis for optimal design of accelerated tests," *Annual Reliability and Maintainability Symposium*, 1990, pp 55–60.

[570] Hansen, M.D., "Human engineering and software maintainability effects on software reliability," *Annual Reliability and Maintainability Symposium*, 1987, pp 374–377.

[571] Harrington, E. , "Cost Efficiency Dynamic Burn-in Definition of the Requirements and Capabilities of Recently Developed Systems," Automatic Testing Deutschland Test and Measurement Exhibition, Wiesbaden, Germany, 1979, pp 1–7.

[572] Härtler, G. , "Parameter estimation for the Arrhenius model," *IEEE Transactions on Reliability*, **R-35** (4), Oct. 1986, pp 414–418.

[573] Head, K.E. and Jones, P.A. , "The use of FRAM to achieve burn-in in G. W. electronic equipment," *Proceedings Joint Conference Automatic Test Systems*, Birmingham, England, 1970, pp 169–186.

[574] Hess, J.A., "Measuring software for its reuse potential," *Annual Reliability and Maintainability Symposium*, 1988, pp 202–207.

[575] Hikami, Toshiya, Obara, Yuichi, Yoshida, Koji and Fuse, Kenichi , "New electric LIF connector with shape memory alloy for burn-in test," *International SAMPE Symposium and Exhibition*, V, Published by *Society for the Advancement of Material and Process Engineering* (SAMPE), 1989, pp 247–258.

[576] Hollander, M. and Proschan, F. , "Testing whether new is better than used," *Annals of Mathematical Statistics*, **43**, 1972, pp 1136-1146.

[577] Hollander, M. and Wolfe, D.A. , *Nonparametric Statistical Methods*, John Wiley & Sons, New York, NY, 1973.

[578] Hu, C., "Reliability Issues of MOS and Bipolar ICs," *Proceedings of IEEE International Conference on Computer Design*, 1989, pp 438–442.

[579] Hudak, Jr., *et al.*, *Development of Standard Methods of Testing and Analyzing Fatigue Crack Growth Rate Data*, Technical Report AFML-TR-78-40, Westinghouse R & D Center, Westinghouse Electric Corporation, Pittsburgh, Pennsylvania 15235, 1978.

[580] Huston, H.H., Wood, M.H., and DePalma, V.M. , "Burn-in effectiveness- theory and measurement, " *Proceedings of International Reliability Physics Symposium,* 1991, pp 271–276.

[581] Jaisingh, L.R., Kolarik, W.J. and Dey, D.K. , "A flexible bathtub hazard model for non-repairable systems with uncensored data," *Microelectronics and Reliability,* **27** (1), 1987, pp 87–103.

[582] Jensen, H.A. and Vairavan, K., "An experimental study of software metrics for real-time software," *IEEE Transactions on Software Engineering,* **SE-11** (2), Feb. 1985, pp 231–234.

[583] Jensen, K.L. and Meeker, W.Q. , "ALTPLAN: microcomputer software for developing and evaluating accelerated life test plans," Paper presented at the Annual ASA Meetings, Anaheim, CA, 1990.

[584] Jensen, F. and Petersen, N.E. , "Burn-in models for non-repairable and repairable equipment based upon bimodal component lifetimes," *Quality Assurance,* **5** (4), 1979, pp 103–107.

[585] Jones, E. and Sheppe, R. , "Alternate approach to traditional burn-In, " *Evaluation Engineering,* Oct. 1991, pp 16–25.

[586] Joyce, W.B., Liou, K-Y, Nash, F.R., Bossard, P.R., and Hartman, R.L. , "Methodology of accelerated aging," *AT&T Technical Journal,* **64**, 1985, pp 717–764.

[587] Kafura, D. and Reddy, G.R., "The use of software complexity metrics in software maintenance," *IEEE Transactions on Software Engineering,* **SE-13** (3), Mar. 1987, pp 335–343.

[588] Kallis, J.M , "Stress screening of electronic modules: investigation of effects of temperature rate of change," *Annual Reliability and Maintainability Symposium,* 1990, pp 59–65.

[589] Keene, S.J. and Keene, K.C., "Reducing the life cycle cost of software through concurrent engineering," *Annual Reliability and Maintainability Symposium,* 1993, pp 274–279.

[590] Keiller, P.A., Littlewood, B., Miller, D.R., and Sofer, A., "Comparison of software reliability predictions," *Proceedings IEEE International Symposium on Fault-Tolerant Computing,* IEEE CS Press, Los Alamitos, CA, 1983, pp 128–134.

[591] Kent, L.L. , "Automotive module burn-in - a means to reliability verification," *Proceedings, International Congress and Exposition,* Detroit, MI, Feb. 23-27, 1987, pp 55–60.

[592] Khuri, A., *Advanced Calculus with Applications in Statistics,* John Wiley & Sons, New York, NY, 1993.

[593] Kimeldorf, G. and Wahba G., "Some results on Tchebycheffian spline functions," *J. Math. Anal. Applic.,* **33**, 1971, pp 82–95.

[594] King, J.C., Chan, W.Y., and Hu, C., "Efficient gate oxide defect screen for VLSI reliability," *IEEE/IEDM,* 1994, pp 597–600.

[595] Kitcheham, B.A., "Metrics in practice," in *Software Reliability: A State of the Art Report,* Sec. 11, A. Bendell and P. Mellor (Editors), Pergamon Infotech Limited, Berkshire, England; 1986, pp 132–144.

[596] Kivenko, K. , "Electronic equipment burn-in for repairable equipment," *Journal of Quality Technology,* **5** (1), 1973, pp 7–10.

[597] Klefsjö, B. and Kumar, U. , "Goodness-of-fit tests for the power-law process based on the TTT-plot," *IEEE Transactions on Reliability,* **41** (4), Dec. 1992, pp 593–598.

[598] Klotz, J., "Spline smooth estimates of survival," in *Survival Analysis* edited by Crowley, J. and Johnson, R.A., 1982, pp 14–25.

[599] Kotz, S. and Johnson, N., (editors), "Burn-in," *Encyclopedia of Statistics,* **1** John Wiley & Sons, Inc., New York, 1982.

[600] Kreyszig, E., *Advanced Engineering Mathematics,* 4th Edition, John Wiley & Sons, New York, NY, 1983.

[601] Lakshmanan, K., Jayaprakash, S., and Sinha, P.K., "Properties of control-flow complexity measures," *IEEE Transactions on Software Engineering,* **SE-17**, Dec. 1991, pp 1289–1295.

[602] Lawford, E. , "Why you should use stress screening," *Test,* **13** (3), Mar. 1991, pp 21, 23.

[603] Lipow, M., "Estimation of Software Package Residual Errors," Correspondence TRW-SS-72-09, TRW Software Series, Redondo Beach, CA, 1972.

[604] Lipow, M., "Some Variations of a Model for Software Time-to-Failure," Correspondence ML-74-2260.1, TRW Systems Group, 1974.

[605] Lipow, M. and Book, E., "Implications of R&M 2000 on Software," *IEEE Transactions on Reliability,* **R-36** (3), Aug. 1987, pp 355–361.

[606] Littlewood, B. and Sofer, A., "A Bayesian modification to the Jelinski-Moranda software reliability growth model," *Software Engineering Journal,* Mar. 1987, pp 30–41.

[607] Lloyd, F.H. , "ESS/burn-in update. The Arrhenius principle in reliability testing," *Evaluation Engineering*, **28** (5), May. 1989, pp 56, 58, 62.

[608] Lochner, R.H. and Basu, A.P., "A Bayesian Approach for Testing Increasing Failure rate," in *The Theory and Applications of Reliability*, **I**, Tsokos, C.P. and Shimi, I.N., Editors., Academic Press, New York, NY, 1977, pp 67–83.

[609] Loranger, J.A. , "The case for component burn-in: the gain is well worth the price," *Electronics*, **48** (2), 1975, pp 73–78.

[610] Louis, T.A. , " Nonparametric analysis of an accelerated failure time model," *Biometrika*, **68**, 1981, pp 381–390.

[611] LuValle, M.J. , "A note on experiment design for accelerated life tests," *Microelectronics and Reliability*, **30**, 1990, pp 591–603.

[612] LuValle, M.J., Welsher, T.L., and Mitchell, J.P. , "A new approach to the extrapolation of accelerated life test data," *The Proceedings of the Fifth International Conference on Reliability and Maintainability*, Biarritz, France, 1986, pp 620–635.

[613] Lynn, N.J. and Singpurwalla, N.D., "Burn-in makes us feel good," GWU/IRRA/Serial T-95/8, 1996.

[614] Lyu, M.R. and Nikora, A., "Applying reliability models more effectively," *IEEE Software*, 1992, pp 43–52.

[615] Lyu, M.R. and Nikora, A., "CASRE– A computer-aided software reliability estimation tool," *Proceedings of the 5th International Workshop on Computer-Aided Software Engineering*, 1992, pp 264–275.

[616] Maly W., Strojwas, A.J., and Director, S.W., "VLSI yield prediction and estimation: a unified framework," *IEEE Transactions on Computer-Aided Design*, **CAD-5** (1), 1986, pp 114-130.

[617] Mann, N.R. , "Tables for obtaining the best linear invariant estimates of parameters of the Weibull distribution," *Technometrics*, **9** (4), 1967, pp 629–645.

[618] Martini, M.R.B., Kanoun, K. and de Souza, J.M., "Software-reliability evaluation of the TROPICO-R switch system," *IEEE Transactions on Reliability*, **39** (3), Aug. 1990, pp 369–379.

[619] McNichols, D.T. and Padgett, W.J. , "Nonparametric estimation from accelerated life tests with random censorship," in *Reliability Theory and Models*, eds. M. S. Abdel-Hameed, E. Cinlar, and J. Quinn, Academic Press, New York, 1984, pp 155–167.

[620] McPherson, J.W. and Baglee, D.A., "Acceleration factors for thin gate oxide stressing," *Proceedings of International Reliability Physics Symposium*, 1985, pp 1–5.

[621] Mead, P.H. , "Reliability growth of electronic equipment," *Microelectronics and Reliability*, **14**, 1975, pp 439–443.

[622] Meeker, W.Q. , *Bibliography on Accelerated Testing*, Available from the Interlibrary Loan Department, Parks Library, Iowa State University, Ames, IA 50011, 1980.

[623] Meeker, W.Q. , "A Comparison of accelerated life test plans for Weibull and lognormal distributions and type I censored data," *Technometrics*, **26**, 1984, pp 157–171.

[624] Meeker, W.Q. and Hahn, G.J. , *How to Plan Accelerated Life Tests: Some Practical Guidelines*, Volume 10 of the ASQC Basic References in Quality Control: Statistical Techniques. Available from the American Society for Quality Control, 310 W. Wisconsin Ave., Milwaukee, WI 53203, 1985.

[625] Meeker, W.Q. and Hahn, G.J. , "How to plan an accelerated life test– some practical guidelines," *Technometrics*, **29**, Dec. 1987, p 491.

[626] Meeker, W.Q. and LuValle, M.J. , "An accelerated life test model based on reliability kinetics," Technical Report, 1991.

[627] Meeker, W.Q. and Nelson, N. , "Optimum accelerated life tests for Weibull and extreme value distributions," *IEEE Transactions on Reliability*, **R-24** (5), Dec. 1975, pp 321–332.

[628] Meeter, C.A. and Meeker, W.Q. , "Optimum Accelerated Life Tests with Nonconstant σ," Technical Report, Department of Statistics, Iowa State University, 1990.

[629] Meindl, J.D., *Microelectronics*, Scientific American, Freeman and Co., San Francisco, 1977, pp 12–23.

[630] Mi, J., *Burn-in*, Ph.D. Dissertation, University of Pittsburgh, Department of Mathematics and Statistics, 1991.

[631] Miller, R. and Nelson, W. , "Optimum simple step-stress plans for accelerated life testing," *IEEE Transactions on Reliability*, **R-32** (1), Apr. 1983, pp 59–65.

[632] Moazzami, R. and Hu, C., "Projecting gate oxide reliability and optimizing reliability screens," *IEEE Transactions on Electron Devices*, **37** (7), 1990, pp 1643–1650.

[633] Moeller, A. , "On the term 'activation energy' in accelerated lifetime tests of plastic encapsulated semiconductor components," *Microelectronics and Reliability*, **20**, 1980, pp 651–664.

[634] Moranda, P.B. , "Prediction of software reliability during debugging," *Annual Reliability and Maintainability Symposium*, 1975, pp 327–332.

[635] Moranda, P.B. , "A failure rate model for burn-in through steady state," Joint National Meeting ORSA/TIMS, Philadelphia, PA, Mar.31-Apr.2, 1976.

[636] Moranda, P.B., "Limits to program testing with random number inputs," *Proc. COMPSAC*, 1978, pp 521–526.

[637] Moranda, P.B., "Event-altered rate models for general reliability analysis," *IEEE Transactions on Reliability*, **R-28** (5), Dec. 1979, pp 376–381.

[638] Mortensen, M.L., *Geometric Modeling*, John Wiley & Sons, New York, NY, 1985.

[639] Mrowiec, T. , "Economic tradeoffs of in-circuit and functional testing," Proceedings International Automatic Test Conference, San Diego, CA, 1978, pp 128–133.

[640] Müller, H. and Wang, J., "Locally adaptive hazard smoothing," *Probability Theory and Related Fields*, **85**, 1990, pp 523–538.

[641] Mulvey, J.M. and Zenios, S.A., *GENO 1.0: A Generalized Network Optimization System*, Report 87-12-03, Decision Science Department, the Wharton School, University of Pensylvania, Philadelphia, 1978.

[642] Musa, J. and Okumoto, K., "A logarithmic Poisson execution time model for software reliability measurement," *Proc. Int'l Conf. Software Eng.*, IEEE CS Press, Los Alamitos, CA, 1984, pp 230–238.

[643] Nakajo, T. and Kume, H., "A case history analysis of software error cause-effect relationships," *IEEE Transactions on Software Engineering*, **SE-17** (8), Aug. 1991, pp 830–838.

[644] Nelson, E.C., "A Statistical Basis for Software Reliability Assessment," TRW Software Series, TRW-SS-73-03, Redondo Beach, CA, 1973.

[645] Nelson, W. , "Two-sample prediction," TIS Report 68-C-404, General Electric Company, Corporate Research and Development, 1968.

[646] Nelson, W. , "A short life test for comparing a sample with previous accelerated test results," *Technometrics*, **14**, 1972, pp 175–186.

[647] Nelson, W. and Meeker, W. , "Theory for optimum censored accelerated life tests for Weibull and extreme value distributions," *Technometrics*, **20**, 1978, pp 171–177.

[648] Nelson, W. , "A survey of methods for planning and analyzing accelerated tests," *IEEE Transactions on Electrical Insulation*, **EI-9** (1), 1973, pp 12–18.

[649] Nelson, W. , "Analysis of Performance Degradation data From Accelerated Tests," *IEEE Transactions on Reliability*, **R-30** (3), Aug. 1981, pp 149–155.

[650] Norris, R.H. , ""Run-in" or "burn-in" of electronic parts: a comprehensive, quantitative basis for choice of temperature, stresses and duration," *Proceedings of the 9th National Symposium on Reliability and Quality Control*, 1963, pp 335–357.

[651] Ost, G., "The practice and economy of burn-in," , *Electronic Engineering*, Aug. 1986, pp 37–43.

[652] Ost, G., "Basics on burn-in/environmental stress screening," *17th Yugoslav Conference on Microelectronics. Proceedings*, **2**, 1989, pp 813–822.

[653] O'Sullivan, F., Yandell, B. and Raynor, W., "Automatic smoothing of regression functions in generalized linear models," *Journal of the American Statistical Association*, **81**, 1986, pp 96–103.

[654] O'Sullivan, P. and Mathewson, A., "Implications of a localized defect model for wafer level reliability measurements of thin dielectrics," *Microelectronics and Reliability*, **33** (11/12), 1993, pp 1679–1685.

[655] Pantic, D. , "Questioning the benefits of burn-in," *Electronic Engineering*, Jul. 1984, pp 45–47.

[656] Parzen, E., "On estimation of a probability density and mode," *Annals of Mathematical Statistics*, **33**, 1962, pp 1065–1076.

[657] Patterson, J.M. , "Electric field equivalency test, a method to burn-in non-package integrated circuits," *Proceedings Advanced Techniques in Failure Analysis Conference*, Los Angeles, CA, 1978, pp 39–41.

[658] Pearson, C.E., *Handbook of Applied Mathematics*, Van Nostrand Reinhold, 1974.

[659] Pecht, M. and Kang, W.C., "A critique of MIL-HDBK-217E reliability prediction methods," *IEEE Transactions on Reliability,* **37** (5), Dec. 1988, pp 453–456 .

[660] Perlstein, D. and Welch, R.L.W., "A Bayesian approach to the analysis of burn-in of mixed populations," *Annual Reliability and Maintainability Symposium,* 1993, pp 417–421.

[661] Phillips, M. , "ESS/burn-in update," *Evaluation Engineering,* **28** (11), Nov. 1989, pp 72–77.

[662] Pleshko, P. , "Aging and burn-in behavior of ?AC plasma panels," *Proceedings SID International Symposium,* Chicago, IL, 1979, pp 56–57.

[663] Prather, R.E. and Myers, Jr., J.P., "The path prefix software testing strategy," *IEEE Transactions on Software Engineering,* **SE-13** (7), Jul. 1987, pp 761–766.

[664] Probert, R.L., "Optimal insertion of software probes in Well-delimited programs," *IEEE Transactions on Software Engineering,* **SE-8** (1), Jan. 1982, pp 34–42.

[665] Proschan, F. , "Theoretical explanation of observed decreasing failure rate," *Technometrics,* **5** (3), 1963, pp 375–383.

[666] Proschan, F. and Singpurwalla, N.D. , "A new approach to inference from accelerated life tests," *IEEE Transactions on Reliability,* **R-29** (3), Aug. 1980, pp 98–102.

[667] Pugacz-Muraszkiewicz, I. , "A methodical approach to accelerated screening of card assemblies, static burn-in case," *Journal of Environmental Sciences,* Mar/Apr 1985, pp 42–47.

[668] Rajarshi, S. and Rajarshi, M.B., "Bathtub distributions: a review," *Communications in Statistics:Theory and Methods,* **17**, 1988, pp 2597–2621.

[669] Rajogopal, A.K. and Teitler, S. , "Parameter compatibility relations for accelerated testing," *IEEE Transactions on Reliability,* **39** (1), Apr. 1990, pp 110–113.

[670] Ramamurthy, B. and Melton, A., "A synthesis of software science measures and the cyclomatic number," *IEEE Transactions on Software Engineering,* **SE-14** (8), Aug. 1988, pp 1116–1121.

[671] Ramsay, J.O. , "Monotone regression splines in action," *Statistical Science,* **3**, 1988, pp 425–461.

[672] Ratnaparkhi, V.M. and Park, W.J. , "Lognormal distribution – model for fatigue life and residual strength of composite materials," *IEEE Transactions on Reliability,* **R-35** (3), Aug. 1986, pp 312–315.

[673] Reda, M.R., Brown, S.G., and Menze, K.L. , "High temperature burn-in and its effects on reliability," *Annual Reliability and Maintainability Symposium,* 1976, pp 72–75.

[674] Redwine, Jr., S.T., "An engineering approach to software test data design," *IEEE Transactions on Software Engineering,* **SE-9** (2), Mar. 1983, pp 191–200.

[675] Reifer, D.J., "Software failure modes and effects analysis," *IEEE Transactions on Reliability,* **R-28** (3), Aug. 1979, pp 247–249.

[676] Romanchik, D. , "Burn-in: still a hot topic," *Test & Measurement World,* Jan. 1992, pp 51–54.

[677] Rosenblatt, M., "Remarks on some nonparametric estimates of a density functions," *Annals of Mathematical Statistics,* **27**, 1959, pp 832-837.

[678] Rosenblatt, M., "Global measures of deviation for kernel and nearest neighbor density estimates," in *Smoothing Techniques for Curve Estimation: Proceedings of Workshop Held in Heidelberg,* April 2-4, 1979, pp 181–190, Gasser, T.A. and Rosenblatt, M., Editors, Springer-Verlag, New York, NY, 1979.

[679] Ross, S.M., "Software reliability: the stopping rule problem," *IEEE Transactions on Software Engineering,* **SE-11**, Dec. 1985, pp 1472–1476.

[680] Rothbart, H.A. , (Editor), *Mechanical Design and Systems Handbook,* Second edition, McGraw-Hill, Inc., New York, NY, 1985.

[681] Rubin, D.B., "Using the SIR algorithm to simulate posterior distributions," *Bayesian Statistics,* **3**, Bernardo, J.M., *et al.*, North-Holland, Amsterdam, 1988, pp 395–402.

[682] Rue, H.D. , "System burn-in for reliability enhancement," *Annual Reliability and Maintainability Symposium,* 1976, pp 336–341.

[683] Runggaldier, W.J., "On stochastic control concepts for sequential burn-in procedures," *Reliability and Decision Making* Barlow, R.E. *et.al* (Editors), Elsevier Applied Science, London, 1993, pp 211–232.

[684] Saraidaridis, C.I. , *System Approach to Burn-in of Digital Systems, Part 1: Factory and Field Device Failure Rates, Economic Analysis,* Bell Labs, Technical Report, 1980.

[685] Saunders, S.C. and Myhre, J.M. , "Maximum likelihood estimation for two-parameter decreasing hazard rate distributions using censored data," *Journal of the American Statistical Association,* **78** (383), 1983, pp 664–673.

[686] Schafer *et al.*, *Validation of Software Reliability Models*, Technical Report RADC-TR-79-147, Rome Air Development Center, Rome, NY, 1979.

[687] Schatz, R., Shooman, M., and Shaw, L., "Application of time dependent stress-strength models of non-electrical and electrical systems," *ASQC Literature Classification System*, pp 540–547.

[688] Schmoyer, R.L. , "Linear interpolation with a nonparametric accelerated failure-time model," *Journal of the American Statistical Association*, **83**, 1988, pp 441–449.

[689] Schmoyer, R.L. , "An exact distribution-free analysis for accelerated life testing at several levels of a single stress," *Technometrics*, **28** (2), May 1986, pp 165–175.

[690] Schnable, G.L. and Swartz, G.A., "In-process voltage stressing to increase reliability of MOS integrated circuits," *Microelectronics and Reliability*, **28** (5), 1988, pp 757–781.

[691] Schneider, B. and Oestergaard, P., "Advanced data compaction approach for test-during burn-in," *International Test Conference*, Sep. 1988, pp 381–390.

[692] Schneidewind, N.F., "Analysis of error process in computer software," *ACM Sigplan Notices*, **10** (6), 1975, pp 337–346.

[693] Schneidewind, N.F., "Application of program graphs and complexity analysis to software development and testing," *IEEE Transactions on Reliability*, **R-28** (3), Aug. 1979, pp 192–198.

[694] Seed, R.B., "Yield, economic, and logistic models for complex digital arrays," *IEEE International Convention Record*, Part 6, Apr. 1967, pp 61–66.

[695] Seed, R.B., "Yield and cost analysis of bipolar LSI," *IEEE International Electron Devices Meeting*, Oct. 1967, pp 12.

[696] Senthilselvan, A., "Penalized likelihood estimation of hazard and intensity functions," *Journal of the Royal Statistical Society*, B, **49** (2), 1987, pp 170–174.

[697] Shanthikumar, J.G., "A general software reliability model for performance prediction," *Microelectronics and Reliability*, **21**, 1981, pp 671–682.

[698] Shanthikumar, J.G., "Software reliability models: a review," *Microelectronics and Reliability*, **23** (5), 1983, pp 903–943.

[699] Shaw, Jr., W.H., *et al.*, "A software science model of compile time," *IEEE Transactions on Software Engineering*, **SE-15** (5), May 1989, pp 543–549.

[700] Shen, V.Y., Conte, S.D., and Dunsmore, H.E., "Software science revisited: a critical analysis of the theory and its empirical support," *IEEE Transactions on Software Engineering*, **SE-9** (2), Mar. 1983, pp 155–165.

[701] Shooman, M.L. , *Quality Programming*, McGraw-Hill, Inc., New York, NY, 1972.

[702] Shooman, M.L., "Probabilistic models for software reliability prediction," *Statistical Computer Performance Evaluation*, Freiberger, W. (Editor), Academic Press, 1972, pp 485–502.

[703] Shooman, M.L., *Software Engineering- Design Reliability and Management*; McGraw-Hill, New York, NY, 1983.

[704] Silverman, B.W., "Density ratios, empirical likelihood and cot death," *Applied Statistics*, **27**, 1978, pp 26–33.

[705] Silverman, B.W., *Density Estimation for Statistics and Data Analysis*, Monographs on Statistics and Applied Probability, Chapman and Hall, New York, NY, 1986.

[706] Singh, A.D., "On wafer burn-in strategies for MCM die," *International Conference and Exhibition on Multichip Modules*, Apr. 1994, pp 255–260.

[707] Singpurwalla, N.D. and Wong, W.H., *Improvement of Kernel Estimators of the Failure Rate Function Using the Generalized Jackknife*, Technical Paper Serial T-415, The George Washington University, Institute for Management Science and Engineering, 1980.

[708] Singpurwalla, N.D. and Wong, W.H.. "Kernel estimators of the failure-rate function and density estimation : an analogy," *Journal of the American Statistical Association*, **78**, 1983, pp 478–481.

[709] Siswadi, Q.C.P. , "Selecting among Weibull, lognormal and gamma distributions using complete and censored samples," *Naval Research Logistics Quarterly*, **29** (4), 1982, pp 557–569.

[710] Slack, L.J. , "Electrified track system helps simplify burn-in test procedures," *Electronic Manufacturing*, **34** (11), Nov. 1988, p 14.

[711] Smith, S.A. and Oren, S. , "Reliability growth of repairable systems," *Naval Research Logistics Quarterly*, **27** (4), 1980, pp 539–547.

[712] Smith, P.L., "Splines as a useful and convenient statistical tool," *The American Statistician*, **33**, 1979, pp 57–62.

[713] Somani, A.K., Palnitkar, S., and Sharma, T. , "Reliability modeling of systems with latent failures using Markov chains," *Annual Reliability and Maintainability Symposium,* 1993, pp 120–125.

[714] Sontz, C. and Wallace, W.E. , "MIL-STD-781," *Annual Reliability and Maintainability Symposium,* 1977, pp 448–453.

[715] Spizzichino, F., "Sequential burn-in procedures," *Journal of Statistical Planning and Inference,* **29**, 1991, pp 187–197.

[716] Spizzichino, F., "A unifying model for the optimal design of life-testing and burn-in," *Reliability and Decision Making* Barlow, R.E. *et.al* (Editors), Elsevier Applied Science, London, 1993, pp 189–210.

[717] Stapper, C.H., "Defect density distribution for LSI yield calculations," *IEEE Transactions on Electron Devices,* **ED-20**, Jul. 1973, pp 655–657.

[718] Stapper, C.H., "The defect-sensitivity effect of memory chips," *IEEE Journal of Solid-State Circuits,* **SC-21**, Feb. 1986, pp 193–198.

[719] Stapper, C.H., "On Murphy's yield Integral," *IEEE Transactions on Semiconductor Manufacturing,* **4** (4), Nov. 1991, pp 294–297.

[720] Stapper, C.H. and Klaasen W., "The evaluation of 16-Mbit memory chips with built-in reliability," *Proceedings of International Reliability Physics Symposium,* 1992, pp 3–7.

[721] Stephens, M.A., "EDF statistics for goodness of fit and some comparisons," *Journal of the American Statistical Association,* **69**, 1974, pp 730–737.

[722] Stitch, M., Johnson, G.M., Kirk, B.P., and Brauer, J.B. , "Microcircuit accelerated testing using high temperature operating tests," *IEEE Transactions on Reliability,* **R-24** (4), Oct. 1975, pp 238–250.

[723] Sukert, A.N., "An investigation of software reliability models," *Annual Reliability and Maintainability Symposium,* 1977, pp 78–84.

[724] Sukyo, A. and Sy, S. , "Development of a burn-in time reduction algorithm using the principles of acceleration factors," *Proceedings of International Reliability Physics Symposium,* 1991, pp 264–270.

[725] Swanson, E.B., "On the user-requisite variety of computer application software," *IEEE Transactions on Reliability,* **R-28** (3), Aug. 1979, pp 221–226.

[726] Tang, S., "New burn-in methodology based on IC attributes, family IC burn-in data, and failure mechanism analysis," *Annual Reliability and Maintainability Symposium,* 1996, pp 185–190.

[727] Tarter, M.E. and Lock, M.D., *Model-Free Curve Estimation,* Monographs on Statistics and Applied Probability 56, Chapman and Hall, New York, NY, 1993.

[728] Taylor, C.H., "Just how reliable are plastic encapsulated semiconductors for military applications and how can the maximum reliability be obtained ?," *Microelectronics and Reliability,* **15**, 1976, pp 131–134.

[729] Terrel, J.A. and McGlone, M.E. , "Evaluation of reliability growth models for electronic engine control systems," AIAA/SAE/ASME 20th Joint Propulsion Conference, June 11-13, Cincinnati, Ohio, 1984, pp 1–7.

[730] Thompson, W.E. , "Software reliability growth modeling," Reliability Growth: Management, Testing and modeling, Seminar Proceedings- Institute of Environmental Sciences, Feb. 27-28, 1978, pp 32–41.

[731] Tomsky, J. , "Regression models for detecting reliability degradation," *Annual Reliability and Maintainability Symposium,* 1982, pp 238–244,

[732] Trindade, D.C., "Can burn-in screen wearout mechanisms?: reliability modeling defective subpopulations- a case study," *Proceedings of International Reliability Physics Symposium,* 1991, pp 260–263.

[733] Truelove, A.J. , "Reliability growth/burn-in: the allocation of testing resources," Proceedings of the 4th Aerospace Testing Seminar, 1978, pp 111–114.

[734] Tuckerman, D.B., *et al.*, "A cost-effective wafer-level burn-in technology," *International Conference and Exibition on Multichip Modules,* Apr. 1994, pp 34–40.

[735] Utréras D., F., "Cross-validation techniques for smoothing spline functions in one or two dimensions," in *Smoothing Techniques for Curve Estimation: Proceedings of Workshop Held in Heidelberg,* April 2-4, 1979, pp 196–231, Gasser, T.A. and Rosenblatt, M., Editors, Springer-Verlag, New York, NY, 1979.

[736] Vander Wiel, S.A. and Meeker, W.Q. , "Accuracy of approximate confidence bounds using censored Weibull regression data from accelerated life tests," *IEEE Transactions on Reliability,* **39** (3), Aug. 1990, pp 346–351.

[737] Vander Wiel, S.A. and Votta, L.G., "Assessing software design using capture-recapture methods," *IEEE Transactions on Software Engineering,* **19** (11), Nov. 1993, pp 1045–1054.

[738] Vaupel, J.W. and Yashin, A.I., "Heterogeneity's rule: some surprising effects of selection on population dynamics," *The American Statistician,* **39**, 1985, pp 176–185.

[739] Viertl, R. , *Statistical Methods for Accelerated Life Testing*, Göttingen: Vandenhoeck and Ruprecht, 1988.

[740] Wager, A.J., Thompson, D.L., and Forcier , "Implications of a model for optimum burn-in," *Proceedings of International Reliability Physics Symposium,* 1983, pp 286–291.

[741] Wagoner, W.L., *The Final Report on a Software Reliability Measurement Study*, Report TOR-0074(41221)-1, The Aerospace Corp., El Segundo, CA, 1973.

[742] Wahba, G., "Partial and interaction splines for the semiparametric estimation of functions of several variables," in Boardman, T.J., ed., *Computer Science and Statistics: Proceedings of the 18th Symposium on the interface*, pp 75–80, Washington, D.C. Amer. Statist. Assoc, 1986.

[743] Wahba, G., *Spline Models for Observational Data*, Society for Industrial and Applied Mathematics, 1990.

[744] Wahba, G. and Wendelberger J., "Some new mathematical methods for variational objective analysis using splines and cross validation," *Monthly Weather Review*, **108**, 1980, pp 1122–1145.

[745] Walters, G.F. and McCall, J.A., "Software quality metrics for life-cycle cost reduction," *IEEE Transactions on Reliability,* **R-28** (3), Aug. 1979, pp 212–220.

[746] Wand, M.P. and Jones, M.C., *Kernel Smoothing*, Chapman & Hall, New York, NY, 1995.

[747] Wang, K.L. , "Demonstrating reliability and reliability growth with environmental stress screening data," *Annual Reliability and Maintainability Symposium,* 1990, pp 47–52.

[748] Watson, G.S. and Leadbetter, M.R., "Hazard analysis II," *Sankya Ser. A*, **26**, 1964, pp 101–116.

[749] Wegman, E.J. and Wright, I.W., "Splines in statistics," *Journal of the American Statistical Association,* **78**, 1983, pp 351–365.

[750] West, M., "Modelling expert opinion," *Bayesian Statistics*, **3**, Bernardo, J.M., *et al.* (Editors), Clarendon, Oxford, 1988, pp 493–508.

[751] Whitbeck, C.W. and Leemis, L.M. , "Component vs system burn-in techniques for electronic equipment," *IEEE Transactions on Reliability,* **38** (2), Jun. 1989, pp 206–209.

[752] Yang, P. and Chern, J., "Design for reliability: the major challenge for VLSI," *Proceedings of the IEEE*, **81** (5), May 1993, pp 730–744.

[753] Yin, X. and Sheng, B. , "Some aspects of accelerated life testing by progressive stress," *IEEE Transactions on Reliability,* **R-36** (1), Apr. 1987, pp 150–155.

[754] Yurkowski, W., Schafer, R.E., and Finkelstein, J.M. , *Accelerated Testing Technology*, Rome Air Development Center Technical Report RADC-TR-67-420, 1967.

A. Notation and Nomenclature

AAE	average absolute error
AC	alternating current
AE	absolute error
AFT	accelerated failure time
AGREE	the Advisory Group on Reliability of Electronic Equipment
ALT	accelerated life test
AMS	agile manufacturing system
AO	asymptotically optimal
AOQL	average outgoing quality level
AQL	acceptable quality level
ARE	average relative error
ASER	accelerated soft error rate
ASIC	application specific integrated circuit
b	subscript to denote burn-in
B^s, B^u_j, B^c_i	per unit burn-in setup cost for system, subsystem j, and component i, respectively
$\mathcal{B}(p,q)$	the Beta function, $\mathcal{B}(p,q) = \frac{\Gamma(p)\Gamma(q)}{\Gamma(p+q)}$
B/I	burn-in
BIB	burn-in board
BIR	built-in reliability
BLIE	best linear invariant estimator
BLUE	best linear unbiased estimator
BPR	Bayesian prediction region
BQLE	Bayes quadratic loss estimator
BT	British Telecommunications
C	burn-in fixed cost
$C_{\max}$	maximum allowable cost
c, u, s	superscript for component, subsystem, and system, respectively
c_f	field repair cost; it denotes the field repair cost for the ith component in jth subsystem if subscripts ij are used; the single subscript j and s represents the field repair cost for the j subsystem

	and for the system, respectively. The same logic used in c_f also applies to c_r, c_s, and c_v, which are defined below.
c_r	redundant cost
c_s	shop repair cost
c_v	burn-in variable cost
$C_T(t)$	total life-cycle cost as a function at t
CAD	computer-aided design
CAM	computer-aided manufacturing
CAPP	computer-aided process planning
CAQC	computer-aided quality control
CAS	computer-aided support system
CASE	computer-aided software engineering
CDF	cumulative distribution function
CDM	charge device model (for ESD tests)
CDR	critical design review
CFR	constant failure rate
CI	confidence interval
CLT	central limit theorem
CML	current mode logic
CMOS	complementary MOS
CP	chip probing
CVD	chemical vapor deposition
DADS	dynamic analysis design system
DBI	dynamic burn-in
DBMS	database management system
DC	direct current
DFR	decreasing failure rate
DFRA	decreasing failure rate average
DIMM	dual in-line memory module
DLBI	die level burn-in
DMRL	decreasing mean residual life
DOD	Department of Defense, U.S.A.
DPRL	decreasing percentile residual life
DRAM	dynamic random access memory
DUT	device under test
$e_{ij,b}$	s-expected fraction of failure during burn-in for component i in subsystem j

$e_{ij,t}$	s-expected fraction of failure at time t for component i in subsystem j
E_a	the activation energy, used in the Arrhenius equation
E-test	electrical test
EB	empirical Bayes
EBE	empirical Bayes estimator
ECL	emitter coupled logic
ECT	expansion coefficient of temperature
EFR	early failure rate
EIAJ	Electronic Industries Association of Japan
EM	electromigration
EOS	electrical over-stress
ESD	electrostatic discharge
ESF	empirical survival function
ESS	environmental stress screening
$F(t)$	CDF at time t
$f(t)$	pdf at time t
FA	failure analysis
FBDP	fair Bayes decision problem
FBLF	fair Bayes loss function
FEA	finite-element analysis
FET	field effect transistor
FIT	failures in unit time (10^9 device-hours)
FMECA	failure-mode effect and criticality analysis
FMS	flexible manufacturing system
FPLD	field-programmable logic devices
FT	final test
$g(\vec{\theta})$	the prior distribution for $\vec{\theta}$
G/L	grid line
Gm	transconductance
GOF	goodness-of-fit
GOI	gate oxide integrity
GT	group technology
GUI	graphical user interface
$H(t)$	cumulative hazard rate function at time t
$h(t)$	hazard rate function at time t
HAST	highly accelerated stress test

HBT	heterojunction bipolar transistor
HBM	human body model (for ESD tests)
HCI	hot carrier injection
HFET	heterojunction FET
HPD	highest posterior density
HTRB	high temperature reverse bias
HTB	high temperature with bias
HTOL	high temperature operating life
HTS	high temperature storage
I/O	input/ output
IC	integrated circuits
IEC	International Electrotechnical Commission
IFR	increasing failure rate
IFRA	increasing failure rate average
IIDE	integrated intelligent design environment
IMD	inter-metallic dielectric
IMRL	increasing mean residual life
IMS	intelligent manufacturing system
iid	independently identical distribution
JEDEC	Joint Electron Device Engineering Council
JFET	Junction FET
JIT	just-in-time
JSE	the James-Stein estimator
kb	kilo-bit (1,000-bit)
KGD	known good die
KME	the Kaplan-Meier estimation (or estimator)
L	subscript denoting the constant failure rate (steady-state condition)
l	ratio used to calculate the cost of the loss of credibility; $l = 0$, low penalty; $l = 1$, medium penalty; $l = 3$, high penalty
$L(\vec{\theta})$	likelihood function of $\vec{\theta}$
L/F	lead frame
L/R	laser repair
LCC	life-cycle cost
LCE	life-cycle engineering
LSAP	logistic support analysis program
LSAR	logistic support analysis records

LSI	large-scale integration
LSTTL	large-scale transistor-transistor logic
LTB	low temperature with bias
LTE	life table estimator
LTOL	low temperature operating life
LTPD	lot tolerance percent defective
Mb	mega-bit (10^6 bits)
MB	mega-byte (10^6 bytes)
MBE	molecular beam epitaxy
MCM	multi-chip module
ME	maximum error
MGF	moment generating function
MLE	maximum likelihood estimation (or estimator)
MM	machine model (for ESD tests)
MOS	metal oxide semiconductor
MPD	maximum percent defective
MRL	mean residual life
MSE	mean square error
$M(t)$	mean value function
MTBF	mean time between failures
MVUE	minimum variance unbiased estimator
MTTF	mean time to failure
MTTR	mean time to repair
N_{ij}	number of components i in subsystem j; note that $N_{v+1,j} = 1$
NBU	new better than used
NBUE	new better than used in expectation
NHPP	nonhomogeneous Poisson process
NMOS	N-channel MOS
NWU	new worse than used
NWUE	new worse than used in expectation
OC	operating characteristic
OODB	object-oriented data base
PAV	pooled-adjacent-violator
PCT	pressure cooker test
pdf	probability density function
PDR	preliminary design review
PEB	parametric empirical Bayes

PED	plastic encapsulated device
PH	proportional hazard
PIND	particle impact noise detector
PLBI	package level burn-in
PLD	programmable logic devices
PLE	product limit estimator
pmf	probability mass function
PMOS	P-channel MOS
ppm/PPM	part per million
PRAT	production reliability acceptance test
PRST	probability ratio sequential test
QA	quality assurance
QBD	charge to breakdown
QC	quality control
QML	qualified manufacturing line
QRA	quality & reliability assurance
Qual.	qualification
R^2	coefficient of determination
RAM	random access memory
RE	relative error
RIE	reaction ion etcher
$R^c_{ij}(t)$	reliability of component i in subsystem j at time t
$R^s(t \mid \mathbf{X}, \mathbf{Y})$	systems reliability at time t, where $\mathbf{X}$, $\mathbf{Y}$ represent the set composed by x^s, x^u_j, x^c_i and y^u_j, respectively; abbreviated as R^s
$R^u_j(t)$	subsystem reliability at time t of subsystem j; can be abbreviated as R^u_j
$R(t)$	reliability function at time t; $R(t) = \overline{F}(t) = 1 - F(t)$
RADC	Roman Air Development Center
RE	reliability engineering
RH	relative humidity
RMSE	root-mean-square error
RQT	reliability qualification test
RRT	routine reliability test
SAPD	strategic approach to product design
SAT	sonic acoustic tomograph
SCA	sneak circuit analysis
SCH	scaled cumulative hazard

SDIP	software development integrity program
SED	statistically experimental design
SLLN	strong law of large number
SOW	statement of work
SQA	software quality assurance
SQC	statistical quality control
SRAM	static random access memory
SSI	small-scale integration
SWEAT	standard wafer level EM acceleration test
T	temperature
t_i	failure time of the ith sample
$t_{i:n}$	the ith smallest failure time in n samples
$t_{ij,b}$	total burn-in time equivalent to burn in at T_P for component i in subsystem j
$t_{ij,L}$	time at which $\lambda_{ij,L}$ is reached; if the superscript $'$ is added, it is adjusted data.
T_p	system ambient temperature (in ^{o}C)
TC	total allowable cost
T/C	temperature cycling
T/S	thermal shock
TDA	test data analysis
TDBI	test during burn-in
TDDB	time-dependent dielectric breakdown
TEG	test element group
THB	temperature/ humidity with bias test
TQM	total quality management
TSS	thermal stress screening
$TT_{i:n}$	the ith smallest (out of n) data after the TTT transformation
TTL	transistor transistor logic
TTT	total time on test
UMA	uniformly most accurate
ULSI	ultra large-scale integration
UMVUE	uniformly minimum variance unbiased estimator
u_j	number of ICs in subsystem j, $u_j = \sum_{i=1}^{v} N_{ij}$
u^s	number of subsystems in the system, $u^s = \sum_{j=1}^{n} y_j^u$

UV	ultraviolet
v	total number of distinct integrated circuits (ICs) in a system
Vt	threshold voltage
VLSI	very large-scale integration
VPE	vapor phase epitaxy
WAT	wafer acceptance test
WBI	wafer burn-in
WLBI	wafer level burn-in
WLR	wafer level reliability
WP	wafer probe
x^s, x_j^u, x_i^c	burn-in time for system, subsystem j and component i, respectively
y_j^u	number of subsystem j used in the system
α	scale parameter of the Weibull distribution ($\alpha > 0$) / type-I error, producer's risk / the significance level
β	shape parameter of the Weibull distribution / type-II error, customer's risk / voltage acceleration multiplier in the voltage acceleration model / a weighting factor for the prior knowledge about a product
Λ	overall compatibility level
λ	hazard rate of the exponential distribution
λ_L	constant failure rate ($\lambda_L > 0$)
θ	MTTF (mean) of the exponential distribution
$\varphi_F(u)$	the TTT transformation of F where $0 \leq u \leq 1$
$\varphi^s, \varphi^u, \varphi^c$	burn-in multipliers for system, subsystem, and component, respectively, on the basis of corresponding burn-in temperatures
ε	compatibility range
ξ_f, ξ_s	repair cost factor for field and shop repair, respectively
$\Gamma(a)$	the gamma function; $\Gamma(a+1) = a\Gamma(a) = a!$
κ	the compatibility factor
η	the time transformation factor
δ	an indicator function; used in likelihood functions
μ	the mean of a distribution / mean defect density
σ	the standard deviation of a distribution
π	the stress factor
$\hat{}$	used to indicate an estimator
$\Phi()$	CDF of N(0,1)

B. FAILURE MODES FOR PARTS

Mechanical Parts

Three categories of a systematic classification are often used to describe mechanical failures: manifestations of a failure (elastic deformation, plastic deformation, rupture or fracture, or material change), failure-inducing agents (force, time, temperature, and reactive environment), and failure locations (body type or surface type). According to Askeland [471], some commonly observed modes of mechanical failure are:

- brinelling,
- brittle fracture,
- buckling,
- corrosion,
- creep,
- ductile rupture,
- fatigue,
- force and/or temperature-induced elastic deformation,
- fretting,
- galling,
- impact,
- radiation damage,
- seizure,
- spalling,
- stress corrosion,
- thermal relaxation,
- thermal shock,
- wear,

- yielding.

Many theories have been proposed for testing various failure modes. Most of them are based on the assumption that failure is predicted to occur when the maximum value of the selected mechanical modulus in the multiaxial state of stress becomes equal to or exceeds the value of the same modulus that produces failure in a simple uniaxial stress test using the same material. Some theories are listed below:

- Distortion energy theory (Huber- Von Mises, and Hencky) for yielding or ductile rupture of isotropic materials.
- Maximum normal strain theory (St. Venant).
- Maximum normal stress theory (Rankine) for brittle fracture in isotropic materials.
- Maximum shearing stress theory (Tresca-Guest).
- Modified Mohr's failure theory for brittle fracture although it exhibits a compressive ultimate strength that is significantly different from the tensile ultimate strength.
- Mohr's failure theory for yielding although it exhibits a compressive yield strength that is significantly different from the tensile yield strength.
- Total strain energy theory (Beltrami).

Several theories are presented for the cumulative damage model, such as the Marco-Starkey, Henry, Gatts, Corten-Dolan, and Marin cumulative damage theory and the Manson double linear damage rule. Rothbart [680, Sec. 17] presents a good review of the aforementioned models.

Electronic and Electrical Parts

Some typical mechanical defects in electronic circuits are:

- cracks in glass or ceramics of package insulators,
- grain bounds segregation at package weld interfaces,
- improper solder joints,
- inadequate bounding between metallization and substrate,
- incomplete cured polymers,
- interface impurities,
- loose conducting particles,
- necked down leads at package,
- overbounded, underbounded, and misplaced wire bounds,

- plating defects,
- poor die attach,
- precipitates and other hard particles within metal films, wires, and interfaces,
- scratches, cracks, and voids in thin film interconnections, dielectrics, chips, and wires,
- thermal-mechanical misfit between circuit materials, and
- thin oxide step coverage.

For selecting package materials, there are always trade-offs between convenient inter-connectability, efficient manufacture, and the long term protection and stability of the circuit structure. Some materials may have useful environmental characteristics but less desirable electrical properties. Some common package-related defects are:

- cracks in glass or other ceramic insulators,
- excessive thermal-mechanical mismatch in package materials,
- improperly applied metallurgical die attachments,
- incompletely cured polymers used for die attach adhesive,
- leakage at metallurgical seals due to surface contamination, incorrect welding or soldering parameters, and grain boundary segregation at weld surfaces,
- loose conducting particles,
- moisture and other chemical agents outgassing from package polymers and metals,
- moisture diffusion through polymers where employed as the seal or else as the package material, and
- trapped moisture and other gases in package.

Defects in solder joints are also commonly seen and result from improper choice of flux, solder technique, temperature-time setting, or solder alloy.

Electrostatic discharge (ESD) results from the interaction of two materials and involves the removal of electrons from the surface atoms of one of the materials. ESD can cause great damage to electronic parts. Some typical ESD-related failure mechanisms are surface breakdown, thermal secondary breakdown, bulk breakdown, metallization melt, dielectric breakdown, and gaseous arc discharge. Various design precautions have been employed to reduce the susceptibility of parts and assemblies to ESD, such as diffused resistors, limiting resistors, Zener diodes, silicon control rectifier (SCR), and FETs. Other measures include applying protective materials and adding protection circuits.

For other types of electronic failure modes, see Fuqua [157, pp 156–159]. In addition, detailed failure modes studies are contained in MIL-STD-883C, and a brief summary is provided by Kuo and Kuo [230].

C. Common Probability Distributions

C.1. Discrete Distributions

Binomial	BIN(n,p)
pdf	$\binom{n}{x} p^x q^{n-x} \quad 0 < p < 1, \quad q = 1-p, \quad x = 0, 1, ..., n$
Mean	np
Variance	npq
MGF	$(pe^t + q)^n$

Bernoulli	BIN(1,p)
pdf	$p^x q^{1-x}$
Mean	p
Variance	pq
MGF	$pe^t + q$

Negative Binomial	NB(r,p)
pdf	$\binom{x-1}{r-1} p^r q^{x-r} \quad 0 < p < 1, \quad r = 1, 2, ..., \quad q = 1-p,$ $x = r, r+1, ..$
Mean	$\frac{r}{p}$
Variance	$\frac{rq}{p^2}$
MGF	$(\frac{pe^t}{1-qe^t})^r$

Geometric	GEO(p)
pdf	$pq^{x-1} \quad 0 < p < 1, \quad q = 1-p, \quad x = 1, 2, ...$
Mean	$\frac{1}{p}$
Variance	$\frac{q}{p^2}$
MGF	$\frac{pe^t}{1-qe^t}$

Hypergeometric HYP(n,M,N)

pdf	$\frac{{}_MC_x \; {}_{(N-M)}C_{(n-x)}}{{}_NC_n}$	$n=1\ldots N,\ x=0\ldots n,\ M=0\ldots N$
Mean	$\frac{nM}{N}$	
Variance	$n\frac{M}{N}(1-\frac{M}{N})\frac{N-n}{N-1}$	

Poisson	POI(μ)	
pdf	$e^{-\mu}\mu^x/x!$	$0<\mu,\quad x=0,1,\ldots$
Mean	μ	
Variance	μ	
MGF	$\exp[\mu(e^t-1)]$	

Discrete Uniform	DU(N)	
pdf	$\frac{1}{N}$	$N=1,2,..,\quad x=1\ldots N$
Mean	$\frac{N+1}{2}$	
Variance	$\frac{N^2-1}{12}$	
MGF	$\frac{1}{N}\frac{e^t-e^{(N+1)^t}}{1-e^t}$	

C.2. Continuous Distributions

Uniform	UNIF(a,b)	
pdf	$1/(a-b)$	$a<b,\quad a<x<b$
Mean	$a+b/2$	
Variance	$(b-a)^2/12$	
MGF	$(e^{bt}-e^{at})/(b-a)t$	

Normal	N(μ,σ^2)	
pdf	$\frac{1}{\sqrt{2\pi\sigma}}\exp[-\frac{(\frac{x-\mu}{\sigma})^2}{2}]$	$0<\sigma^2$
Mean	μ	
Variance	σ^2	
MGF	$e^{\mu+\frac{\sigma^2 t^2}{2}}$	

Lognormal	LN(μ,σ^2)	
pdf	$\frac{1}{\sqrt{2\pi\sigma x}}\exp[-\frac{1}{2}(\frac{ln\ x-\mu}{\sigma})^2]$	
Mean	$\exp[\mu+\frac{\sigma^2}{2}]$	
Variance	$[\exp(\sigma^2)-1]\exp(2\mu+\sigma^2)$	
MGF	does not exist	

Exponential	EXP(θ)	
pdf	$\frac{1}{\theta}e^{-\frac{x}{\theta}}$	$\theta > 0, \quad x > 0$
Mean	θ	
Variance	θ^2	
MGF	$\frac{1}{1-\theta t}$	
Double Exponential	DE(θ, η)	
pdf	$\frac{1}{2\theta}e^{-\frac{\lvert x-\eta \rvert}{\theta}}$	$\theta > 0$
Mean	η	
Variance	$2\theta^2$	
MGF	$\frac{e^{\eta t}}{1-\theta^2 t^2}$	
Beta	BETA(p, q)	
pdf	$\frac{\Gamma(p+q)}{\Gamma(p)\Gamma(q)}x^{p-1}(1-x)^{q-1}$	$0 < x < 1$
Mean	$\frac{p}{p+q}$	
Variance	$\frac{pq}{(p+q)^2(p+q+1)}$	
MGF	$1 + \sum_{k=1}^{\infty}(\prod_{r=0}^{k-1}\frac{p+r}{p+q+r})\frac{t^k}{k!}$	
Gamma	GAM(α, β)	
pdf	$\frac{\beta^\alpha}{\Gamma(\alpha)}x^{\alpha-1}\exp(-\beta x)$	$\alpha > 0, \quad \beta > 0, \quad x > 0$
Mean	α/β	
Variance	α/β^2	
MGF	$(\frac{1}{1-t/\beta})^\alpha$	
Weibull	WEI(α, β)	
pdf	$\frac{\beta}{\alpha^\beta}x^{\beta-1}\exp[-(x/\alpha)^\beta]$	$\alpha > 0, \quad \beta > 0, \quad x > 0$
Mean	$\alpha\Gamma(1+\frac{1}{\beta})$	
Variance	$\alpha^2[\Gamma(1+\frac{2}{\beta}) - \Gamma^2(1+\frac{1}{\beta})]$	
MGF	exists only for $\alpha \geq 1$	
Extreme Value	EV(θ, η)	
pdf	$\frac{1}{\theta}\exp[\frac{x-\eta}{\theta} - e^{\frac{x-\eta}{\theta}}]$	$\theta > 0$
Mean	$\eta - \gamma\theta$	$\gamma \doteq 0.5772$
Variance	$\frac{\pi^2\theta^2}{6}$	

MGF	$e^{\eta t}\Gamma(1+\theta t)$	
Cauchy	$\mathrm{CAU}(\theta,\eta)$	
pdf	$\frac{1}{\theta\pi[1+(\frac{x-\eta}{\theta})^2]}$	$\theta > 0$
Mean	does not exist	
Variance	does not exist	
MGF	does not exist	
Pareto	$\mathrm{PAR}(\theta,\kappa)$	
pdf	$\kappa/[\theta(1+\frac{x}{\theta})^{\kappa+1}]$	$\theta > 0, \quad \kappa > 0, \quad x > 0$
Mean	$\theta/(\kappa-1)$	
Variance	$\frac{\theta^2\kappa}{(\kappa-2)(\kappa-1)^2}$	
MGF	does not exist	
Chi-square	$\mathrm{CHI}(\nu)$	
pdf	$\frac{1}{2^{\frac{\nu}{2}}\Gamma(\frac{\nu}{2})}x^{\frac{\nu}{2}-1}e^{-\frac{x}{2}}$	$\nu = 1,2,...,\quad x > 0$
Mean	ν	
Variance	2ν	
MGF	$(\frac{1}{1-2t})^{\frac{\nu}{2}}$	
Logistic	$\mathrm{LG}(\mu,\sigma)$	
pdf	$\frac{\exp(\frac{x-\mu}{\sigma})}{\sigma[1+\exp(\frac{x-\mu}{\sigma})]^2}$	$-\infty < \mu < \infty, \quad \sigma > 0, \quad -\infty < x < -\infty$
Mean	μ	
Variance	$\pi^2\sigma^2/3$	
MGF	$\exp(\mu t)\Gamma(1-\sigma t)\Gamma(1+\sigma t), \quad \lvert t\rvert < \frac{1}{\sigma}$	

D. Simulation for U-shaped Hazard Rate Curves

D.1. Generating U-shaped Hazard Rate Curves

The area under the curve in $h(t) - t$ plotting, i.e., $H(t)$, can be obtained by summing all the small "strips" decomposed under the curve. T^D and T^C as well as the parameters of the Weibull distributions assumed for DFR and IFR have to be specified. The h^C, the hazard rate of the constant part of a U-shaped hazard rate function, can be expressed as

$$h^C = \lambda^D \beta^D (\lambda^D \, T^D)^{\beta^D - 1}. \tag{D.1}$$

The hazard rate in the wearout section is first determined by Eq. (3.16) under different t and then shifted upward by h^C as determined by Eq. (D.1).

D.2. Simulation

Time is divided into small segments with time length Δt. There are $T_{\text{SIM}}/\Delta t$ testing periods if T_{SIM} represents the total simulating time. Only the failure times between $(k-1)\Delta t$ and $k\Delta t$ $(k = 1, 2, 3, \ldots, N)$ are recorded. Assume that N_k components are put in test (at risk) in segment k, then the hazard rate in the kth segment is expressed as

$$h_k(t) = \frac{(\text{number failed in the k}^{\text{th}} \text{ testing period})}{(\text{total number in test})(\text{duration})} = \frac{f_k}{N_k \Delta t}. \tag{D.2}$$

A random number U $(0 \leq U \leq 1)$ is generated and the cdf of the component with Weibull failure rate $\beta \neq 1$ (or nonexponential) is obtained by

$$U = \frac{F(t) - F(k\Delta t)}{1 - F(k\Delta t)}, \quad (k-1)\Delta t \leq t < k\Delta t, \quad k = 2, 3, \ldots, N. \tag{D.3}$$

A failure time t of the Weibull distribution with parameters λ and β is derived by inverse transformation technique

$$t = \frac{1}{\lambda}(-\ln\, U)^{\frac{1}{\beta}} \tag{D.4}$$

The failure times of the nonexponential components in the DFR $(k\Delta t < T^D)$ region are derived by Eqs. (D.3) and (D.4). If

$$T^D \leq k\Delta t < (T^D + T^C),$$

the component is in the CFR region in which the component possesses the memoryless property. Thus, every failure time would begin from $k\Delta t$; i.e.,

$$\text{A failure time in the CFR region} = k\Delta t - \frac{\ln U}{h^C} \tag{D.5}$$

When $k\Delta t \geq (T^D + T^C)$, the component is in the wearout region. Since the hazard rate in the IFR region has been shifted upward by an amount h^C when we simulate the wearout portion, it can be treated as two sub-components in series, where one has the Weibull distribution with parameter i (λ^I, β^I) and the other exponential with parameter h^C. Two failure times are then derived by Eqs. (D.3), (D.4) and (D.5), respectively; the smaller one is treated as the failure time of the component in IFR (since these two sub-components are in series). Without loss of generality, assume that T^D and T^C are chosen to be the multiplications of Δt to avoid complexity at the changing points. The procedure is summarized as the next pseudo code:

```
For (k = 1; k≤TestingPeriod; k++)
   For (j = 1; j≤NumberInPeriod; j++)
      FailFlag = 0;
      minFailTime = kΔt + 1;
      For (i = 1; i ≤ CurveNumber; i++)
         switch(WhichPart)
            DFR: FailTime = WeibullFail(λ_i^D, β_i^D);
            CFR: FailTime = ExponentialFail(h_i^C) + kΔt
            IFR: FailTime = min[WeibullFail(λ_i^I, β_i^I) + T^D + T^C,
                      ExponentialFail(h_i^C) + kΔt];
         If (kΔt ≤ FailTime < (k + 1) Δt)
            minFailTime = FailTime;
            FailFlag = 1;
      If (FailFlag = 1)
         FailNumber[k]++;
         Message "System fails at minFailTime";
```

E. Sample Programs

E.1. A Sample GINO Program

```
MODEL:
1) MIN=     (   (3.583 -h1)^2+
                (1.667 -h2)^2+
                (1.333 -h3)^2+
                (0.875 -h4)^2+
                (0.667 -h5)^2+
                (1.250 -h6)^2+
                (0.500 -h7)^2+
                (0.250 -h8)^2+
                (1.250 -h9)^2+
                (0.500 -h10)^2+
                (1.000 -h11)^2+
                (1.000 -h12)^2+
                (0.500 -h13)^2+
                (0.750 -h14)^2+
                (0.750 -h15)^2+
                (0.500 -h16)^2+
                (1.000 -h17)^2+
                (0.500 -h18)^2+
                (0.750 -h19)^2+
                (0.750 -h20)^2+
                (0.0   -h21)^2+
                (1.000 -h22)^2+
                (0.750 -h23)^2+
                (1.000 -h24)^2+
                (0.500 -h25)^2) ;
  2)  h5 -4.0*h4 +6.0*h3 -4.0*h2 +h1 >0.0 ;
  3)  h6 -4.0*h5 +6.0*h4 -4.0*h3 +h2 >0.0 ;
  4)  h7 -4.0*h6 +6.0*h5 -4.0*h4 +h3 >0.0 ;
  5)  h8 -4.0*h7 +6.0*h6 -4.0*h5 +h4 >0.0 ;
  6)  h9 -4.0*h8 +6.0*h7 -4.0*h6 +h5 >0.0 ;
  7)  h10-4.0*h9 +6.0*h8 -4.0*h7 +h6 >0.0 ;
  8)  h11-4.0*h10+6.0*h9 -4.0*h8 +h7 >0.0 ;
  9)  h12-4.0*h11+6.0*h10-4.0*h9 +h8 >0.0 ;
 10)  h13-4.0*h12+6.0*h11-4.0*h10+h9 >0.0 ;
 11)  h14-4.0*h13+6.0*h12-4.0*h11+h10>0.0 ;
 12)  h15-4.0*h14+6.0*h13-4.0*h12+h11>0.0 ;
 13)  h16-4.0*h15+6.0*h14-4.0*h13+h12>0.0 ;
 14)  h17-4.0*h16+6.0*h15-4.0*h14+h13>0.0 ;
```

```
  15)  h18-4.0*h17+6.0*h16-4.0*h15+h14>0.0 ;
  16)  h19-4.0*h18+6.0*h17-4.0*h16+h15>0.0 ;
  17)  h20-4.0*h19+6.0*h18-4.0*h17+h16>0.0 ;
  18)  h21-4.0*h20+6.0*h19-4.0*h18+h17>0.0 ;
  19)  h22-4.0*h21+6.0*h20-4.0*h19+h18>0.0 ;
  20)  h23-4.0*h22+6.0*h21-4.0*h20+h19>0.0 ;
  21)  h24-4.0*h23+6.0*h22-4.0*h21+h20>0.0 ;
  22)  h25-4.0*h24+6.0*h23-4.0*h22+h21>0.0 ;
  23)  h26-4.0*h25+6.0*h24-4.0*h23+h22>0.0 ;
  24)  h27-4.0*h26+6.0*h25-4.0*h24+h23>0.0 ;
  25)  h28-4.0*h27+6.0*h26-4.0*h25+h24>0.0 ;
  26)  h29-4.0*h28+6.0*h27-4.0*h26+h25>0.0 ;
  27)  h29>0.0 ;
  28)  h28-h29>0.0 ;
  29)  h29-2.0*h28+h27>0.0 ;
  30)  h29-3.0*h28+3.0*h27-h26<0.0 ;
 END

LEAVE
```

E.2. A Sample GAMS program

```
$TITLE  A trial model on completely monotone
$OFFUPPER
*

 SETS    K         TIME PERIODS      /1*25/
         J         TOTAL TIME SPAN   /1*29/ ;

 SCALAR  SMALLD    order            /4/ ;

 PARAMETER empirical hazard rates
   HAT(K)
  / 1  3.583
    2  1.667
    3  1.333
    4  0.875
    5  0.667
    6  1.250
    7  0.500
    8  0.250
    9  1.250
   10  0.500
   11  1.000
   12  1.000
   13  0.500
   14  0.750
   15  0.750
   16  0.500
```

```
  17  1.000
  18  0.500
  19  0.750
  20  0.750
  21  0.0
  22  1.000
  23  0.750
  24  1.000
  25  0.500  /  ;

PARAMETER W(K) weighting factors ;
          W(K) = 1 ;

VARIABLES H(J) smoothed hazard rates
          D total deviation ;

POSITIVE VARIABLE H ;

EQUATIONS Dd1
          Dd2
          Dd3
          Dd4
          Dd5
          Dd6
          Dd7
          Dd8
          Dd9
          Dd10
          Dd11
          Dd12
          Dd13
          Dd14
          Dd15
          Dd16
          Dd17
          Dd18
          Dd19
          Dd20
          Dd21
          Dd22
          Dd23
          Dd24
          Dd25
          Dlastb
          Dlastc
          Dlastd
          OBJD ;
*----------------------------------------------------------------------*
Dlastb..  H("28")-H("29") =G= 0.0 ;
Dlastc..  H("29")-2.0*H("28")+H("27") =G= 0.0 ;
Dlastd..  H("29")-3.0*H("28")+3.0*H("27")-H("26") =L= 0.0 ;
```

```
Dd1..  H("5") -4.0*H("4") +6.0*H("3") -4.0*H("2") +H("1")  =G=  0.0 ;
Dd2..  H("6") -4.0*H("5") +6.0*H("4") -4.0*H("3") +H("2")  =G=  0.0 ;
Dd3..  H("7") -4.0*H("6") +6.0*H("5") -4.0*H("4") +H("3")  =G=  0.0 ;
Dd4..  H("8") -4.0*H("7") +6.0*H("6") -4.0*H("5") +H("4")  =G=  0.0 ;
Dd5..  H("9") -4.0*H("8") +6.0*H("7") -4.0*H("6") +H("5")  =G=  0.0 ;
Dd6..  H("10")-4.0*H("9") +6.0*H("8") -4.0*H("7") +H("6")  =G=  0.0 ;
Dd7..  H("11")-4.0*H("10")+6.0*H("9") -4.0*H("8") +H("7")  =G=  0.0 ;
Dd8..  H("12")-4.0*H("11")+6.0*H("10")-4.0*H("9") +H("8")  =G=  0.0 ;
Dd9..  H("13")-4.0*H("12")+6.0*H("11")-4.0*H("10")+H("9")  =G=  0.0 ;
Dd10.. H("14")-4.0*H("13")+6.0*H("12")-4.0*H("11")+H("10") =G=  0.0 ;
Dd11.. H("15")-4.0*H("14")+6.0*H("13")-4.0*H("12")+H("11") =G=  0.0 ;
Dd12.. H("16")-4.0*H("15")+6.0*H("14")-4.0*H("13")+H("12") =G=  0.0 ;
Dd13.. H("17")-4.0*H("16")+6.0*H("15")-4.0*H("14")+H("13") =G=  0.0 ;
Dd14.. H("18")-4.0*H("17")+6.0*H("16")-4.0*H("15")+H("14") =G=  0.0 ;
Dd15.. H("19")-4.0*H("18")+6.0*H("17")-4.0*H("16")+H("15") =G=  0.0 ;
Dd16.. H("20")-4.0*H("19")+6.0*H("18")-4.0*H("17")+H("16") =G=  0.0 ;
Dd17.. H("21")-4.0*H("20")+6.0*H("19")-4.0*H("18")+H("17") =G=  0.0 ;
Dd18.. H("22")-4.0*H("21")+6.0*H("20")-4.0*H("19")+H("18") =G=  0.0 ;
Dd19.. H("23")-4.0*H("22")+6.0*H("21")-4.0*H("20")+H("19") =G=  0.0 ;
Dd20.. H("24")-4.0*H("23")+6.0*H("22")-4.0*H("21")+H("20") =G=  0.0 ;
Dd21.. H("25")-4.0*H("24")+6.0*H("23")-4.0*H("22")+H("21") =G=  0.0 ;
Dd22.. H("26")-4.0*H("25")+6.0*H("24")-4.0*H("23")+H("22") =G=  0.0 ;
Dd23.. H("27")-4.0*H("26")+6.0*H("25")-4.0*H("24")+H("23") =G=  0.0 ;
Dd24.. H("28")-4.0*H("27")+6.0*H("26")-4.0*H("25")+H("24") =G=  0.0 ;
Dd25.. H("29")-4.0*H("28")+6.0*H("27")-4.0*H("26")+H("25") =G=  0.0 ;

OBJD..               D =E=   W("1")* (H("1")- HAT("1"))**2 )+
                             W("2")* (H("2")- HAT("2"))**2 )+
                             W("3")* (H("3")- HAT("3"))**2 )+
                             W("4")* (H("4")- HAT("4"))**2 )+
                             W("5")* (H("5")- HAT("5"))**2 )+
                             W("6")* (H("6")- HAT("6"))**2 )+
                             W("7")* (H("7")- HAT("7"))**2 )+
                             W("8")* (H("8")- HAT("8"))**2 )+
                             W("9")* (H("9")- HAT("9"))**2 )+
                            W("10")*(H("10")-HAT("10"))**2 )+
                            W("11")*(H("11")-HAT("11"))**2 )+
                            W("12")*(H("12")-HAT("12"))**2 )+
                            W("13")*(H("13")-HAT("13"))**2 )+
                            W("14")*(H("14")-HAT("14"))**2 )+
                            W("15")*(H("15")-HAT("15"))**2 )+
                            W("16")*(H("16")-HAT("16"))**2 )+
                            W("17")*(H("17")-HAT("17"))**2 )+
                            W("18")*(H("18")-HAT("18"))**2 )+
                            W("19")*(H("19")-HAT("19"))**2 )+
                            W("20")*(H("20")-HAT("20"))**2 )+
                            W("21")*(H("21")-HAT("21"))**2 )+
                            W("22")*(H("22")-HAT("22"))**2 )+
                            W("23")*(H("23")-HAT("23"))**2 )+
                            W("24")*(H("24")-HAT("24"))**2 )+
```

```
                          W("25")*(H("25")-HAT("25"))**2 ) ) ;
MODEL  MONO1   / H, D / ;
*----------------------------------------------------------------------
SOLVE MONO1 USING NLP MINIMIZING D ;
```

INDEX

FSC
www.fsc.org
MIX
Papier | Fördert
gute Waldnutzung
FSC® C083411

Kolibri 360 GmbH
Ferdinand-Jühlke-Straße 7
99095 Erfurt, Deutschland
produktsicherheit@kolibri360.de